Responses of Plants to Environmental Stresses

Responses of Plants to Environmental Stresses

Editor

Renata Szymańska

MDPI • Basel • Beijing • Wuhan • Barcelona • Belgrade • Manchester • Tokyo • Cluj • Tianjin

Editor
Renata Szymańska
AGH University of Science and
Technology
Poland

Editorial Office
MDPI
St. Alban-Anlage 66
4052 Basel, Switzerland

This is a reprint of articles from the Special Issue published online in the open access journal *Plants* (ISSN 2223-7747) (available at: https://www.mdpi.com/journal/plants/special_issues/Responses_of_plants).

For citation purposes, cite each article independently as indicated on the article page online and as indicated below:

LastName, A.A.; LastName, B.B.; LastName, C.C. Article Title. *Journal Name* **Year**, *Volume Number*, Page Range.

ISBN 978-3-0365-0830-6 (Hbk)
ISBN 978-3-0365-0831-3 (PDF)

Contents

About the Editor

Renata Szymańska is an assistant professor at the Faculty of Physics and Applied Computer Science, AGH University of Science and Technology, Krakow, Poland. She works in the Molecular Biophysics and Bioenergetics Group. She acquired her Ph.D. in plant biochemistry at Jagiellonian University in 2011. After her graduation, she worked as an assistant in the Department of Plant Physiology and Biochemistry, Jagiellonian University (Kraków, Poland) till 2013. Following that, she got an opportunity to work for several months as a post-doctoral research fellow in the Plant Physiology Max Planck Research Group. Her areas of expertise are plant biochemistry and physiology. She focuses on plant responses to environmental stresses, reactive oxygen species and oxidative stress, as well as metabolism of plant-derived antioxidants. She has over 40 peer-reviewed research publications in leading international journals.

Preface to "Responses of Plants to Environmental Stresses"

Plants are constantly exposed to unfavorable environmental conditions, which play a major role in determining the productivity of crops worldwide. Abiotic factors also determine the geographical distribution of plant species across Earth's different zones. Among the abiotic stresses with which plants must cope, we distinguish: water availability, extreme temperatures, nutrient deficiency, excess light, heavy metals, and salinity. The plant's success depends on the adaptation and acclimation mechanisms, which are directly or indirectly related to photosynthesis as well as growth and development processes.

The plant response to environmental stresses involves multiple processes, which include complex mechanisms on all organismal levels: whole-plant, physiological, cellular, biochemical, and molecular. The abiotic stress response activates multidirectional interactions and crosstalk between these levels. One of the greatest challenges in modern plant biology is identifying these complex interactions underlying abiotic stress responses using all available tools (e.g., genetic engineering and bioinformatics) and experimental approaches (e.g., physiological, biochemical, molecular, and omics studies).

Today, a large amount of new information is available in the dynamic and expanding field of knowledge on plants' environmental stress responses. The data provided in this book fill a large gap in our understanding of these processes and integrate a large part of the scientific knowledge spectrum of plants' abiotic-stress-related mechanisms.

The present book encompasses 15 articles: 1 review, 1 brief report and communication, as well as 13 original papers. Together, they demonstrate the complexity of the plant response to environmental factors, which engages different mechanisms on all organism levels. All original reports provide a methodological approaches for studying the plant stress response. Research was conducted on different species, including model plant *Arabidopsis*, crops (i.e., crabapple, bottle gourd, cotton, wheat, moso bamboo, rice, barley, eggplant, and ramie), ornamental plants (Portulaca ad Dendrobium), as well as species growing in severe conditions (*Elymus sibiericus* and *Colobanthus quitensis*). Eleven articles present information concerning changes in physiological parameters in plants exposed to heat, drought, salinity, or nitrogen, and CO_2 deficiency. The emphasis is placed, amongst others, on phenotypic, biochemical, and molecular traits, such as: photosynthetic activity and chlorophyll fluorescence (Wang et al.; Gomez-Espinoza et al.), root system architecture (Rafael et al.), rhizome integration (Jing et al.), antioxidants activity (Borsai et al.; Toth et al.; Huang et al.; Lei et al.), cellulose fiber quality (Ayele et al.), stress responsive gene expression (Huang et al.), proteomic analyses (Lei et al.; Yoo et al.), and secondary metabolites (Faralli et al.). Three articles describe genome-wide approaches, which provide a holistic overview of the adaptation/acclimation molecular mechanisms of plant growing under different stress conditions. These papers include studies on ATP-binding cassette transporters (Zhang et al.), heat shock factors (Wang et al.), and expression of nitrogen-related genes (Tan et al.). In addition, the review by Li et al. provides a comprehensive information on the *WKRY* gene family, which includes the plant-specific transcription factors playing a important roles in different abiotic stress-response pathways.

We think that this book will inspire further studies by researchers worldwide, which will help us to understand the mechanisms of plants' environmental stress responses and allow the application of this knowledge to practical use. We would like to express our gratitude to all authors, reviewers, and contributors for their support and creation of this book. l be answered in this Special Issue that focuses on one of the most studied and relevant food-associated mycotoxins.

Renata Szymańska
Editor

Article

Plasticity of the Root System Architecture and Leaf Gas Exchange Parameters Are Important for Maintaining Bottle Gourd Responses under Water Deficit

Dinoclaudio Zacarias Rafael [1], Osvin Arriagada [2], Guillermo Toro [3], Jacob Mashilo [4], Freddy Mora-Poblete [1] and Rodrigo Iván Contreras-Soto [5,*]

[1] Institute of Biological Science, University of Talca, Talca 3460000, Chile; mosembaba@gmail.com (D.Z.R.); morapoblete@gmail.com (F.M.-P.)
[2] Departamento de Ciencias Vegetales, Facultad de Agronomía e Ingeniería Forestal, Pontificia Universidad Católica de Chile, Santiago 306-22, Chile; arriagada.lagos.o@gmail.com
[3] Plant Stress Physiology Laboratory, Centro de Estudios Avanzados en Fruticultura (CEAF), Rengo 2940000, Chile; gtoro@ceaf.cl
[4] Limpopo Department of Agriculture and Rural, Bela-Bela 0480, South Africa; jacobmashilo@yahoo.com
[5] Instituto de Ciencias Agroalimentarias, Animales y Ambientales, Universidad de O'Higgins, San Fernando 3070000, Chile
* Correspondence: rodrigo.contreras@uoh.cl

Received: 28 September 2020; Accepted: 18 November 2020; Published: 3 December 2020

Abstract: The evaluation of root system architecture (RSA) development and the physiological responses of crop plants grown under water-limited conditions are of great importance. The purpose of this study was to examine the short-term variation of the morphological and physiological plasticity of *Lagenaria siceraria* genotypes under water deficit, evaluating the changes in the relationship between the root system architecture and leaf physiological responses. Bottle gourd genotypes were grown in rhizoboxes under well-watered and water deficit conditions. Significant genotype-water regime interactions were observed for several RSA traits and physiological parameters. Biplot analyses confirmed that the drought-tolerant genotypes (BG-48 and GC) showed a high net CO_2 assimilation rate, stomatal conductance, transpiration rates with a smaller length, and a reduced root length density of second-order lateral roots, whereas the genotypes BG-67 and Osorno were identified as drought-sensitive and showed greater values for average root length and the density of second-order lateral roots. Consequently, a reduced length and density of lateral roots in bottle gourd should constitute a response to water deficit. The root traits studied here can be used to evaluate bottle gourd performance under novel water management strategies and as criteria for breeding selection.

Keywords: rhizoboxes; gaseous exchange; sub-Saharan Africa; root length density

1. Introduction

Drought is widely recognized as one of the most significant agricultural constraints in many regions worldwide, accounting for more than 80% of crop damage and losses [1]. In Mediterranean regions, for instance, the increase in annual average temperatures and the lower-than-average precipitation affect food production and sustainability in various agricultural systems [2]. In the context of climate change, it is highly probable that drought stress intensity will increase in the future as a result of more variable and unpredictable precipitation patterns. In the Mediterranean-like climate of central Chile, which is the main region for fruit and vegetable production in Chile, this phenomenon could potentially induce economic losses in agricultural systems. In fact, a recent study has indicated that

Central Chile will likely experience detrimental effects on water availability and vegetation changes that will have social and economic impacts [3].

Chile is one of the major contributors to fruit and vegetable production in South America. In addition, Central Chile plays an important role and has positioned itself as a leading exporter of diverse agricultural products. Vegetable crop production in this region is dominated by small-scale farmers whose lands are vulnerable to climate change [4,5]. The increasing probability of drought occurrences coupled with the increasing demand for food for the growing human population indicate the need to develop crop management strategies that improve water-use efficiency and productivity and increase crop yield outputs, especially under water-restricted agricultural systems [6].

Drought tolerance in plants is associated with the modification of various morphological and physiological responses. These responses improve the adaptation and production of crops grown under water-limited conditions. The most common physiological parameters associated with drought tolerance in the short-term include enhanced net CO_2 assimilation by the control of stomatal conductance and reduced transpiration rates for water conservation [7]. The maintenance of these physiological responses is widely associated with sustainable crop production in water-stressed environments [8]. Among the various plant organs, root development/morphology plays an important role associated with water-extraction from the soil profile, especially when water is limited [9,10]. The root system has great potential for improving plant adaptation and production under drought stress conditions [10–12]. In this context, Lynch [6] proposed that reduced root development would be advantageous for drought resistance in high-input agroecosystems. Root traits that improve water capture include fewer axial roots, a reduced density of lateral roots, and a greater loss of roots that do not contribute to water capture [6].

Several studies have reported a significant correlation between root and shoot traits, suggesting a coordinated strategy between below- and above-ground plant organs in response to water deficit [10,13]. These findings have enabled the selection of both root and shoot traits to improve drought tolerance and increase yield potential in several plant species, such as common bean [14], tomato [15], and quercus [16]. Moreover, Hund et al. [17] found that tolerant maize genotypes developed longer crown roots, which increased transpiration, stomatal conductance, and relative water content. Another study, also involving maize cultivars, supported the assumption that water stress reduces the production of crown roots, and lines with fewer crown roots had substantially deeper rooting and a greater capture rate of subsoil water and, consequently, improved the plant water status, stomatal conductance, leaf and canopy photosynthesis, biomass, and seed yield [18]. These results indicate that both root and physiological traits confer drought adaptation and should be useful for screening and selection for breeding purposes.

Bottle gourd (*Lagenaria siceraria* (Mol. Standl)) is an important cucurbit crop that is often grown under rainfed conditions in arid and semi-arid ecosystems. In semi-arid regions of sub-Saharan Africa, for instance, genetically diverse landraces of bottle gourd are commonly cultivated by local farmers in water-restricted conditions, yielding reasonable fruit production as a consequence of several years of selection and cultivation [8,19]. In this sense, the investigation of wild species or landraces from different gene pools could be useful to identify the morpho-physiological traits related to drought tolerance [20]. In addition, in genotypes of South African bottle gourd, Mashilo et al. [8] found that enhanced instantaneous water-use and intrinsic water-use efficiencies linked to high net CO_2 assimilation (An), stomatal conductance (gs), and transpiration (E) rates were significantly associated with drought tolerance. In the present study, we hypothesized that, in the initial development of bottle gourd, enhanced physiological performance could be associated with changes in root phenes due to water reduction. In fact, there is a lack of information regarding the relationships that may exist between root system architecture (RSA) traits and physiological responses in bottle gourd. In light of this, the objective of this study was to examine the short-term variation of the morphological and physiological plasticity of *Lagenaria siceraria* genotypes under water deficit, evaluating the changes in the relationship between the root system architecture and leaf physiological responses.

2. Results

2.1. Differences in Water Consumption of Bottle Gourd Genotypes

The plot of the normalized transpiration of bottle gourd genotypes against the fraction of transpirable soil water (FTSW) is shown in Figure 1. In most of the genotypes, except for BG-48 (Figure 1E), a relatively high FTSW was observed with normalized transpiration (NTR) values of ~1. Illapel and BG-67 decreased the FTSW below a critical threshold value, and there was a marked linear decrease in NTR in response to further declines in FTSW. Segmented regression indicated that the threshold value for transpiration occurred at an FTSW that ranged from 0.80 (±0.1) for Chepica to an FTSW of 0.37 (±0.03) for BG-67 (Figure 1, Table 1). Osorno, Chepica, Aurora, and BG-48 genotypes showed high FTSW threshold values of 0.77, 0.80, 0.76, and 0.82, respectively, compared with the relatively low FTSW threshold values recorded for Illapel (0.47) and BG-67 (0.37).

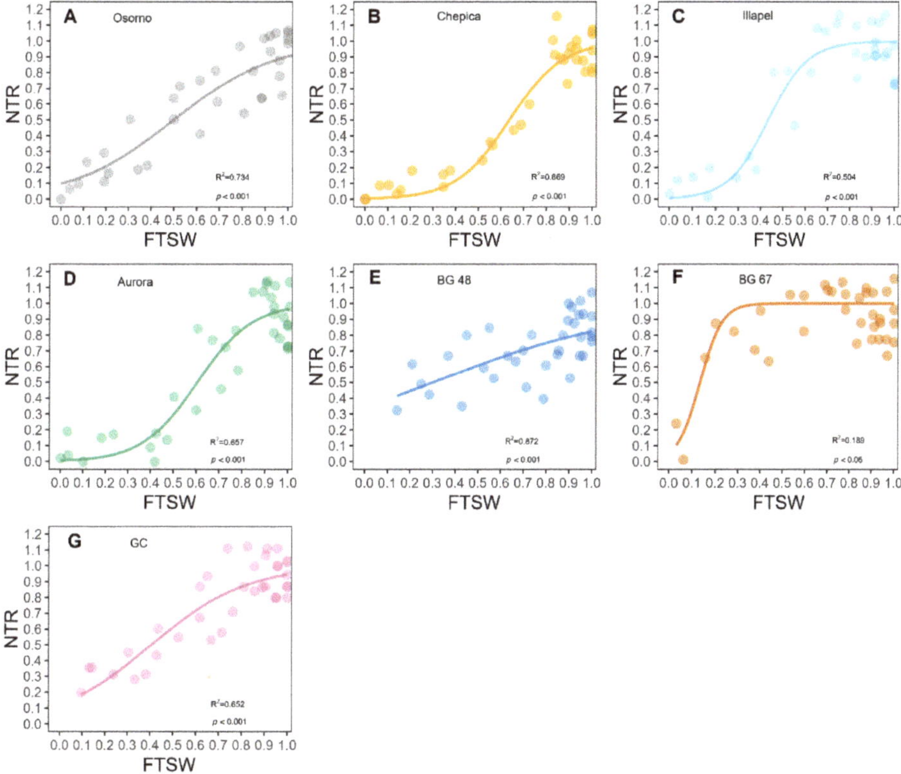

Figure 1. Normalized transpiration (NTR) response to fractions of transpirable soil water (FTSW) of seven genotypes of bottle gourd. Segmented regression indicated a threshold FTSW value above which there was a linear plateau of ~1.0 and below which there was a linear decline of NTR in response to decreasing FTSW. The genotypes are named on every figure.

Table 1. Comparison of fractions of transpirable soil water threshold ($FTSW_t$) values of seven genotypes of bottle gourd.

Genotype	$FTSW_t$	S.E	*
BG-48	0.824	0.08	a
Chepica	0.800	0.02	b
Osorno	0.777	0.10	c
Aurora	0.760	0.04	c
GC	0.696	0.10	d
Illapel	0.474	0.02	e
BG-67	0.368	0.03	f

S.E: standard error; * lowercase letters compare FTSWt between genotypes (Tukey test, $p < 0.05$).

2.2. Analysis of Variance and Mean Comparison for Physiological Parameters, Biomass, and Root System Architecture Traits

An analysis of variance (ANOVA) showed a highly significant ($p < 0.001$) effect of the genotype-water regime interaction for only stomatal conductance (gs), transpiration rate (E), and intrinsic water-use efficiency (WUEi) (Table 2). In most of the tested bottle gourd genotypes, the contrasting means in the comparison of the physiological traits under both water regimes showed that water deficit significantly reduced all traits (Tables S2–S4). In addition, BG-48 and GC genotypes showed significant differences between well-watered (WW) and water deficit (WD) treatments for stomatal conductance (gs) and transpiration rate (E) (Tables S2 and S3). On the other hand, for WUEi, the same genotypes showed an increment under the WD condition, whereas non-significant differences were observed for WUEins (Tables S4 and S5). For the same physiological traits, other genotypes showed a reduction in WUEi under the WD condition (Table S4). For RSA parameters, the genotype-water regime interaction effect was non-significant for the average root length of tap and basal roots (ARL), root angle of the first-order lateral of the tap and basal roots (ARA_1), and root length density of the lateral of the tap and basal roots (RLD_L). However, RSA traits measured in first-order and second-order lateral tap and basal roots—ARL_1, ARL_2, ARA_2, and RLD_{L1}—were influenced by the genotype-water regime interaction (Table 2). BG-48 showed a significant reduction in the length of lateral roots (i.e., ARL_1 and ARL_2) and a reduced density of lateral roots. Contrastingly, GC also showed a significant increment for both the length and density of lateral roots. Non-significant differences were observed for the genotypes Osorno, Chepica, and BG-67 for the same RSA traits (Table 3).

Table 2. Results of analysis of variance for physiological parameters and root system architecture traits evaluated in seven bottle gourd genotypes under well-watered and water deficit conditions.

Source of Variation	Significance (Physiological Traits)							
	An	gs	E	Ci	WUEi	WUEins	WUEwp	Biomass
Genotype (G)	**	**	**	ns	ns	ns	ns	ns
Water regime (W)	**	**	**	ns	ns	ns	**	**
G*W	ns	**	**	ns	**	ns	ns	ns
CV (%)	35.8	60.7	40.5	18.0	34.1	30.7	15.4	7.5
	Significance (root system architecture traits)							
	RLD_L	ARL	ARA	RLD_{L1}	ARL_1	ARA_1	ARL_2	ARA_2
Genotype (G)	**	**	**	ns	ns	**	ns	ns
Water regime (W)	ns	ns	ns	ns	ns	**	ns	ns
G*W	ns	ns	**	**	**	ns	**	**
CV (%)	16.2	16.2	5.3	14.9	14.9	7.2	19.8	7.5

CV (%): coefficient of variation in percentage; net CO_2 assimilation rate (An), stomatal conductance (gs), transpiration rate (E), intercellular CO_2 concentration (Ci), intrinsic water-use efficiency (WUEi), instantaneous water-use efficiency (WUEins), and whole plant water-use efficiency (WUEwp); average root length of tap and basal roots (ARL), root length of the first-order lateral of the tap and basal roots (ARL_1), and root length of the second-order lateral of the tap and basal roots (ARL_2); average root angle for the tap and basal roots (ARA), root angle of the first-order lateral of the tap and basal roots (ARA_1), root angle of the second-order lateral of the tap and basal roots (ARA_2), root length density of tap and basal roots (RLD), and root length density of the first-order lateral of the tap and basal roots (RLD_{L1}). ns, non-significant; **, significant at 1% probability by the F-test, respectively.

Table 3. Results of orthogonal contrasting tests for the difference in mean values between water deficit (WD) and well-watered (WW) conditions for the average root length of the first-order and second-order of the lateral of tap and basal roots (ARL_1 and ARL_2), the average root angle of the second-order of the lateral of tap and basal roots (ARA_2), and the root length density of the lateral tap and basal roots (RLD_{L1}).

Genotype	ARL_1 (cm)	ARL_2 (cm)	RLD_{L1} (cm/cm³)	ARA_2 (°)
Osorno	−27.9 ns	0.07 **	−0.01 ns	1.91 ns
Chepica	86.3 ns	2.5 ns	0.02 ns	3.04 ns
Illapel	2.85 ns	−0.27 ns	0 ns	−0.43 ns
Aurora	−62.5 ns	−4.83 ns	−0.01 ns	5.86 *
BG-48	44.1 **	−7.92 ***	0.01 *	1.11 *
BG-67	40.4 ns	−3.84 ns	0.01 ns	−1.06 ns
GC	−47.3 **	7.7 ns	−0.01 *	−0.01 ns

ns: non-significant; * significant at 5%; ** significant at 1%; *** significant at 0.1%.

2.3. Correlations between Physiological and Root System Architecture Traits under Well-Watered and Water Deficit Conditions

Pearson correlation coefficients between physiological and RSA traits among the evaluated bottle gourd landraces under WW and WD conditions are presented in Figure 2. Negative and significant associations were observed for several physiological and RSA traits and biomass production under water deficit conditions. The net CO_2 assimilation (An), stomatal conductance (gs), and transpiration rate (E) were negatively and significantly correlated with ARL_2, ARA_2, and biomass. On the other hand, Ci values were positively and significantly correlated with ARL_2, ARA_2, and biomass, but negatively correlated with leaf gas exchange parameters (An, gs, and E). Intrinsic and instantaneous water use-efficiencies were negatively and significantly correlated with ARL_2, whereas WUEwp was negatively and significantly correlated with ARA_2, but positively correlated with ARL_1 and RLD_{L1}.

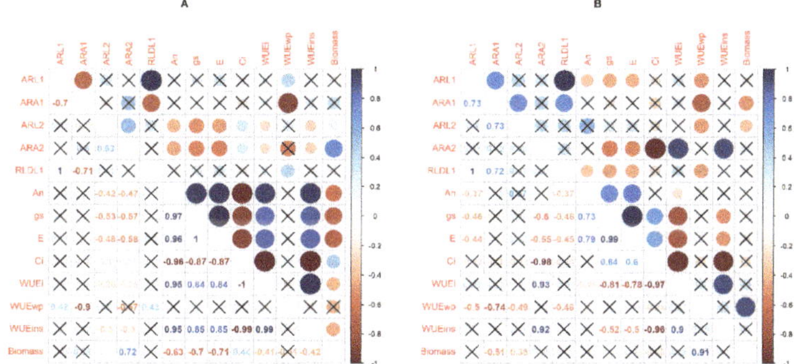

Figure 2. Pearson correlation coefficients among physiological and root system architecture (RSA) traits assessed in genotypes of bottle gourd in water deficit (**A**) and well-watered (**B**) conditions. Net CO_2 assimilation rate (An); stomatal conductance (gs); intercellular CO_2 concentration (Ci); transpiration rate (E); instantaneous water-use efficiency (WUEins); intrinsic water-use efficiency (WUEi); and whole plant water use efficiency (WUEwp). Average root length of the first-order lateral of the tap and basal roots (ARL_1) and second-order lateral of the tap and basal roots (ARL_2); average root angle of the first-order lateral of the tap and basal roots (ARA_1) and second-order lateral of the tap and basal roots (ARA_2); and root length density of the first-order lateral of the tap and basal roots (RLD_{L1}). Positive correlations are displayed in blue and negative correlations in red. The color intensity and the size of the circle are proportional to the correlation coefficients. On the right side of the correlogram, the color legend shows the correlation coefficients and the corresponding colors.

Under the WW condition, biomass was negatively correlated with some RSA traits (ARA_1 and ARL_2) and positively correlated with WUEwp. ARA_2 was negatively and significantly correlated with gs and E and positively correlated with WUEi and WUEins. WUEins and WUEi were both negatively correlated with gs, E, and Ci. Furthermore, WUEwp was negatively correlated with RSA traits (ARL_1, ARA_1, and ARL_2) (Figure 2).

2.4. Principal Component Analysis for the Differentiation of Drought-Tolerant and Sensitive Bottle Gourd Genotypes

Principal component analyses of physiological and RSA parameters measured under water deficit and well-watered conditions are presented in Table 4. Under the WD condition, the total variability of the three-dimensional space was efficiently summarized by the two principal components, which accounted for 51% and 26% of the variability, respectively. The first component consisted of high positive loadings for leaf gas exchange parameters as well as An, gs, E, WUEi, and WUEins and negative loadings for some RSA traits (ARA_2, ARA_2, and ARL_2), biomass, and Ci. In contrast, the second component consisted of high positive and negative loadings of root traits such as RLD_{L1}, ARL_1, ARA_1, ARL_2, and ARA_2. Under the WW condition, the first component consisted of negative loadings of RSA traits and biomass, while the leaf gas exchange parameters consisted of positive loadings (An, gs, E, and Ci) that accounted for 43% of the total variation. On the other hand, the second component consisted of negative loadings for most of the leaf gas exchange parameters (An, gs, E, and Ci) and root traits (RLD_{L1}, ARL_1, ARA_2, and ARL_2), which accounted for 29% of the total variation.

Table 4. Principal component analysis showing eigenvectors, eigenvalues, and percentage of variance of physiological and root system architecture traits of seven bottle gourd genotypes under water deficit and well-watered conditions.

Traits	Water Deficit (Eigenvectors)			Well-Watered (Eigenvectors)		
	PC1	PC2	PC3	PC1	PC2	PC3
RLD_{L1}	0.09	−0.42	0.38	−0.25	−0.32	−0.21
ARL_1	0.09	−0.42	0.38	−0.24	−0.33	−0.22
ARA_1	−0.02	0.53	0.12	−0.19	−0.44	0.09
ARL_2	−0.17	−0.19	0.40	−0.07	−0.37	0.37
ARA_2	−0.22	0.19	0.52	−0.38	0.03	0.28
An	0.39	0.08	0.05	0.16	−0.10	0.56
gs	0.38	0.02	−0.04	0.37	−0.08	0.27
E	0.39	−0.03	−0.01	0.35	−0.12	0.29
Ci	−0.36	−0.12	−0.19	0.37	−0.09	−0.26
WUEi	0.35	0.11	0.20	−0.40	0.14	0.11
WUEwp	0.01	−0.46	−0.24	−0.01	0.48	0.11
WUEins	0.36	0.13	0.17	−0.33	0.14	0.33
Biomass	−0.27	0.17	0.31	−0.06	0.38	0.03
Eigenvalues	2.56	1.86	1.35	2.36	1.92	1.57
Proportion of total variance (%)	0.50	0.26	0.14	0.43	0.29	0.19
Cumulative variance (%)	0.50	0.76	0.91	0.43	0.72	0.91

A principal component biplot (PC1 and PC2) was used to visualize the relationships between bottle gourd genotypes based on physiological and RSA parameters (Figure 3). In this biplot, smaller angles with the same direction among the vectors represented the most informative and correlated physiological and/or root traits, identifying groups of genotypes based on the assessed traits. The genotypes that were closed or in the same direction as the vectors were plotted as associated with an increase or reduction of these traits. Under the WW condition, genotypes Aurora and BG-48 were grouped with high values of ARA_1, ARL_1, ARL_2, and RLD_{L1}. Osorno and Illapel were differentiated by high values of WUEwp. On the other hand, reduced values of the leaf gas exchange parameters of An, gs, E, and Ci were associated with BG-67 and GC. Under the WD condition, Aurora was grouped with high values of ARA_1. Osorno, Illapel, and BG-67 were grouped as expressing high ARA_2, ARL_2, and Ci. On the contrary, BG-48 was differentiated by high An, gs, E, WUEins, and WUEi values and a reduction in the length and density of lateral roots (ARL_1, ARL_1, and RLD_{L1}). GC possessed high Ci, gs, and E (Figure 3).

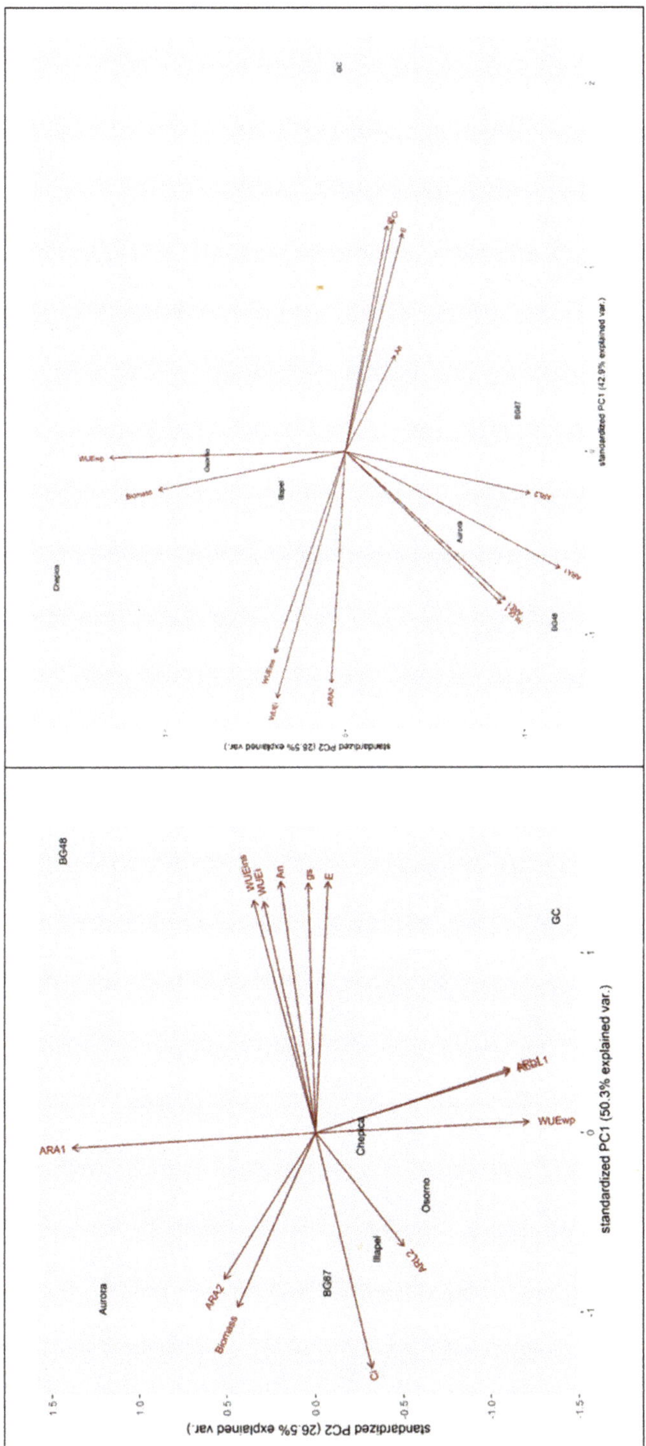

Figure 3. Principal component (PC) biplot showing the percentage of variance explained by PC1 and PC2, and grouping of bottle gourd genotypes based on physiological and root system architecture traits under water deficit and well-watered conditions.

2.5. Morphological and Physiological Plasticity

Among the seven bottle gourd genotypes, significant differences were observed in the relative distance plasticity index (RDPI) in physiological and morphological traits (biomass and RSA) (Figures 4 and 5). In general, low plasticity was observed for physiological and morphological traits. No significant differences were observed among genotypes for RDPI in Ci, WUEi, and WUEins. GC and Illapel showed higher RDPI values for leaf gas exchange (Gs and E) than Osorno and Chepica. Furthermore, GC and BG-48 showed higher RDPI values for biomass when compared with the other five genotypes (Figure 4). Regarding the root morphological traits, no significant differences were observed among genotypes for RDPI in ARL and RLD_L. BG-48 showed higher plasticity for the second-order lateral of the tap and basal roots (ARA_2 and ARL_2) and the root length density of the first-order lateral of the tap and basal roots when compared with the other six genotypes (Figure 5).

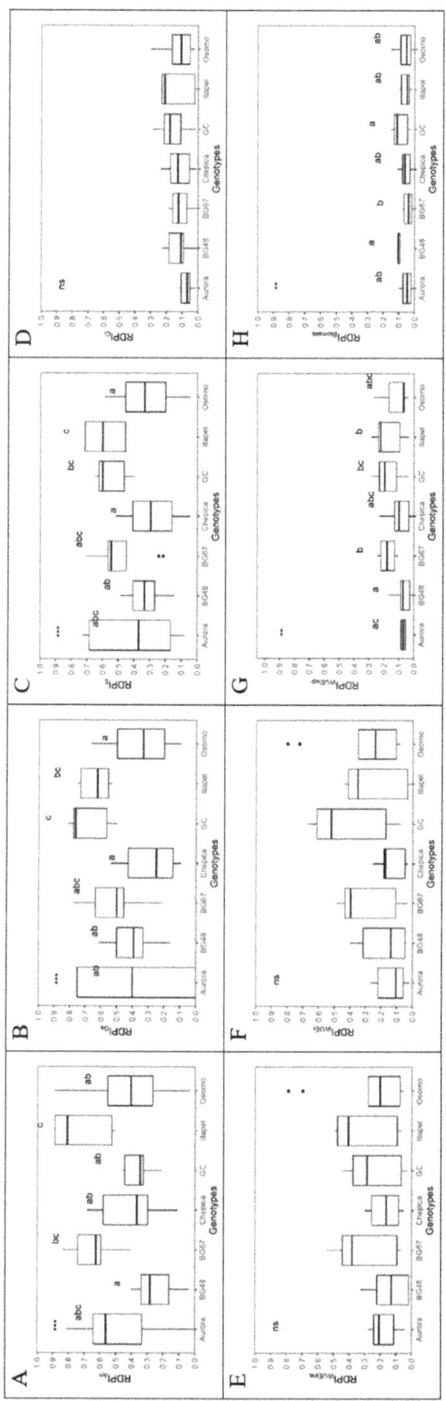

Figure 4. Relative distance plasticity index (RDPI) considering physiological traits (**A**–**G**) and biomass (**H**) in seven bottle gourd genotypes subjected to the water deficit condition. *p*-value for difference between genotypes ($p < 0.001$, Tukey's test). Lowercase letters compare the RDPI in each genotype (Tukey test, $p < 0.001$). ns = non-significant analysis of variance (ANOVA); ** significant at 1%; *** significant at 0.1%.

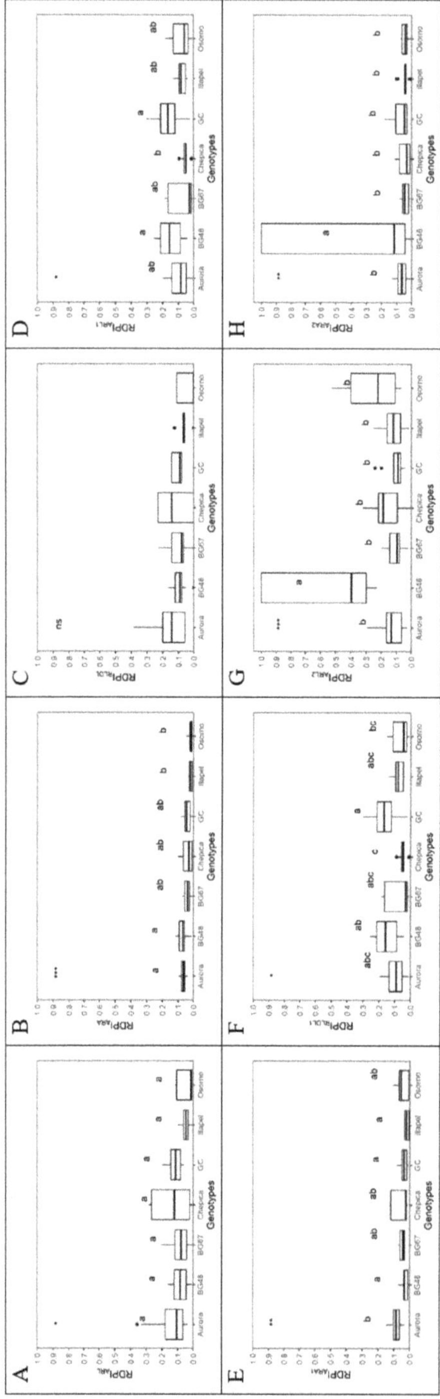

Figure 5. Relative distance plasticity index (RDPI) considering morphological traits (**A–H**) in seven bottle gourd genotypes subjected to the water deficit condition. *p*-value for difference between genotypes (*p* < 0.001, Tukey's test). Lowercase letters compare the RDPI in each genotype (Tukey test, *p* < 0.001). ns = non-significant ANOVA; * significant at 5%; ** significant at 1%; *** significant at 0.1%.

3. Discussion

When studying plant responses to water deficit, several morphological and physiological traits have been evaluated and reported [20,21]. Such a large dataset of numbers and variables makes it difficult to form an overall idea of how water deficit affects plants and how plants respond to such a limiting condition [22]. The approach of the relative distance plasticity index used here has been used to study plant adaptations under different conditions or environments and to evaluate growth responses under stressful conditions [22–24]. Our results for the RDPI showed that most of the traits studied showed some level of plasticity in response to water reduction, even though the plasticity presented here was relatively low (RDPI < 0.5); in addition, there was also some evidence that not all traits that contributed significantly and highly to variation presented higher plasticity indices in response to water reduction.

In the present study, the root morphological and physiological plasticity of drought-related traits and the negative correlation between leaf gas exchange parameters with lateral tap and basal roots allowed us to characterize the response to water reduction of bottle gourd. In different plant species, previous studies have also reported traits that were responsible for plastic responses with the aim of obtaining an integrative index related to the sensitivity to drought stress of various genotypes [16,22,24,25]. For instance, in our study, high variability and a genotype-dependent relative plasticity index were observed between bottle gourd genotypes; in particular, the Osorno genotype showed the lowest physiological and morphological plasticity index, whereas Illapel and BG-48 showed higher physiological and morphological plasticity indices, respectively. Furthermore, these results confirmed that, in some bottle gourd genotypes, the leaf gas exchange parameters were positively influenced by drought stress [8,26] and, consequently, could be used as drought-related traits. On the other hand, this study also shows that conclusions regarding the response of bottle gourd to water reduction are a result of different strategies associated with root morphological drought-related traits.

Plant responses, soil water availability, and the water uptake capacity from shallow or deep soils have been widely studied as important key factors to assess the tolerance degree to water deficit of different plant species [27–29]. In general, two strategies have been described to explain the behavior of plants to face water deficit: a "productive" strategy, which attempts to maintain open stomata, assuming water losses, but increasing net CO_2 assimilation to yield biomass; and a "conservative" strategy, which ensures water conservation in the soil and promotes early stomatal closure in response to water deficit [30]. In this study, we found a variability in the FTSW threshold between South African and Chilean genotypes, highlighting BG-48, Chépica, Osorno, Aurora, and GC genotypes as exhibiting "conservative" behavior, while Illapel and BG-67 showed "productive" behavior.

Some physiological traits (An, Gs, and E) revealed that Chepica and Osorno were more sensitive to water deficit than GC and Illapel. Specifically, Osorno, BG-67, and Chepica showed a severe reduction of some leaf gas exchange parameters (mean values of stomatal conductance, photosynthetic rate, and transpiration) as a result of water stress when compared with BG-48 and GC. BG-67 and Osorno genotypes recorded reductions of 91% and 84% in photosynthetic rates, 88% and 84% reductions in stomatal conductance, and 84% and 81% reductions in transpiration, respectively. Similarly, a reduction in stomatal conductance and the CO_2 assimilation rate under water deficit has been reported in different plant species including watermelon [31], squash [32], and quercus [29]. In previous studies based on the physiological performance of *L. siceraria*, Mashilo et al. [8] classified BG-48 and GC as drought-tolerant genotypes. Our study also revealed contrasting abilities to tolerate water stress, where bottle gourd genotypes that originated from arid and semi-arid environments (i.e., BG-48 and GC) showed better tolerance compared with Chilean genotypes grown in temperate or cold environments (i.e., Osorno and Chepica).

The BG-48 and GC genotypes, which are tolerant to water deficit [8], recorded a decreased intercellular CO_2 concentration due to water stress, although this tendency was not significant (Table S6). These findings may confirm that, under water deficit conditions, the stomatal closure reduces the internal CO_2 concentration of the leaf, as proposed by Cornic [33], Zhang et al. [34], and Flexas et al. [35].

However, there are contradicting reports on the mechanism responsible for stomatal closure. Some studies endorse the view that chemical signals are responsible for stomatal closure, while others support the idea that hydraulic signals are responsible [7]. This report probably supported the contrasting results previously reported for bottle gourd by Mashilo et al. [8], which revealed that drought-stressed genotypes (BG-48 and GC) showed an increased intercellular CO_2 concentration irrespective of reduced stomatal conductance, photosynthetic, and transpiration rates. Although similar results that reported increased CO_2 concentration were observed under water stress in cowpea [36], maize [37], and wheat [38], we suggest that more research is necessary on stomatal closure as a response to water deficit in bottle gourd.

Regarding the morphological plasticity indices, BG-48 and GC genotypes presented higher plasticity than the other five genotypes, which was based on ARL_1, ARL_2, ARA_2, RLD_{L1}, and biomass. As BG-48 and GC genotypes presented a greater biomass than the other five genotypes, we may argue that the secondary growth and ability to maintain or increase root length and the density of lateral tap and basal roots under water deficit may be related to the good growth and yield performance of bottle gourd under drought conditions. It is important to note that ARL_2 and ARA_2 showed relatively moderate plasticity in comparison with the other RDPIs. In addition, under the water deficit condition, BG-48 had specific phene states as the reduced length (ARL_1 and ARL_2) and density of lateral tap and basal roots (RLD_{L1}) permit greater resource allocation to deeper roots. In cassava and maize, some authors noted that genotypes with high yield potential under drought are characterized by having a more intensive and extensive fine root system, which enables the acquisition of more water from larger and deeper volumes of soil [24,39,40]. In fact, Lynch [6], in a revision of root phenotypes for drought resistance, proposed that specific root phenes such as fewer axial roots and a reduced density of lateral roots may contribute to improving water capture in dry topsoil.

In addition to the morpho-physiological plasticity index, principal component analysis was conducted to discriminate tolerant and susceptible bottle gourd genotypes based on their physiological and RSA traits. In particular, the PCA was able to reduce and group physiological and root morphological traits into components according to their ability to describe the variability among bottle gourd genotypes under the water deficit condition. Plotting the bottle gourd genotypes by means of their component scores, PC1 separated BG-48 with positive values of An, gs, E, WUEins, and WUEi and a reduction in the length and density of lateral roots (ARL_1, ARL_2, and RLD_{L1}). This finding indicated that water reduction led to fewer axial roots and a reduced density of lateral roots, which may contribute to improving water capture in dry topsoil. On the other hand, the genotype BG-67 showed a reduction in leaf gas exchange parameters with some increment in the length of lateral roots, which may be considered another strategy associated with "productive" behavior.

4. Material and Methods

4.1. Plant Material

The plant material used in this study consisted of seven bottle gourd genotypes. Three were commercial varieties sourced from the Limpopo Department of Agriculture and Rural Development (Towoomba Research Station) of South Africa, one was a commercial variety from Chile, and the rest were accessions collected from three regions of Chile. Breeding varieties from South Africa were identified with a high level of drought tolerance and cultivated under dryland conditions with limited agricultural inputs (i.e., fertilization and irrigation) [8]. Details of the bottle gourd genotypes are shown in Table S7.

4.2. Experimental Design and Growing Conditions

Bottle gourd seeds were sterilized by immersion in 2% (*v/v*) sodium hypochlorite in water for 10 min, rinsed twice with deionized water for 10 min, and germinated for 5–7 seven days at 20–25 °C in 7 cm × 7 cm × 8 cm (0.23 L) pots with peat and sand substrate in an equal ratio of 1:1. Plants with the first fully expanded true leaf and with an absence of damage or disease were considered as criteria for transplantation to the rhizobox. For root system architecture phenotyping, experiments were conducted in rhizoboxes (length × width × height = 60 × 2 × 40 cm), which were boxes with transparent plexiglass plates and covered by a non-transparent plastic box on the outside (Figure 6). Rhizoboxes were inclined by 45° to the horizontal plane with the plexiglass plate on the underside, so that roots could grow along the surface (Figure 6B). Each rhizobox was filled with ~2 kg of substrate (1:1 peat/sand *v/v*). Fertilizer was not applied during the entire experiment to avoid a confusion of the applied stress.

Figure 6. Rhizobox structure and dimensions used to evaluate root system architecture traits (**A**); angle and position of rhizobox in the field experiment (**B**); rhizobox (2) placed horizontally on a black surface (3) at a distance of ~134 cm from the digital camera (1) to avoid the effect of light (4) (**C**).

Bottle gourd plants were grown under field conditions, where the average air temperature was 23.8 ± 2.7 °C with a relative humidity of 54% and solar radiation level of 27 Mj/m^2. The experiment was conducted in the 2019–2020 growing season in the field condition using a shade net cover (Raschel sun-shading net with 50% light transmittance). A completely randomized design with a 7 × 2 factorial arrangement and three replicates was used. Factors consisted of seven bottle gourd genotypes and two water regimes (well-watered and water deficit conditions).

4.3. Water Deficit Treatment, Fractions of Transpirable Soil Water, and Transpiration Rate

Twenty days after sprouting, the plants of each genotype were transplanted to the rhizobox. At this time, plants were subjected to two water availability irrigation conditions: well-watered (WW) and water deficit (WD). Plants under the WW condition were irrigated three times per week, adding water to reach the corresponding 100% of the substrate water content of each rhizobox during the period of the experiment (28 days). In contrast, the WD condition was induced by suspending the irrigation supply for 28 days, followed by weighting each rhizobox three times per week to determine the amount of water consumed by each plant for the assessed genotypes. The fraction of transpirable soil water (FTSW) relative to well-watered treatments, which represented the portion of remaining volumetric soil water available for transpiration on each day of the experiment, was used as the indicator of stress [41]. The FTSW for each day of the experiment was calculated using Equation (1):

$$\text{FTSW} = [\text{Pot weight day n} - \text{Final pot weight}]/[\text{Initial pot weight} - \text{Final pot weight}] \qquad (1)$$

The normalized transpiration rates (NTRs) of WW and WD plants were determined by dividing the daily transpiration rate (gravimetrically) of each replication in each treatment of WD plants by the transpiration rate of WW plants. The NTR and FTSW were calculated for each rhizobox in the WD treatment using rhizobox weights recorded three times per week. For plants growing under the WD condition, the NTR of each bottle gourd genotype was plotted against the FTSW by fitting a segmented non-linear regression to determine the FTSW threshold value at which the NTR began to decline. The non-linear regression was fitted using R 4.0 [R Core Development Team, 2020].

4.4. Physiological Parameters and Biomass Production

Gas exchange parameters including the stomatal conductance (gs), transpiration rate (E), intercellular CO_2 concentration (Ci), and net CO_2 assimilation rate (An) were measured once per week for four weeks using a CIRAS-2 portable IRGA photosynthesis system (PPSystem, Hitchin, UK) with a controlled environment CIRAS PLC cuvette (broad windows 2.5 cm^2). The CO_2 concentration and photosynthetically active radiation inside the cuvette were adjusted to 400 μmol mol^{-1} and 1500 μmol m^{-2} s^{-1}, respectively. The measurements were all carried out between 09:00 and 14:00 on clear days on the fifth and fully-expanded leaves of the plants. Intrinsic water-use efficiency (WUEi) was calculated as the ratio between An and gs, and instantaneous water-use efficiency (WUEins) as the ratio between An and E. To calculate the whole-plant water-use efficiency (WUEwp), three plants per genotype and water treatments were harvested at the end of the experiment. Leaves, shoots, and roots for each plant were separated and dried in an oven at 60 °C to obtain dry weights. The total biomass increase during the experiment was estimated as the difference between the whole-plant dry weights at the beginning and end of the experiment. Plant water consumed over the four-week period was estimated from the sum of the daily water consumption. WUEwp was determined according to Medrano et al. [42] using Equation (2):

$$\text{WUE}_{wp}\left(\text{gL}^{-1}\right) = \frac{(\text{dry weight of final biomass} - \text{dry weight of initial biomass})}{\text{total water consumed}} \tag{2}$$

Finally, to determine the dry-mass (biomass) of each genotype in WD and WW conditions, the stem and roots were put in an oven for a minimum of 48 h at 70 °C, and then the mass in grams was measured.

4.5. Root Parameters and Image Processing

To characterize the root system architecture (RSA) of plants grown under WW and WD conditions, the rhizoboxes were photographed once per week with a high-resolution Nikon digital camera (Nikon D3500) fitted with a Nikkor AF-P 18–55 mm 1:3.5–5.6 G lens. For standard imaging, the focus of the camera was placed vertically, which was also done to avoid the effect of light on the acrylic of the rhizoboxes. The rhizoboxes were placed horizontally on a black surface at a distance of ~134 cm from the camera to obtain the best focus of fine roots (Figure 6C). The focus of the camera was adjusted manually and remained fixed for all images of the rhizoboxes.

The CI-690 RootSnap was used to measure the root traits, and the RSA traits or root classes based on the site of origin were classified as proposed by Zobel and Waiser [43]. In our study, the tap and basal roots were used to calculate the average root length (ARL) and average root angle (ARA). Furthermore, the RSA was classified and measured as the average root length of the first-(ARL_1) and second-order (ARL_2) lateral tap and basal roots and the average root angle of the first-(ARA_1) and second-order (ARA_2) lateral tap and basal roots.

Other RSA parameters, including root length density (RLD), which was expressed as the total length of root per unit of volume of soil (RLD_L), were calculated according to Johnson et al. [44] using Equation (3):

$$RLD_L = L/(A \times D) \tag{3}$$

where

- RLD_L: root length density, based on the length of roots (cm root cm^3 soil);
- L: total length of root observed under the rhizobox (cm);
- A: framework area observed in the rhizobox ($60 \times 40 = 2400$ cm^2);
- D: depth of the rhizobox (2 cm).

Two measurements of RLD_L were calculated as the RLD of tap and basal roots (RLD_L) and the first-order lateral tap and basal roots (RLD_{L1}).

4.6. Morphological and Physiological Plasticity Index

The relative distance plasticity index (RDPI) was calculated for morphological and physiological traits following Valladares et al. [23] and Marchiori et al. [22]. The data obtained at 28 days after transplanting were used to calculate the morphological and physiological plasticity, which indicated the relative phenotypic distance between individuals of the same genotype exposed to different treatments (WW and WD). Briefly, for each bottle gourd genotype, a 2×3 matrix of each morphological and physiological parameter was constructed, where the rows (i) represented the treatments and the columns represented the bottle gourd individuals (j) (i.e., the replicate for each treatment). We considered two water regimes ($i = 1, 2$) and three individuals of each bottle gourd genotype ($j = 1, 2, 3$). The phenotypic plasticity for a given variable x can be related to the difference of x between two individuals of the same genotype grown under different water treatments. The phenotypic plasticity was described by the absolute distance between two selected individuals (j and j') of the same genotype grown under distinct water conditions (i and i'). Regarding this assumption for the whole data set, we computed pairwise distances across all individuals and water conditions. For a given variable x, the distance among values ($dij{\rightarrow}i'j'$) was the difference $xi'j' - xij$, and the relative distances ($rdij{\rightarrow}i'j'$) were defined as $dij{\rightarrow}i'j'/(xi'j' + xij)$ for all pairs of individuals of a given genotype grown under different water availabilities. Finally, RDPI was calculated as $\sum(rdij{\rightarrow}i'j')/n$, where n represented the number of distances. Detailed descriptions of the relative distance plasticity index and its bases are given in Valladares et al. [21] and Marchiori et al. [22].

The RDPI differences between genotypes were evaluated with a one-way ANOVA and post-hoc Tukey mean comparison test ($p < 0.05$) using the packages 'ggpubr', 'plry', and 'multicompView' and considering 'bottle gourd genotypes' as a factor.

4.7. Data Analysis

An analysis of variance (ANOVA) was performed after testing the homogeneity of variances and normality of the residuals using Bartlett and Shapiro–Wilk tests.

A two-way ANOVA was performed for physiological and RSA traits. For the multiple comparison analysis test, orthogonal contrasts were performed to compare the mean values of genotypes by the water regime interaction effect. Statistical analyses were performed using the PROC GLM procedure of SAS software (SAS version 9.3).

The mean values of the studied RSA traits and physiological parameters for each condition (WW and WD) were used to compute the Pearson's linear correlation coefficients to describe the pattern of association between physiological and RSA traits in the R-package, using the "corrplot" function. Significance tests for the correlation coefficients were determined using Student's *t*-test.

A principal component analysis (PCA) based on the correlation matrix was performed using the "princomp" function in R. The eigenvectors derived from the PCA were used to identify the variables that had a strong relationship with a specific principal component. The PC biplot was then generated using the "ggbiplot" package in R to describe and group bottle gourds for their level of drought tolerance according to Shah et al. [45].

5. Conclusions

In conclusion, our results provided evidence that most of the traits studied showed some level of plasticity in response to water reduction. Some RSA traits, such as a reduced length and density of lateral roots (RLD_{L1}, ARL_2, ARA_2, and ARL_1), were able to improve the morphological plasticity of root biomass production in bottle gourd under the water deficit condition. These findings may contribute to a better understanding of the drought-tolerant mechanisms conferred by root system architecture traits and the physiological responses of bottle gourd, leading to efficient selection criteria and enhancements of the drought adaptation and phenotypic plasticity in this vegetable crop.

Supplementary Materials: The following are available online at http://www.mdpi.com/2223-7747/9/12/1697/s1, Table S1: Increment and reduction of CO_2 assimilation (An), stomatal conductance (gs), and intrinsic water-use efficiency (An/gs) between initial and final mean values (experimental time period) of stress treatments in seven bottle gourd landraces. Table S2: Results of contrast tests comparing the mean values difference between well-watered (WW) and water deficit (WD) conditions for stomatal conductance (gs). Table S3: Results of contrast test comparing the mean value differences between well-watered (WW) and water deficit (WD) conditions for transpiration (E). Table S4: Results of contrast tests comparing the mean value differences between well-watered (WW) and water deficit (WD) conditions for intrinsic water use efficiency (WUEi). Table S5: Results of contrast tests comparing the mean value differences between well-watered (WW) and water deficit (WD) conditions for instantaneous water use efficiency (WUEins). Table S6: Results of contrast tests comparing the mean value differences between well-watered (WW) and water deficit (WD) conditions for intercellular CO_2 concentration (Ci). Table S7: Origin and geographical coordinates of seven bottle gourd genotypes evaluated under well-watered and water deficit conditions.

Author Contributions: D.Z.R.: Data curation and formal analysis. O.A.: Data curation and root analyses. G.T.: Measurement and validation of physiological parameters and writing—review. J.M.: Conceptualization, landraces support, and writing—review & editing. F.M.-P.: Supervision and writing—review & editing. R.I.C.-S.: Conceptualization, methodology, resources, supervision, and writing—draft & editing. All authors have read and agreed to the published version of the manuscript.

Funding: This work was supported by the National Commission for Scientific and Technological Research (CONICYT, Chile) FONDECYT (grant number 11180278).

Acknowledgments: D.Z.R. thanks the Chilean Agency for the International Cooperation for Development (AGCIDChile) Nelson Mandela Scholarship. R.I.C.-S. thanks CEAF (Centro de Estudios Avanzados en Fruticultura) for their support of the project, specifically Ariel Salvatierra for plant propagation, Paula Pimentel for Lab facilitation. R.I.C.-S. also thanks to Jonathan Lynch for the comments to improve the manuscript.

Conflicts of Interest: The authors declare no conflict of interest.

References

1. FAO. *The Impact of Disasters on Agriculture and Food Security*; FAO: Rome, Italy, 2015; p. 76.
2. Dong, C.; MacDonald, G.; Okin, G.S.; Gillespie, T.W. Quantifying drought sensitivity of mediterranean climate vegetation to recent warming: A case study in Southern California. *Remote Sens.* **2019**, *11*, 2902. [CrossRef]
3. Garreaud, R.D.; Boisier, J.P.; Rondanelli, R.; Montecinos, A.; Sepúlveda, H.H.; Veloso-Aguila, D. The Central Chile Mega Drought (2010–2018): A climate dynamics perspective. *Int. J. Climatol.* **2020**, *40*, 421–439. [CrossRef]
4. Aldunce, P.; Araya, D.; Sapiain, R.; Ramos, I.; Lillo, G.; Urquiza, A.; Garreaud, R. Local perception of drought impacts in a changing climate: The mega-drought in central Chile. *Sustainability* **2017**, *9*, 2053. [CrossRef]
5. Roco, L.; Engler, A.; Bravo-Ureta, B.E.; Jara-Rojas, R. Farmers' perception of climate change in mediterranean Chile. *Reg. Environ. Chang.* **2015**, *15*, 867–879. [CrossRef]
6. Lynch, J.P. Rightsizing root phenotypes for drought resistance. *J. Exp. Bot.* **2018**, *69*, 3279–3292. [CrossRef]
7. Parkash, V.; Singh, S. A review on potential plant-based water stress indicators for vegetable crops. *Sustainability* **2020**, *12*, 3945. [CrossRef]
8. Mashilo, J.; Odindo, A.O.; Shimelis, H.A.; Musenge, P.; Tesfay, S.Z.; Magwaza, L.S. Drought tolerance of selected bottle gourd [*Lagenaria siceraria* (Molina) Standl.] landraces assessed by leaf gas exchange and photosynthetic efficiency. *Plant Physiol. Biochem.* **2017**, *120*, 75–87. [CrossRef]
9. Djanaguiraman, M.; Prasad, P.V.V.; Kumari, J.; Rengel, Z. Root length and root lipid composition contribute to drought tolerance of winter and spring wheat. *Plant Soil.* **2019**, *439*, 57–73. [CrossRef]
10. Khasanova, A.; Lovell, J.T.; Bonnette, J.; Weng, X.; Jenkins, J.; Yoshinaga, Y.; Schmutz, J.; Juenger, T.E. The genetic architecture of shoot and root trait divergence between mesic and xeric ecotypes of a perennial grass. *Front. Plant Sci.* **2019**, *10*, 1–10. [CrossRef]
11. Ahmadi, J.; Pour-Aboughadareh, A.; Fabriki-Ourang, S.; Mehrabi, A.A.; Siddique, K.H.M. Screening wheat germplasm for seedling root architectural traits under contrasting water regimes: Potential sources of variability for drought adaptation. *Arch. Agron. Soil Sci.* **2018**, *64*, 1351–1365. [CrossRef]

12. Cochavi, A.; Rachmilevitch, S.; Bel, G. The effect of irrigation regimes on plum (*Prunus cerasifera*) root system development dynamics. *Plant Biosyst.* **2019**, *153*, 529–537. [CrossRef]

13. Price, A.H.; Steele, K.A.; Gorham, J.; Bridges, J.M.; Moore, B.J.; Evans, J.L.; Richardson, P.; Jones, R.G.W. Upland rice grown in soil-filled chambers and exposed to contrasting water-deficit regimes. I. Root distribution, water use and plant water status. *Field Crop Res.* **2002**, *76*, 11–24. [CrossRef]

14. Polania, J.; Rao, I.M.; Cajiao, C.; Grajales, M.; Rivera, M.; Velasquez, F.; Raatz, B.; Beebe, S.E. Shoot and root traits contribute to drought resistance in recombinant inbred lines of MD 23–24 × SEA 5 of common bean. *Front. Plant Sci.* **2017**, *8*. [CrossRef] [PubMed]

15. Hernandez-Espinoza, L.H.; Barrios-Masias, F.H. Physiological and anatomical changes in tomato roots in response to low water stress. *Sci. Hortic.* **2020**, *265*, 109208. [CrossRef]

16. Di Iorio, A.; Montagnoli, A.; Scippa, G.S.; Chiatante, D. Fine root growth of *Quercus pubescens* seedlings after drought stress and fire disturbance. *Environ. Exp. Bot.* **2011**, *74*, 272–279. [CrossRef]

17. Hund, A.; Ruta, N.; Liedgens, M. Rooting depth and water use efficiency of tropical maize inbred lines, differing in drought tolerance. *Plant Soil* **2009**, *318*, 311–325. [CrossRef]

18. Gao, Y.; Lynch, J.P. Reduced crown root number improves water acquisition under water deficit stress in maize (*Zea mays* L.). *J. Exp. Bot.* **2016**, *67*, 4545–4557. [CrossRef]

19. Sithole, N.; Modi, A.T. Responses of selected bottle gourd [*Lagenaria siceraria* (Molina Standly)] landraces to water stress. *Acta Agric. Scand. Sect. B Soil Plant Sci.* **2015**, *65*, 350–356. [CrossRef]

20. Abenavoli, M.R.; Leone, M.; Sunseri, F.; Bacchi, M.; Sorgonà, A. Root Phenotyping for Drought Tolerance in Bean Landraces from Calabria (Italy). *J. Agron. Crop Sci.* **2016**, *202*, 1–12. [CrossRef]

21. Valladares, F.; Sanchez-Gomez, D.; Zavala, M.A. Quantitative estimation of phenotypic plasticity: Bridging the gap between the evolutionary concept and its ecological applications. *J. Ecol.* **2006**, *94*, 1103–1116. [CrossRef]

22. Marchiori, P.E.R.; Machado, E.C.; Sales, C.R.G.; Espinoza-Núñez, E.; Magalhães Filho, J.R.; Souza, G.M.; Pires, R.C.M.; Ribeiro, R.V. Physiological plasticity is important for maintaining sugarcane growth under water deficit. *Front. Plant Sci.* **2017**, *8*, 1–12. [CrossRef] [PubMed]

23. Valladares, F.; Sánchez-Gómez, D. Ecophysiological traits associated with drought in Mediterranean tree seedlings: Individual responses versus interspecific trends in eleven species. *Plant Biol.* **2006**, *8*, 688–697. [CrossRef] [PubMed]

24. Adu, M.O. Causal shoot and root system traits to variability and plasticity in juvenile cassava (*Manihot esculenta* Crantz) plants in response to reduced soil moisture. *Physiol. Mol. Biol. Plants* **2020**, *26*, 1799–1814. [CrossRef] [PubMed]

25. Romano, A.; Sorgonà, A.; Lupini, A.; Araniti, F.; Stevanato, P.; Cacco, G.; Abenavoli, R.M. Morpho-physiological responses of sugar beet (*Beta vulgaris* L.) genotypes to drought stress. *Acta Physiol. Plant.* **2013**, *35*, 853–865. [CrossRef]

26. Mashilo, J.; Shimelis, H.; Odindo, A. Yield-based selection indices for drought tolerance evaluation in selected bottle gourd [*Lagenaria siceraria* (Molina) Standl.] landraces. *Acta Agric. Scand. Sect. B Soil Plant Sci.* **2017**, *67*. [CrossRef]

27. Adiredjo, A.L.; Navaud, O.; Muños, S.; Langlade, N.B.; Lamaze, T.; Grieu, P. Genetic control of water use efficiency and leaf carbon isotope discrimination in sunflower (*Helianthus annuus* L.) subjected to two drought scenarios. *PLoS ONE* **2014**, *9*, e101218. [CrossRef]

28. Casadebaig, P.; Debaeke, P.; Lecoeur, J. Thresholds for leaf expansion and transpiration response to soil water deficit in a range of sunflower genotypes. *Eur. J. Agron.* **2008**, *28*, 646–654. [CrossRef]

29. Chiatante, D.; Tognetti, R.; Scippa, G.S.; Congiu, T.; Baesso, B.; Terzaghi, M.; Montagnoli, A. Interspecific variation in functional traits of oak seedlings (*Quercus ilex, Quercus trojana, Quercus virgiliana*) grown under artificial drought and fire conditions. *J. Plant Res.* **2015**, *128*, 595–611. [CrossRef]

30. Opazo, I.; Toro, G.; Solis, S.; Salvatierra, A.; Franck, N.; Albornoz, F.; Pimentel, P. Late reduction on transpiration is an important trait for water deficit tolerance in interspecific Prunus rootstock hybrids. *Theor. Exp. Plant Physiol.* **2019**, *31*, 493–506. [CrossRef]

31. Mo, Y.; Yang, R.; Liu, L.; Gu, X.; Yang, X.; Wang, Y.; Zhang, X.; Li, H. Growth, photosynthesis and adaptive responses of wild and domesticated watermelon genotypes to drought stress and subsequent re-watering. *Plant Growth Regul.* **2016**, *79*, 229–241. [CrossRef]

32. Ors, S.; Ekinci, M.; Yildirim, E.; Sahin, U. Changes in gas exchange capacity and selected physiological properties of squash seedlings (*Cucurbita pepo* L.) under well-watered and drought stress conditions. *Arch. Agron. Soil Sci.* **2016**, *62*, 1700–1710. [CrossRef]

33. Cornic, G. Drought stress inhibits photosynthesis by decreasing stomatal aperture—Not by affecting ATP synthesis. *Trends Plant Sci.* **2000**. [CrossRef]

34. Zhang, L.; Zhang, L.; Sun, J.; Zhang, Z.; Ren, H.; Sui, X. Rubisco gene expression and photosynthetic characteristics of cucumber seedlings in response to water deficit. *Sci. Hortic.* **2013**, *161*, 81–87. [CrossRef]

35. Flexas, J.; Bota, J.; Loreto, F.; Cornic, G.; Sharkey, T.D. Diffusive and metabolic limitations to photosynthesis under drought and salinity in C3 plants. *Plant Biol.* **2004**, *6*, 269–279. [CrossRef]

36. Singh, S.K.; Raja Reddy, K. Regulation of photosynthesis, fluorescence, stomatal conductance and water-use efficiency of cowpea (*Vigna unguiculata* [L.] Walp.) under drought. *J. Photochem. Photobiol. B Biol.* **2011**, *105*, 40–50. [CrossRef]

37. De Carvalho, R.C.; Cunha, A.; da Silva, J.M. Photosynthesis by six Portuguese maize cultivars during drought stress and recovery. *Acta Physiol. Plant.* **2011**, *33*, 359–374. [CrossRef]

38. Guan, X.K.; Song, L.; Wang, T.C.; Turner, N.C.; Li, F.M. Effect of Drought on the Gas Exchange, Chlorophyll Fluorescence and Yield of Six Different-Era Spring Wheat Cultivars. *J. Agron. Crop Sci.* **2015**, *201*, 253–266. [CrossRef]

39. El-Sharkawy, M.A.; Cock, J.H. Response of cassava to water stress. *Plant Soil* **1987**, *100*, 345–360. [CrossRef]

40. York, L.M.; Nord, E.A.; Lynch, J.P. Integration of root phenes for soil resource acquisition. *Front. Plant Sci.* **2013**, *4*. [CrossRef]

41. Ritchie, J. Soil water availability. *Plant Soil* **1981**, *58*, 327–338. [CrossRef]

42. Medrano, H.; Tomás, M.; Martorell, S.; Flexas, J.; Hernández, E.; Rosselló, J.; Pou, A.; Escalona, J.-M.; Bota, J. From leaf to whole-plant water use efficiency (WUE) in complex canopies: Limitations of leaf WUE as a selection target. *Crop J.* **2015**, *3*, 220–228. [CrossRef]

43. Zobel, R.W.; Waisel, Y. A plant root system architectural taxonomy: A framework for root nomenclature. *Plant Biosyst.* **2010**, *144*, 507–512. [CrossRef]

44. Johnson, M.G.; Tingey, D.T.; Phillips, D.L.; Storm, M.J. Advancing fine root research with minirhizotrons. *Environ. Exp. Bot.* **2001**, *45*, 263–289. [CrossRef]

45. Shah, T.M.; Imran, M.; Atta, B.M.; Ashraf, M.Y.; Hameed, A.; Waqar, I.; Shafiq, M.; Hussain, H.; Naveed, M.A.; Maqbool, M.A. Selection and screening of drought tolerant high yielding chickpea genotypes based on physio-biochemical indices and multi-environmental yield trials. *BMC Plant Biol.* **2020**, *20*, 1–16. [CrossRef] [PubMed]

Publisher's Note: MDPI stays neutral with regard to jurisdictional claims in published maps and institutional affiliations.

Article

Shoot Characterization of Isoprene and Ocimene-Emitting Transgenic *Arabidopsis* Plants under Contrasting Environmental Conditions

Michele Faralli, Mingai Li and Claudio Varotto *

Department of Biodiversity and Molecular Ecology, Research and Innovation Centre, Fondazione Edmund Mach, via Mach 1, 38010 San Michele all'Adige (TN), Italy; michele.faralli@fmach.it (M.F.); mingai.li@fmach.it (M.L.)
* Correspondence: claudio.varotto@fmach.it

Received: 16 March 2020; Accepted: 6 April 2020; Published: 9 April 2020

Abstract: Isoprenoids are among the most abundant biogenic volatile compounds (VOCs) emitted by plants, and mediate both biotic and abiotic stress responses. Here, we provide for the first time a comparative analysis of transgenic *Arabidopsis* lines constitutively emitting isoprene and ocimene. Transgenic lines and Columbia-0 (Col-0) *Arabidopsis* were characterized under optimal, water stress, and heat stress conditions. Under optimal conditions, the projected leaf area (PLA), relative growth rate, and final dry weight were generally higher in transgenics than Col-0. These traits were associated to a larger photosynthetic capacity and CO_2 assimilation rate at saturating light. Isoprene and ocimene emitters displayed a moderately higher stress tolerance than Col-0, showing higher PLA and gas-exchange traits throughout the experiments. Contrasting behaviors were recorded for the two overexpressors under water stress, with isoprene emitters showing earlier stomatal closure (conservative behavior) than ocimene emitters (non-conservative behavior), which might suggest different induced strategies for water conservation and stress adaptation. Our work indicates that (i) isoprene and ocimene emitters resulted in enhanced PLA and biomass under optimal and control conditions and that (ii) a moderate stress tolerance is induced when isoprene and ocimene are constitutively emitted in *Arabidopsis*, thus providing evidence of their role as a potential preferable trait for crop improvement.

Keywords: isoprene; ocimene; heat stress; water stress

1. Introduction

A large number of plants constitutively emit volatile organic compounds (VOCs) and it has been shown that 36% of the total photosynthetic assimilates produced by terrestrial plants are destined for VOCs' biosynthesis [1]. The involvement of specific VOCs in a wide range of physiological processes has been largely reported [2]. Plant defense against insects, pollinator attraction, plant–plant communication, plant–pathogen interactions, reactive oxygen species scavenging, thermo-tolerance, and environmental stress adaptation are some of the most relevant ecological functions of VOCs [2].

Isoprene (2-methyl-1,3-butadiene, C_5H_8) is the most abundant naturally emitted biogenic VOC [3]. In plants, isoprene biosynthesis is catalyzed in chloroplasts by isoprene synthase (IspS) from dimethylallyl diphosphate anion (DMADP), which is formed by the 2-C-methyl-D-erythritol 4-phosphate (MEP) pathway [4]. Isoprene emission rates depend on the activity of IspS and the pool size of DMADP [5,6], which are in turn influenced by many factors, such as the endogenous developmental stage of a leaf [7,8] and several environmental stimuli and constraints [9,10]. Phylogenetic analyses show that the isoprene biosynthesis capacity was lost in *Glycine max* probably during the domestication process [11] while it was present in ancestral lines, including *Glycine soja*. Monson et al. [12] reported that isoprene emission is likely ancestral within the family Fabaceae, but several independent evolutionary events led to at least 16 losses and 10- gains in the isoprene biosynthesis capacity. The elevated frequency in gaining and losing the trait has been explained by the relatively few mutations necessary to produce or lose the IspS gene coupled with the evidence that isoprene emission is advantageous in a narrow range of environments. Recent phylogenetic reconstruction indicates that *Arundo donax* IspS (AdoIspS) and dicots IspS most likely originated by parallel evolution from Terpene Synthase b (TPS-b) monoterpene synthases, suggesting potentially different physiological roles of the two VOCs (isoprene and ocimene) under environmental stresses [13]. Therefore, understanding how isoprene affects plant growth and physiology and comparing the induced protection under abiotic stresses of isoprene and other monoterpenes will allow determination of whether isoprene emission is a beneficial trait to be reintroduced to plants, especially for the purpose of crop improvement.

Indeed, although the MEP pathway is ubiquitous in plants, only a small portion of plants emit isoprene due to the lack of the *IspS* gene [14]. Since the biosynthesis of isoprene is a cost in terms of carbon [15,16], the great investment of energy into isoprene of some species must have relevant functional reasons. In particular, isoprene and other monoterpenes are believed to play a protective role against thermal and oxidative stresses, possibly because of the capacity of this molecule to stabilize thylakoid membranes [17,18], or to remove reactive oxygen within the mesophyll [19,20]. However, there is evidence that more stable monoterpenoids replace isoprene emission, allowing plant adaptation to more xeric environments, while isoprene emission is maintained in fast-growing plants potentially adapted to a high water availability and subjected to short and moderate stress conditions [21]. For instance, alien species of Hawaii emit more monoterpenes than native ones, and this has been suggested to be an indication of greater evolutionary success of alien species since monoterpene emission is associated with higher stress resistance [22]. More specifically, it was shown that isoprene biosynthesis evolved as an ancestral mechanism in plants to cope with transient oxidative stresses during their water-to-land transition [23]. Indeed, fast-growing hygrophilous *Quercus* species, such as most North American and some European oaks (e.g., *Quercus robur*), emit isoprene, whereas isoprene is replaced by monoterpenes in xeric oaks, such as *Q. ilex* [23,24]. In particular, it has been shown that ocimene is a commonly emitted monoterpene under stress conditions, in particular under heat stress [24]. For instance, leaves of *Quercus ilex* emit high levels of ocimene and the emission is temperature dependent and maximal at 35°C [24]. This suggests that (1) environmental conditions seem to shape isoprenoids' emission capacity and (2) isoprene and ocimene-emitting plants may display different potential responses in stress tolerance.

To compare the role of isoprene and ocimene on environmental stress tolerance, we used some transgenic *Arabidopsis* produced in [13]. Wild-type plants and two transgenic lines per type of emitted VOC were compared in three independent experiments for their growth and stress tolerance by using a series of non-invasive shoot phenotyping techniques. This work provided for the first time a comparative analysis of *Arabidopsis* plants constitutively emitting isoprene and monoterpene regarding their growth under optimal and stress conditions and showed potential contrasting physiological behavior under disadvantageous environments.

2. Materials and Methods

2.1. Plant Materials and Growth Conditions

Arabidopsis thaliana L. ecotype Columbia-0 (Col-0) was used for all experiments as a wild-type control while AdoIspS-44 and AdoIspS-79 lines emitting high levels of isoprene (~300 parts per billion volume) and AdoIspS_m1-8 and AdoIspS_m1-73 lines (F310A mutation: Phenylalanine at position 310 replaced with alanine) emitting high levels of ocimene and small amounts of isoprene, were selected from a previous work [13]. All the transgenic lines overexpress the transgene under the constitutive 35S promoter in the *Arabidopsis* Col-0 background and emit the respective VOCs constitutively. Seeds of wild-type and all other lines were previously harvested from plants grown on pots in a mixture of soil (48%, 48%, and 4% of Flora gard special mixture, Einheits Erde Classic, and perlite, respectively),

23 °C temperature, 75 μmol m^{-2} s^{-1} photosynthetically active radiation (PAR), and 16/8-h light/dark photoperiod. For phenotypic and physiological characterization, seeds were germinated either on agar plates under sterile conditions (Experiment 1) or in pots (Experiment 2 and 3) in similar growing conditions to the plants used for seed collection. After seeding, agar plates or pots were stratified for 3 days in the dark at 4 °C and grown under long day conditions at 23 °C, light intensity of 75 μmol m^{-2} s^{-1} photosynthetically active radiation (PAR), and 50% relative humidity. After the full cotyledons' emergence, seedlings were transplanted in pots containing soil and used for physiological characterization. Details on the growth conditions for each specific experiment are provided in the respective method section below.

2.2. Experimental Design and Stress Application

Three independent experiments were carried out in growth chambers (KBV400, BINDER GmbH, Tuttlingen, Germany). Experiment 1 was a factorial 5 × 2 experiment in a randomized block design with lines (Col-0, AdoIspS-44 and AdoIspS-79 isoprene emitters, AdoIspS_m1-8 and AdoIspS_m1-73 ocimene emitters) and watering regime (well-watered, WW and water stressed, WS) as factors in 10 blocks ($n = 10$). The experiment was set up in two identical growth chambers (KBV400, BINDER GmbH, Tuttlingen, Germany) and pots were placed in trays, with each tray containing 10 pots and treated as a block. Experiment 2 design was equal to Experiment 1, but it consisted of 12 blocks ($n = 12$) and after stress application, plants were subjected to a 7-day recovery period (fully re-watering to WS plants) and subsequent biomass harvesting. In both experiments, germinated seedlings at the two-cotyledon stage were transplanted to plastic 8x8x8 cm pots with a very similar amount of soil (~130 g of a 48%-48%-4% of Flora gard special mixture, Einheits Erde Classic, and perlite, respectively), with two per pot. Plants were subsequently thinned as one per pot according to uniform growth before the stress application. After transplanting, pots were transferred to growth chambers set at 23/22 °C daytime/nighttime temperature, an average 60% relative humidity (i.e., optimal vapor pressure deficit of ~1.1 kPa), and ~80 μmol m^{-2} s^{-1} PAR on average at the rosette level. The photoperiod was 12/12 h day/night in Experiment 1 and 10/14 h day/night in Experiment 2. The shorter photoperiod in Experiment 2 compared to Experiment 1 allowed a higher stress intensity before the onset of reproductive stages (i.e., flowering) owing to the longer vegetative phase and, therefore, total pot water loss. Pots were watered every two days to saturation to avoid soil moisture deficit. Twenty-three days after sowing, the selected pots were subjected to WS by withholding watering in both experiments. The available water content of the pots was expressed as a fraction of the transpirable soil water as FTSW= $(P_g - P_d)$/TTSW, where (i) the total transpirable soil water (TTSW) was the difference between the pot weights at a 100% water holding capacity (WHC) (pot weight ~230 g including the plant and plastic pot) and when the transpiration rate of the stressed plants decreased to 10% of the control plants (~90 g), (ii) P_g was the actual pot weight on a given date, and (iii) P_d was the pot weight at the time when the transpiration rate of the stressed plants was 10% of the control plants (~90 g of pot weight). Pot weight was assessed every day with a balance (Pioneer PA2102C, Ohaus, Parsippany, NJ, USA).

Experiment 3 had a factorial 5 × 2 design in a randomized block with lines (as above) and temperature (control temperature (CT) and heat stress (HS)) as factors in 12 blocks ($n = 12$). For this experiment, the blocks were split in two chambers, one at the CT temperature and one at the HS temperature. Plants were transferred in pots and grown as in Experiment 2. Stress was applied 22 days after sowing and by increasing the temperature to 29/28 °C day/nighttime (standard temperature for the heat stress experiments in *Arabidopsis*, e.g., [25]) in the selected chamber devoted to HS. During the HS application, pots were watered daily to avoid a soil moisture deficit.

2.3. Gravimetric Assessment of Daily Transpiration

In Experiment 1 and 2 (i.e., when WS was applied), pots ($n = 12$ or 10) were weighed daily in the morning and within a 30-min time frame and from the date after treatment application (DAT) 1 on. In Experiment 3, $n = 3$ pots were used to pot FTSW daily and re-watering was carried out to avoid a soil moisture deficit. In Experiment 1 and 2, WW pots were re-watered daily to a target weight reflecting approximately 0.8–0.9 FTSW while no water was added to WS pots. The pot weights P_1 and P_2 of two consecutive days were used to calculate the water use of the plant over 24 h. Since soil evaporation was not minimized, empty pots were placed randomly in the growth chamber and at different FTSW to estimate the average daily evaporation, which was subtracted from the total plant water use and calculated daily plant transpiration (TR, mL day^{-1}).

2.4. Imaging Projected Leaf Area

For all the experiments, the projected leaf area (PLA, cm^2) was taken for all the pots starting from DAT-1 ($n = 10$ or $n = 12$) and every two days after gravimetric assessment. Images were collected with a Samsung Galaxy A20 camera and analyzed with Easy Leaf Area software as described in [26]. Briefly, the selected pot was quickly taken from the growth chamber and placed on a table next to a red calibration area of 4 cm^2 fixed at the top of an 8x8x8 pot (i.e., the distance between the camera and the plant/reference was identical). The picture was taken from the top at a distance of 40 cm and the camera was always positioned parallel to the plant. Image segmentation and PLA measurement was immediately carried out with the Easy Leaf Area free app and the value recorded. The PLAs P_1 and P_2 of two consecutive days were used to calculate the relative shoot growth rate (RGR) (% d^{-1}) according to the equation: RGR = 100 × 1/t × ln (P_2/P_1), where t is the days between P_2 and P_1 (i.e., two days).

2.5. Leaf number, Leaf Emergence Rate, and Phenology

In Experiment 2 and 3, the dynamic of the leaf number was characterized after shoot imaging by visually counting visible leaves ($n = 12$). Subsequently, the leaf emergence rate was calculated as the maximum slope of the linear relationship between the leaf number (LN) and time (t) during the experimental period (i.e., during the linear phase of plant growth). Plants were also visually inspected for phenological stages according to [27] and the date of inflorescence emergence (GS 5.10) and first flower opening (GS 6.0) were recorded.

2.6. Gas-Exchange Measurements

Gas-exchange measurements ($n = 4$ to 5) were performed for Experiment 1, 2, and 3 with a Li-Cor 6400 (Li-Cor, Lincoln, NE, USA) using an integrated fluorescence leaf cuvette (LI-6400-40; Li-Cor) between 0900 and 1400. To minimize the potential leaf position and developmental stage effects, all the gas-exchange measurements were taken on the sixth fully expanded leaf of four to five randomly selected plants for each treatment. When needed, the leaf area was recalculated by imaging the portion used for gas exchange. In the Li-Cor cuvette, all the parameters (leaf CO_2 assimilation at saturating light, A; and stomatal conductance, g_s) were collected at 400 ppm CO_2. Leaf temperature was maintained at 23 °C, a VPD between 0.9 and 1.3 kPa, and PAR was 600 µmol m^{-2} s^{-1} (saturating PAR for *Arabidopsis* previously evaluated by light curves (Figure S1)), with a 10:90 blue:red light and a flow rate of 400 µmol s^{-1}. In Experiment 3, the block temperature was maintained either at 23 or 29 °C depending on the plant treatment (i.e., CT or HS plants). In Experiment 1, data were collected at DAT 15 (i.e., mild water stress); in Experiment 2, at DAT 22 (severe water stress); and in Experiment 3, at DAT 18.

2.7. A/Ci Analysis

During Experiment 2 and 3, control plants were used for further gas-exchange characterization in both Col-0 wild-type and transgenic lines ($n = 4$). Measurements of the response of A to sub-stomatal CO_2 concentrations (C_i) were performed using a Li-Cor 6400 (Li-Cor, Lincoln, NE, USA) and a 2-cm^2 leaf cuvette with an integral blue–red light-emitting diode (LED) light source as described above. Cuvette conditions were maintained as described in the previous section for WS experiments. When steady-state conditions were achieved, the CO_2 concentration was sequentially decreased to 300, 200, 150, 75, and 50 µmol mol^{-1} before returning to the initial concentration of 400 µmol mol^{-1}. This was followed by a sequential increase to 500, 700, 900, 1100, 1300, and 1500 µmol mol^{-1}. Readings were recorded when A had stabilized to the new conditions. The maximum velocity of Rubisco for carboxylation (V_{cmax}) and the maximum rate of electron transport demand for Ribulose 1,5-bisphosphate (RuBP) regeneration (J_{max}) were derived by curve fitting, as described by Sharkey et al. [16].

2.8. Final Biomass Assessment

In Experiment 2 and 3, on DAT 28 and 26, respectively, the shoot biomass was destructively assessed by harvesting the plants (n = 12). The fresh weight (FW, g) of the shoot was recorded immediately after harvest with a precision balance and samples were immediately placed inside an oven (BD115, BINDER GmbH, Tuttlingen, Germany) at 60 °C for four days. Shoot dry weight (DW, g) was then recorded by weighing the dried samples.

2.9. Statistical Analysis

Statistical analyses were conducted using STATISTICA 13th (Dell Software) and RStudio (R Core Team, 2017). Randomization and experimental design were produced in RStudio with the *Agricolae* package. All the data were subjected to repeated measurement and subsequent two-way analysis of variance (ANOVA) for each DAT when line and stress factors were present (e.g., DW, PLA). Single-factor analysis was carried out with one-way ANOVA. Shapiro–Wilk and Levene's tests were used to test data for normality and homogeneity of variance, respectively. Fisher's least significant difference test was used for multiple comparisons. Estimation of TR$_{break}$ was carried out with segmented regression by plotting TR and FTSW and estimated as the intersection between the two linear segments as in Faralli et al. [28].

3. Results

3.1. Stress Application and Phenology

In this work, we aimed to characterize the transgenic lines under semi-realistic environmental conditions and several stressors. In Experiment 1, WS was slowly (slope linear fitting −0.053) applied for 16 days until FTSW was 0.14 on average, therefore mimicking a relatively mild WS environment (Figure 1A). Conversely, in Experiment 2, a relatively more severe stress was applied (FTSW below 0.1 for four days) followed by a re-watering period to saturation (FTSW 0.8) (Figure 1B). In Experiment 3, HS was applied by increasing the chamber temperature to 29 °C while maintaining saturating levels of the soil moisture to avoid confounding factors from WS (Figure 1C). Phenological assessments suggest that WS did not trigger an escape strategy in Experiment 2, at least at the stress conditions applied (p = 0.613), with similar days to bolting between WW and WS plants. Similarly, no significant differences were detected between lines in both WW and WS conditions (p = 0.617) (Figure 2A,B). The leaf emergence rate was around 0.85 leaf day^{-1} on average under WW conditions and this was not significantly different between lines (p = 0.086). Under WS, the leaf emergence rate was significantly (p < 0.001) lower than WW conditions for all the lines (Figure 2C,D). HS application significantly reduced the days for bolting compared with CT conditions (p < 0.001) (Figure 2E,F) and decreased the leaf emergence rate (p < 0.001) (Figure 2G,H), while no significant differences were found between lines (p = 0.169 and p = 0.621).

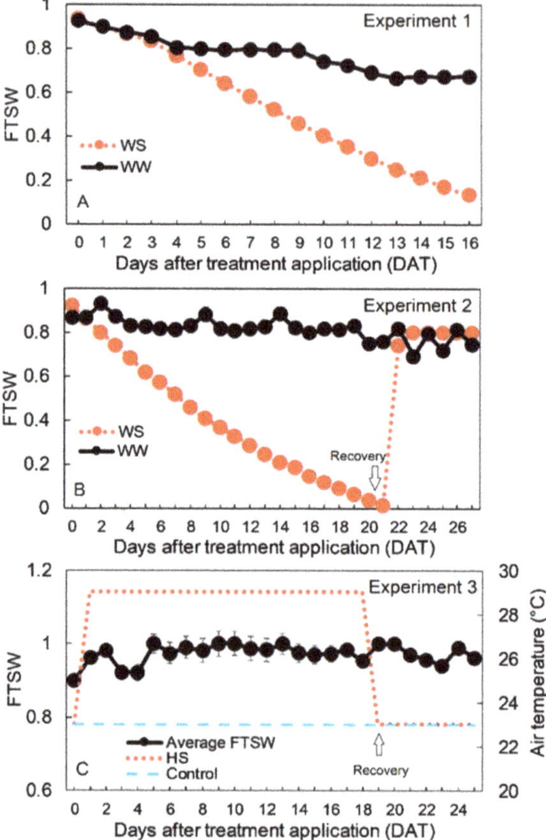

Figure 1. Environmental conditions for Experiment 1, 2, and 3 (**A–C**, respectively). In **A**, black dots represent fraction of transpirable soil water (FTSW) of well-watered (WW) plants while red dots represent FTSW for water stress (WS) plants over a 16-day experimental period ($n = 50$). In **B**, black dots represent FTSW of WW plants while red dots represent FTSW for WS plants over a 27-day experimental period ($n = 60$). From days after treatment application (DAT) 22, WS plants were subjected to a recovery period. In **C**, black dots represent the average FTSW evaluated on three average pots per treatment ($n = 30$), the red dotted line represents the average day-time air temperature of the heat stress (HS) chamber and the blue line represents the average day-time temperature of the control temperature (CT) chamber.

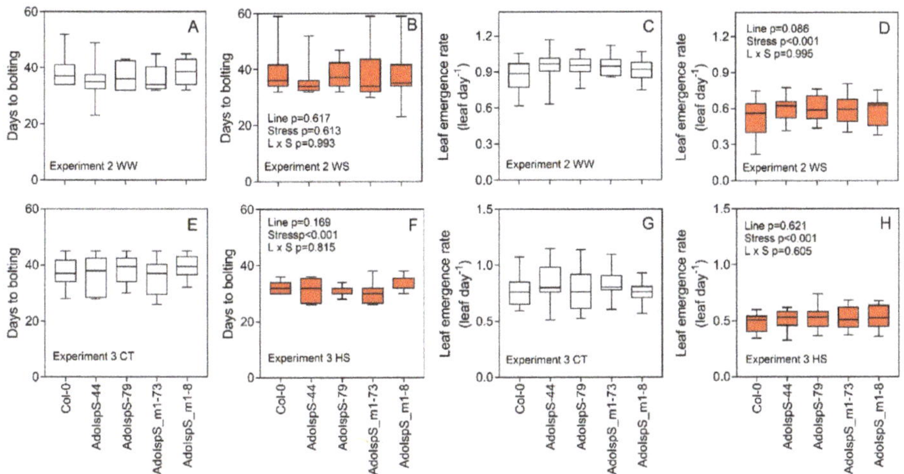

Figure 2. Days to bolting and the leaf emergence rate assessed in Experiment 2 (**A–D**) and 3 (**E–H**). For all the graphs, black bars represent the control (either WW or CT) conditions, whereas red bars represent plants subjected to stress (either WS or HS) (*n* = 12). Data were analyzed with two-way ANOVA and the output is included in the graph. Means separation was carried out with Fisher's test (since no differences are present between lines, the respective letters were omitted for simplicity).

3.2. Shoot Biomass Assessment

Under WW conditions and in Experiment 2 (50 days after seeding), Col-0 showed lower shoot dry weight biomass than AdoIspS-44 and AdoIspS_m1-8 (*p* = 0.036) (Figure 3A). WS conditions significantly (*p* < 0.001) reduced the shoot dry weight compared with the WW plants on average and for all the lines. No statistically significant differences were found between lines under WS conditions (Figure 3B). In Experiment 3 (45 days after seeding), CT Col-0 showed a reduced shoot dry weight compared to AdoIspS-44 and AdoIspS_m1-73 (*p* = 0.006) (Figure 3C). HS significantly (*p* < 0.001) reduced the shoot dry weight biomass for all the lines and by ~45% on average. When compared with Col-0 under HS, the transgenic lines showed a significantly higher dry weight biomass (*p* = 0.006) (Figure 3D)

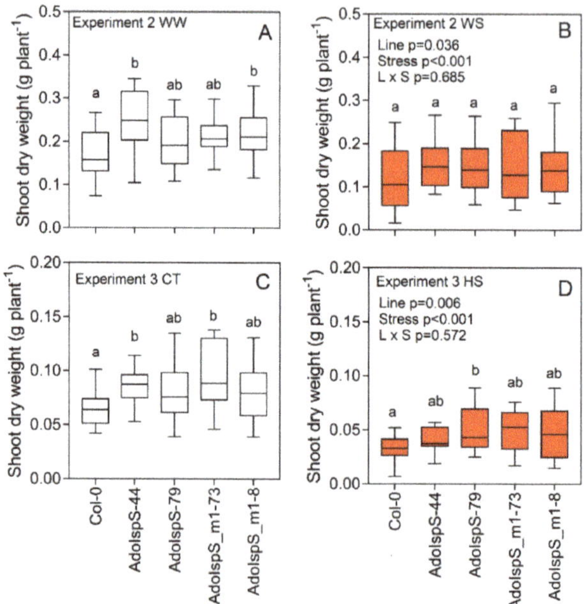

Figure 3. Shoot dry weight for Col-0, isoprene emitters (AdoIspS-44 and AdoIspS-79), and ocimene emitters (AdoIspS_m1-73 and AdoIspS_m1-8) *Arabidopsis* lines grown under control and stress conditions. In Experiment 2 (**A,B**), plants were grown under control (**A**) and water stress conditions (**B**) while in Experiment 3 (**C,D**), plants were grown under control (**C**, 23 °C temperature) and heat stress conditions (**D**, 29 °C temperature) (*n* = 12). The two-way ANOVA output is shown in the graph. Different letters represent significant differences according to Fisher's test.

3.3. Water-Use Strategies under Reduced Water Availability

Under WS conditions, the reduction in transpiration for Col-0 started at FTSW of 0.35 and 0.38 for Experiment 1 and 2, respectively (Figure 4A,B). When compared with Col-0, isoprene-emitting lines (i.e., AdoIspS-44 and 79) displayed a more pronounced water conservation strategy, with a TR_{break} ranging between 0.43 and 0.47 FTSW in both experiments ($p < 0.001$). Conversely, ocimene-emitting plants showed reduced transpiration at lower FTSW compared with both Col-0 and AdoIspS lines, with a TR_{break} between 0.28 and 0.31 in both experiments ($p < 0.001$).

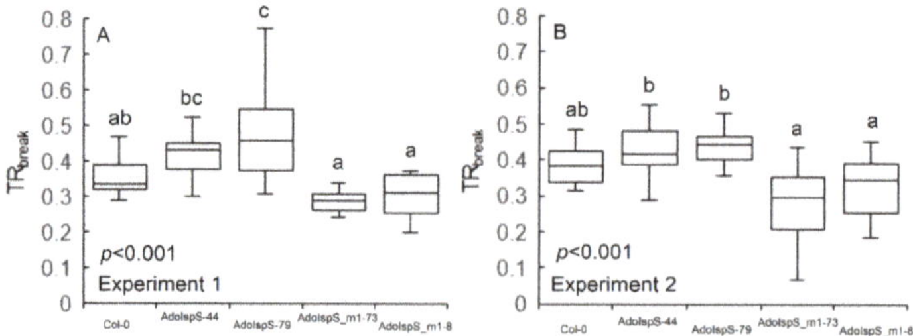

Figure 4. Breakpoint of plant transpiration to reduced soil water availability (TR_{break}). Data were gravimetrically collected in Experiment 1 (**A**) and 2 (**B**) and daily plant transpiration was plotted against FTSW curves and subjected to segmented regression (*n* = 10 in A and *n* = 12 in B). Data were analyzed with one-way ANOVA and different letters represent significant differences according to Fisher's test.

3.4. Projected Leaf Area and Relative Growth Rate

Correlations between PLA and shoot dry weight ($p < 0.001$, $R^2 = 0.84$, data not shown) confirmed the reliability of the dynamic estimation of the leaf area over the experimental periods. For all the experiments under WW or CT conditions, all the transgenic lines showed larger PLA compared with the wild-type Col-0 (Figure 5A,E,I). The higher PLA relative to Col-0 was statistically significant by up to 30% for AdoIspS_m1-73 in Experiment 1 (DAT 2-6), by up to 20%–25% for AdoIspS-44 in Experiment 2 (DAT 21-28), and by up to 20% in Experiment 3 (DAT 6-12). The limitation of water availability (i.e., Experiment 1 and 2 Figure 5B,F) caused a pronounced growth retardation for all the lines (from DAT 14 both experiments), as shown by evident reductions in PLA and RGR. However, while under moderate water stress (Experiment 1), no significant differences were observed among the lines, and significantly higher PLA values for the emitters were found in Experiment 2 (severe water stress) when compared with Col-0 and during the recovery period. In particular, AdoIspS-44 and AdoIspS_m1-8 showed significantly higher PLA than Col-0 in Experiment 2, with a much sharper recovery period compared with the wild type. HS severely affected plants' growth, with a significant reduction in PLA and RGR since DAT 6 compared with WW conditions for the lines tested. Significantly lower PLA values for Col-0 were recorded compared with emitters (e.g., AdoIspS_m1-8 DAT 25, AdoIspS_m1-73 and AdoIspS-79 DAT 21, AdoIspS_m1-73, and AdoIspS-44 DAT 8).

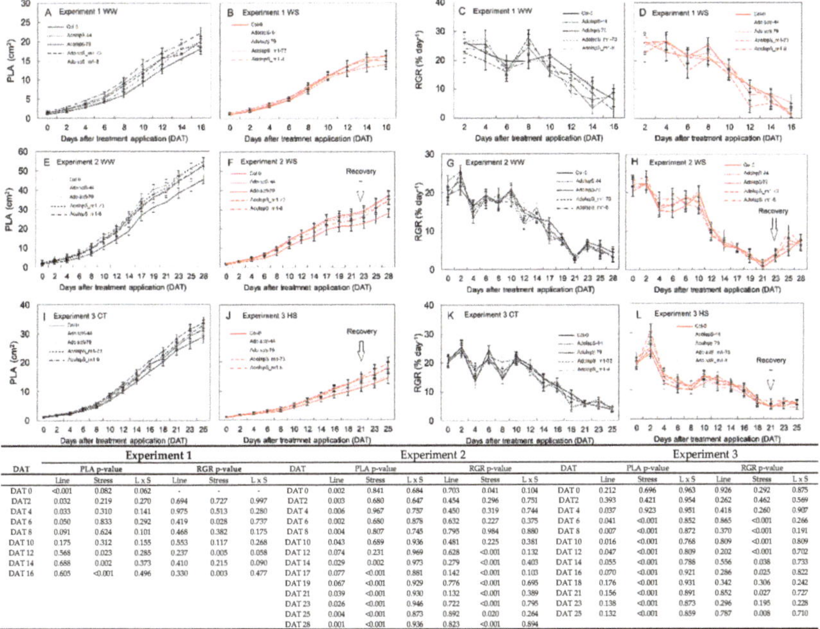

	Experiment 1							Experiment 2							Experiment 3					
DAT	PLA p-value			RGR p-value			DAT	PLA p-value			RGR p-value			DAT	PLA p-value			RGR p-value		
	Line	Stress	L x S	Line	Stress	L x S		Line	Stress	L x S	Line	Stress	L x S		Line	Stress	L x S	Line	Stress	L x S
DAT 0	<0.001	0.082	0.062	·	·	·	DAT 0	0.002	0.841	0.684	0.703	0.041	0.104	DAT 0	0.212	0.696	0.963	0.926	0.292	0.875
DAT2	0.032	0.219	0.270	0.694	0.727	0.997	DAT2	0.003	0.680	0.647	0.454	0.296	0.751	DAT2	0.393	0.421	0.954	0.262	0.462	0.569
DAT 4	0.033	0.310	0.141	0.975	0.513	0.280	DAT 4	0.006	0.967	0.737	0.450	0.319	0.744	DAT 4	0.037	0.923	0.951	0.418	0.260	0.907
DAT 6	0.050	0.833	0.292	0.419	0.028	0.737	DAT 6	0.002	0.680	0.878	0.632	0.227	0.375	DAT 6	0.041	<0.001	0.852	0.865	<0.001	0.266
DAT 8	0.091	0.624	0.101	0.468	0.382	0.175	DAT 8	0.004	0.807	0.745	0.795	0.984	0.880	DAT 8	0.007	<0.001	0.872	0.370	<0.001	0.191
DAT 10	0.175	0.312	0.155	0.553	0.117	0.268	DAT 10	0.043	0.689	0.936	0.481	0.225	0.381	DAT 10	0.016	<0.001	0.768	0.809	<0.001	0.809
DAT 12	0.568	0.023	0.285	0.237	0.005	0.058	DAT 12	0.074	0.231	0.969	0.628	<0.001	0.132	DAT 12	0.047	<0.001	0.809	0.202	<0.001	0.702
DAT 14	0.688	0.002	0.373	0.410	0.215	0.090	DAT 14	0.029	0.002	0.973	0.279	<0.001	0.403	DAT 14	0.055	<0.001	0.788	0.556	0.038	0.733
DAT 16	0.605	<0.001	0.496	0.330	0.003	0.477	DAT 17	0.077	<0.001	0.881	0.142	<0.001	0.103	DAT 16	0.070	<0.001	0.921	0.286	0.025	0.822
							DAT 19	0.067	<0.001	0.929	0.776	<0.001	0.695	DAT 18	0.176	<0.001	0.931	0.342	0.306	0.242
							DAT 21	0.039	<0.001	0.930	0.132	<0.001	0.389	DAT 21	0.156	<0.001	0.891	0.852	0.027	0.727
							DAT 23	0.026	<0.001	0.946	0.722	<0.001	0.795	DAT 23	0.138	<0.001	0.873	0.296	0.195	0.228
							DAT 25	0.004	<0.001	0.873	0.892	0.020	0.264	DAT 25	0.132	<0.001	0.859	0.787	0.008	0.710
							DAT 28	0.001	<0.001	0.936	0.823	<0.001	0.894							

Figure 5. Projected leaf area (PLA) and relative growth rate (RGR) of Col-0 and isoprene- or ocimene-emitting lines. Data were collected over three experiments and during the entire experimental treatment. In (**A,E,I**), continuous assessment of PLA for control plants (WW in Experiment 1 and 2 or CT in Experiment 3) and in (**C,G,K**), the calculated RGR is shown. In (**B,F,J**), continuous assessment of PLA for stressed plants (WS in Experiment 1 and 2 or HS in Experiment 3) and in (**D,H,L**), the calculated RGR is shown. Values are means ± standard error of the means ($n = 12$ while $n = 10$ in Experiment 1). Data were subjected to repeated measurements analysis ($p < 0.001$) and two-way ANOVA and the output is shown in the table for each experiment. A multiple comparisons test (Fisher's test) was carried out for each day after treatment (DAT) and is shown in Supplementary 2 for simplicity.

3.5. Gas-Exchange, A/Ci Analysis

Saturating A for Col-0 was ~8 µmol m^{-2} s^{-1} on average, which was significantly ($p < 0.001$ and $p = 0.003$, respectively, Figure 6A,E,I) lower than transgenic lines in Experiment 1 and 3 under optimal conditions (WW and CT). Indeed, the A/C_i analysis supports the in situ gas-exchange measurements, with both isoprene- and ocimene-emitting lines displaying higher A ($p = 0.045$) and J_{max} ($p = 0.041$) while no significant differences were found for V_{cmax} (Table 1). As expected, WS severely reduced A and g_s and the reduction was very similar for all the lines ($p < 0.001$) (Figure 6B,F). However, in some lines (e.g., AdoIspS-44 exp 1 and ISPs 79 exp 2), higher A values compared with Col-0 were present. Under HS, g_s was not negatively affected for all the lines, and in some cases (e.g., AdoIspS-79 and AdoIspS_m1-73), higher g_s values were obtained compared with the CT (Figure 6N). On the contrary, HS reduced A by 20% in Col-0 ($p = 0.016$) while no significant differences compared with the WW control were recorded for ocimene emitters and AdoIspS-79 (Figure 6L).

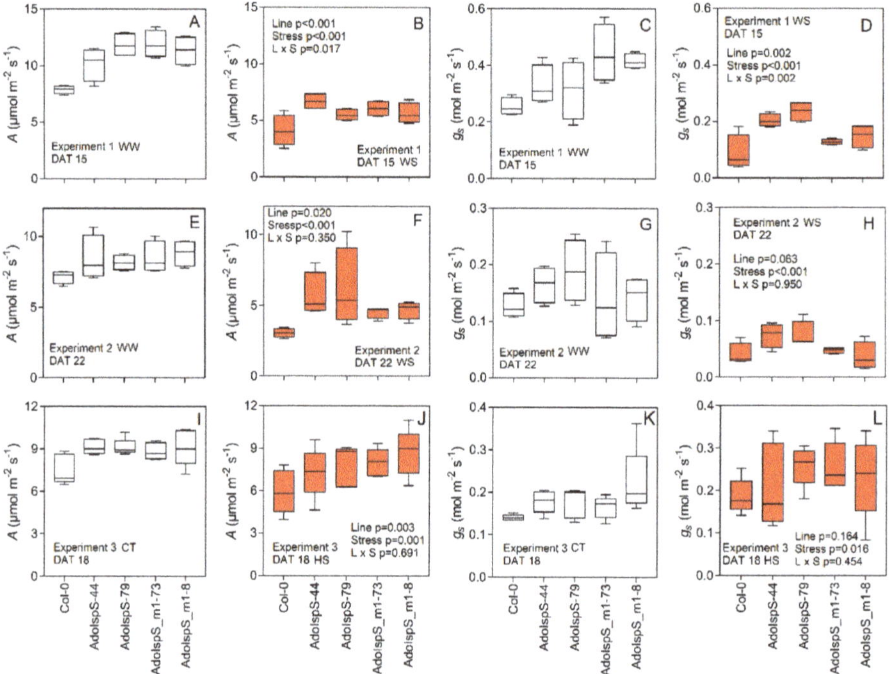

Figure 6. CO$_2$ assimilation rate (A) and stomatal conductance (g_s) collected in Experiment 1 (**A–D**), Experiment 2 (**E–H**), and Experiment 3 (**I–L**). Values are means ($n = 4$ to 5) and the two-way ANOVA output is shown in each graph. Measurements ($n = 4$ to 5) were performed at 600 µmol m^{-2} s^{-1} PAR, 23 °C leaf temperature, and 400 µmol mol^{-1} [CO$_2$] in Experiment 1 and 2 (water stress experiments) while in Experiment 3 (heat stress experiment), the block temperature was 23 and 29 °C for CT and HS, respectively.

Table 1. Photosynthetic to sub-stomatal CO_2 concentration response curve (A/C_i) output for the wild-type Col-0 and transgenic lines. Data were collected over Experiments 2 and 3 on control plants with a Licor 6400XT. Parameter estimation was carried out as described by [16]. Values are means $\pm$ standard error of the means (n = 4 to 5) and analyzed with one-way ANOVA while means separation was carried out by Fisher's test.

Line	A	V_{cmax}	J_{max}
Col-0	8.0 [a]	45.2	94.4 [a]
AdoIspS-79	9.4 [b]	49.1	102.3 [a,b]
AdoIspS-44	9.6 [b]	56.6	119.6 [b]
AdoIspS_m1-8	9.8 [b]	55.9	114.5 [b]
AdoIspS_m1-73	10.2 [b]	53.5	114.4 [b]
		p-value	
	0.045	0.321	0.041

4. Discussion

Transgenic approaches have been largely used to engineer isoprene emission in non-emitter species [13,19,25,29,30]. This has led to a large amount of information regarding the role of isoprene in plant growth, stress tolerance, and signaling. As of today, however, much less effort has been devoted to the comparative dissection of the differences between the biological functions of isoprene and monoterpenes. In this work, *Arabidopsis* plants transformed to emit constitutively isoprene or ocimene were compared for the first time and a comprehensive shoot characterization was carried out in order to assess their potential role on stress tolerance and plant growth. The comparative approach used in our study has two major advantages with respect to similar studies carried out in the past in different species [31]. First, it compared in the same genetic background the physiological effects of isoprene and ocimene emission, thus normalizing the effect of the starting pool of metabolites, which is known to vary among species and affect emissions [32,33]. Second, it employed two enzymes differing by only one amino acid [13], thus minimizing among transgenic emitters any confounding effects on plant growth due to the protein length or translational efficiency. Leveraging on similar emission levels from selected transgenic lines, in this study, we thus characterized the physiological effects of isoprene and ocimene under optimal conditions as well as two main abiotic stresses, temperature excess and water limitation.

4.1. Hemi- and Mono-Terpene Emission Improves Plant Growth under Optimal Conditions

Under optimal conditions, isoprenoids' emission is a metabolically expensive trait, with high energy and photosynthetic carbon requirements [15]. However, independent studies demonstrated that the emission of isoprenoids led to an increased biomass, leaf area, and pigment content in several species (e.g., [25,34]), which is consistent with our data, suggesting the existence of a tight but complex relationship between isoprene/monoterpene emission and growth. In Zuo et al. [34], *Arabidopsis* plants transformed with a *Eucalyptus globulus IspS* gene had a higher leaf area, leaf number, and final dry weight than the wild-type Col-0, consistent with our data on both isoprene and ocimene emitters. Similarly, in Loivamäki et al. [25], *Arabidopsis* lines transformed with an *IspS* gene from gray poplar had higher growth rates under optimal growth conditions. The role of isoprene as a signaling molecule has recently been shown, with a significant upregulation in the expression of genes belonging to signaling networks or associated with specific growth regulators (e.g., gibberellic acid, cytokinins, and jasmonic acid) in *Arabidopsis* engineered to emit isoprene [34]. In particular, greater accumulation of gibberellic acid, potentially through an enhanced expression level of genes encoding for zinc fingers proteins (e.g., *TZF5*), has been suggested as a potential explanation of these phenotypes with an enhanced leaf area. Additionally, isoprene appeared to enhance cytokinin levels mainly through changes to the expression levels of genes associated to cytokinin signaling. In our work, isoprene-emitting lines showed a higher PLA and final dry weight than Col-0 under optimal conditions, corroborating the hypothesis that isoprene and monoterpene emission might be involved in enhancing or modulating the gene network and signaling of plant growth.

An interesting output of our work is the enhanced photosynthetic capacity and CO_2 assimilation per unit of leaf area (A) in emitters compared with Col-0 under optimal conditions. It was previously reported that isoprene and monoterpene emission might increase the chlorophyll content in leaves, and potentially enhance A [25,34]. This increase in A can also partially explain the higher PLA and biomass of the emitters compared with Col-0, suggesting a higher carbon availability that can sustain growth. Intraspecific variation within the *Arundo* tribe and some dicots for A and isoprene emission revealed a positive and significant correlation between isoprene emission and photosynthesis [7,35]. Morfopoulos et al. [35] proposed a mechanism by which the isoprene emission rate is directly proportional to the excess of reducing power (nicotinamide adenine dinucleotide phosphate, NADPH) generated by the linear electron flow and unused by photosynthesis. Indeed, in our experiment, J_{max} was significantly higher in emitters than Col-0, suggesting that a fraction of the total electron flux generated by the photosystem II might be allocated to isoprenoid biosynthesis.

4.2. Isoprene and Ocimene Emission Resulted in a Moderate Tolerance to Environmental Stresses

In our work, albeit at the boundaries of significance and partly inconsistent between lines transformed with the same construct, both isoprene and ocimene emission resulted in marginally higher A and PLA under water and heat stress than Col-0. Conversely, significant positive effects were recorded for dry weight under heat stress only. Indeed, our data are in line with most of the literature showing the efficacy of hemi- and monoterpenes at protecting the photosynthetic apparatus under high temperatures [25,30,34]. Somehow surprisingly, however, these results indicate relatively minor phenotypic variations consequent to VOCs' emission under the stress conditions tested. Since the heat stress regime applied in this work was milder and indubitably closer to physiological conditions than in some previous reports [30], further characterization is needed to better evaluate potentially different degrees of responses under broader environmental conditions.

However, isoprene and ocimene emitters showed an opposite behavior concerning their water-use under reduced water availability. Ocimene emitters reduced their transpiration at a very low value of FTSW, suggesting a non-conservative water-use behavior. On the contrary, isoprene emitters showed a highly conservative water-use strategy, with early stomatal closure and an elevated sensitivity of transpiration to soil drying. While dryland agriculture might benefit from conservative genotypes [28,36], a non-conservative strategy is advantageous for maximizing nutrient capture and for successful colonization of dry habitats with extreme fluctuations in resource availability [37]. Under short resource fluctuations, fast nutrient and water uptake can take over resource utilization by slower neighbors, thus providing a competitive advantage in disadvantageous epiphytic habitats [37,38]. This might indicate why monoterpene-emitting species are more common in xeric habitat than isoprene-emitting species. It was already shown that hygrophilic isoprene emitters showed elevated stomatal sensitivity to soil water stress, mainly to avoid tissue dehydration [39,40], which is consistent with our work. We speculate that the conservative behavior of the isoprene emitters analyzed in this study might also suggest a strategy to increase internal the isoprene concentration (owing to its high volatility) under stress conditions to enhance its potential beneficial effect, which is minimized under low concentrations [41]. The two contrasting strategies therefore, although they did not produce a higher dry weight biomass than Col-0 under water stress, led, for different reasons, to similar phenotypes between the two constitutively emitting lines. Further investigation on this is required in order to understand the physiological basis of this behavior and exploit the potential advantages of these responses under different magnitudes of water stress.

4.3. Agricultural and Evolutionary Relevance

The role of isoprenoids in plant defense strategies against biotic and abiotic stresses and their potential applications to agriculture are increasingly being appreciated [42]. Our results highlight both similarities and differences in abiotic stress tolerance among isoprene and ocimene emitters, which on the one hand determined their evolution in natural environments and on the other hand will affect the possibility of applying them to agricultural settings. Agro-ecosystems, in fact, represent simplified environments in which human beings buffer environmental conditions to provide steady and sufficient amounts of water, light, and nutrients to crops [43]. The consistently higher PLA and *A* we observed for the transgenic lines compared with Col-0 under optimal conditions recorded in this work suggest that enhanced growth is a resulting phenotype in isoprenoid-emitting plants (supported by other literature, e.g., [34]) and might be a preferable trait, at least in biomass crops. In general, irrespective of the VOC emitted, heat tolerance was generally enhanced compared with Col-0. This is relevant considering that in the future, climate change will increase the frequency of extreme weather events [44], and further suggests that enhanced isoprenoids' emission could be a viable strategy to be used in crop improvement. For instance, under stressful conditions, the induced stress tolerance (e.g., reactive oxygen species scavenging, membrane stability, gas exchange and dry weight maintenance) and the synergistic effects between isoprenoids, secondary metabolites (e.g., carotenoids), and hormones (i.e., cytokinins) [42] is of major interest, in particular to assess whether a potential delayed in senescence is induced in isoprenoid-emitting plants under thermal stress (a favorable trait in several crops, e.g., cereals [45]). However, which terpene should be better suited to this task is still a matter of debate, as the evidence in favor of either of them is still fragmentary and partly conflicting. The presence of isoprene emission in wild soybean (*Glycine soja*) and its lack in cultivated soybean (*Glycine max*) suggests that isoprene emission in *Fabaceae* could have been counter-selected during the domestication process in favor of monoterpene emission [46]. Given the naturally occurring multiple losses and gains of isoprene emission during the course of evolution in *Fabaceae* [12], however, it is still doubtful whether IspS pseudogenization in cultivated soybean is simply a by-product or the result of domestication analogously to the loss of resistance toward pathogens observed in several other crops [47,48]. On the other hand, our results indicate that isoprene emission could be preferable over ocimene emission in the long run under the current climate change scenario, as it provides conservative water-use behavior and thus potentially higher sustainability over time [49]. From a broader evolutionary perspective, the increased sensitivity to dehydration of isoprene-emitting plants compared with ocimene emitters provides a rationale for the observation that isoprene is usually associated to a perennial lifestyle, where dehydration avoidance rather than drought escape is advantageous [50]. These results are in line with previous work suggesting that while isoprene evolved in plants adapted to high water availability and subjected to short stresses, it was replaced by monoterpenes or more stable isoprenoids in xeric environments [21].

5. Conclusions

To our knowledge, this is the first study where a comprehensive characterization of *Arabidopsis* lines constitutively emitting isoprene and ocimene has been carried out. Our data support the most recent literature on hemi- and monoterpene plant biosynthesis suggesting a positive effect of the emission on both growth and stress tolerance and corroborating the idea of their potential usefulness in crop improvement. Given the differences found in water-use strategies followed by contrasting stomatal sensitivity to water limitation, the potential application of isoprene emission for perennial crops and of monoterpene emission for annual crops will need to be assessed further. In particular, the dissection of the possible differences in the signaling cascade in isoprene and monoterpene emitters by transcriptomic approaches holds the promise to improve our understanding of the role of VOCs as signaling molecules for stress priming in plants.

Supplementary Materials: The following are available online at http://www.mdpi.com/2223-7747/9/4/477/s1, Figure S1: Curve of *A* to photosynthetic active radiation ($n = 4$) carried out in Col-0. Photosynthesis was fully saturated at 600 μmol m^{-2} s^{-1} PAR, Table S1: Details of statistical tests related to PLA and RGR.

Author Contributions: Conceptualization, M.L. and C.V.; Methodology, M.F., M.L. and C.V.; Software, M.F.; Validation, M.F., M.L. and C.V.; Formal Analysis, M.F.; Resources, M.L.; Data Curation, M.F.; Writing—Original

Draft Preparation, M.F.; Writing—Review and Editing, M.L. and C.V.; Supervision, C.V.; Project Administration, C.V.; Funding Acquisition, C.V. All authors have read and agreed to the published version of the manuscript.

Acknowledgments: This work was supported by the Autonomous Province of Trento through core funding. The authors would like to thank Damiano Gianelle for the use of the Licor 6400XT instrument.

Conflicts of Interest: The authors declare no conflict of interest.

References

1. Kesselmeier, J.; Ciccioli, P.; Kühn, U.; Stefani, P.; Biesenthal, T.; Rottenberger, S.; Wolf, A.; Vitullo, M.; Valentini, R.; Nobre, A.; et al. Volatile organic compound emissions in relation to plant carbon fixation and the terrestrial carbon budget. *Glob. Biogeochem. Cycles* **2002**, *16*, 73-1. [CrossRef]

2. Spinelli, F.; Cellini, A.; Marchetti, L.; Mudigere, K.; Piovene, C. Emission and function of volatile organic compounds in response to abiotic stress. In *Abiotic Stress in Plants—Mechanisms and Adaptations*; IntechOpen: London, UK, 2011; pp. 367–394.

3. Guenther, A.; Karl, T.; Harley, P.; Wiedinmyer, C.; Palmer, P.I.; Geron, C. Estimates of global terrestrial isoprene emissions using MEGAN (Model of Emissions of Gases and Aerosols from Nature). *Atmos. Chem. Phys. Discuss.* **2006**, *6*, 107–173. [CrossRef]

4. Schwender, J.; Zeidler, J.; Gröner, R.; Müller, C.; Focke, M.; Braun, S.; Lichtenthaler, F.W.; Lichtenthaler, H.K. Incorporation of 1-deoxy-D-xylulose into isoprene and phytol by higher plants and algae. *FEBS Lett.* **1997**, *414*, 129–134. [CrossRef]

5. Wiberley, A.E.; Donohue, A.R.; Westphal, M.M.; Sharkey, T.D. Regulation of isoprene emission from poplar leaves throughout a day. *Plant Cell Environ.* **2009**, *32*, 939–947. [CrossRef] [PubMed]

6. Rasulov, B.; Hüve, K.; Bichele, I.; Laisk, A.; Niinemets, Ü. Temperature response of isoprene emission in vivo reflects a combined effect of substrate limitations and isoprene synthase activity: A kinetic analysis1. *Plant Physiol.* **2010**, *154*, 1558–1570. [CrossRef]

7. Ahrar, M.; Doneva, D.; Koleva, D.; Romano, A.; Rodeghiero, M.; Tsonev, T.; Biasioli, F.; Stefanova, M.; Peeva, V.; Wohlfahrt, G.; et al. Isoprene emission in the monocot Arundineae tribe in relation to functional and structural organization of the photosynthetic apparatus. *Environ. Exp. Bot.* **2015**, *119*, 87–95. [CrossRef]

8. Niinemets, Ü.; Sun, Z.; Talts, E. Controls of the quantum yield and saturation light of isoprene emission in different-aged aspen leaves. *Plant Cell Environ.* **2015**, *38*, 2707–2720. [CrossRef]

9. Loreto, F.; Schnitzler, J.-P.; Schnitzler, J.-P. Abiotic stresses and induced BVOCs. *Trends Plant Sci.* **2010**, *15*, 154–166. [CrossRef]

10. Sharkey, T.D.; Monson, R.K. Isoprene research—60 years later, the biology is still enigmatic. *Plant Cell Environ.* **2017**, *40*, 1671–1678. [CrossRef]

11. Sharkey, T.D.; Gray, D.W.; Pell, H.K.; Breneman, S.R.; Topper, L. Isoprene synthase genes form a monophyletic clade of acyclic terpene synthases in the tps-b terpene synthase family. *Evolution* **2012**, *67*, 1026–1040. [CrossRef]

12. Monson, R.K.; Jones, R.T.; Rosenstiel, T.N.; Schnitzler, J.-P.; Schnitzler, J.-P. Why only some plants emit isoprene. *Plant Cell Environ.* **2012**, *36*, 503–516. [CrossRef] [PubMed]

13. Li, M.; Xu, J.; Alarcon, A.A.; Carlin, S.; Barbaro, E.; Cappellin, L.; Velikova, V.; Vrhovsek, U.; Loreto, F.; Varotto, C. In planta recapitulation of isoprene synthase evolution from ocimene synthases. *Mol. Boil. Evol.* **2017**, *34*, 2583–2599. [CrossRef] [PubMed]

14. Loreto, F.; Fineschi, S. Reconciling functions and evolution of isoprene emission in higher plants. *New Phytol.* **2014**, *206*, 578–582. [CrossRef]

15. Sharkey, T.D.; Wiberley, A.E.; Donohue, A.R. Isoprene emission from plants: Why and how. *Ann. Bot.* **2007**, *101*, 5–18. [CrossRef] [PubMed]

16. Sharkey, T.D.; Bernacchi, C.; Farquhar, G.D.; Singsaas, E. Fitting photosynthetic carbon dioxide response curves for C3leaves. *Plant Cell Environ.* **2007**, *30*, 1035–1040. [CrossRef] [PubMed]

17. Singsaas, L.; Lerdau, M.; Winter, K.; Sharkey, T.D. Isoprene Increases thermotolerance of isoprene-emitting species. *Plant Physiol.* **1997**, *115*, 1413–1420. [CrossRef] [PubMed]

18. Velikova, V.; Várkonyi, Z.; Szabo, M.; Maslenkova, L.; Nogues, I.; Kovács, L.; Peeva, V.; Busheva, M.; Garab, G.; Sharkey, T.D.; et al. Increased thermostability of thylakoid membranes in isoprene-emitting leaves probed with three biophysical techniques. *Plant Physiol.* **2011**, *157*, 905–916. [CrossRef]

19. Vickers, C.E.; Possell, M.; Cojocariu, C.I.; Velikova, V.; Laothawornkitkul, J.; Ryan, A.; Mullineaux, P.M.; Hewitt, C.N. Isoprene synthesis protects transgenic tobacco plants from oxidative stress. *Plant Cell Environ.* **2009**, *32*, 520–531. [CrossRef]

20. Loreto, F.; Velikova, V. Isoprene produced by leaves protects the photosynthetic apparatus against ozone damage, quenches ozone products, and reduces lipid peroxidation of cellular membranes. *Plant Physiol.* **2001**, *127*, 1781–1787. [CrossRef]

21. Loreto, F.; Dicke, M.; Schnitzler, J.-P.; Turlings, T.C.J. Plant volatiles and the environment. *Plant Cell Environ.* **2014**, *37*, 1905–1908. [CrossRef]

22. Llusià, J.; Peñuelas, J.; Sardans, J.; Owen, S.M.; Niinemets, Ü. Measurement of volatile terpene emissions in 70 dominant vascular plant species in Hawaii: Aliens emit more than natives. *Glob. Ecol. Biogeogr.* **2010**, *19*, 863–874. [CrossRef]

23. Fini, A.; Brunetti, C.; Loreto, F.; Centritto, M.; Ferrini, F.; Tattini, M. Isoprene responses and functions in plants challenged by environmental pressures associated to climate change. *Front. Plant Sci.* **2017**, *8*, 1–8. [CrossRef] [PubMed]

24. Loreto, F.; Förster, A.; Dürr, M.; Csiky, O.; Seufert, G. On the monoterpene emission under heat stress and on the increased thermotolerance of leaves of Quercus ilex L. fumigated with selected monoterpenes. *Plant Cell Environ.* **1998**, *21*, 101–107. [CrossRef]

25. Loivamäki, M.; Gilmer, F.; Fischbach, R.J.; Sörgel, C.; Bachl, A.; Walter, A.; Schnitzler, J.-P. Arabidopsis, a model to study biological functions of isoprene emission? *Plant Physiol.* **2007**, *144*, 1066–1078. [CrossRef] [PubMed]

26. Bloom, A.J. Easy leaf area: Automated digital image analysis for rapid and accurate measurement of leaf area. *Appl. Plant Sci.* **2014**, *2*, 1400033.

27. Boyes, U.C.; Zayed, A.M.; Ascenzi, R.; McCaskill, A.J.; Hoffman, N.E.; Davis, K.; Görlach, J. Growth stage –based phenotypic analysis of arabidopsis. *Plant Cell* **2001**, *13*, 1499–1510. [CrossRef]

28. Faralli, M.; Williams, K.S.; Han, J.; Corke, F.M.K.; Doonan, J.H.; Kettlewell, P.S. Water-saving traits can protect wheat grain number under progressive soil drying at the meiotic stage: A phenotyping approach. *J. Plant Growth Regul.* **2019**, *38*, 1562–1573. [CrossRef]

29. Sharkey, T.D.; Yeh, S.; Wiberley, A.E.; Falbel, T.G.; Gong, D.; Fernandez, D.E. Evolution of the isoprene biosynthetic pathway in Kudzu1[w]. *Plant Physiol.* **2005**, *137*, 700–712. [CrossRef]

30. Sasaki, K.; Saito, T.; Lämsä, M.; Oksman-Caldentey, K.-M.; Suzuki, M.; Ohyama, K.; Muranaka, T.; Ohara, K.; Yazaki, K. Plants utilize isoprene emission as a thermotolerance mechanism. *Plant Cell Physiol.* **2007**, *48*, 1254–1262. [CrossRef]

31. Feng, Z.; Yuan, X.; Fares, S.; Loreto, F.; Li, P.; Hoshika, Y.; Paoletti, E. Isoprene is more affected by climate drivers than monoterpenes: A meta-analytic review on plant isoprenoid emissions. *Plant Cell Environ.* **2019**, *42*, 1939–1949. [CrossRef]

32. Nogues, I.; Brilli, F.; Loreto, F. Dimethylallyl diphosphate and geranyl diphosphate pools of plant species characterized by different isoprenoid emissions1. *Plant Physiol.* **2006**, *141*, 721–730. [CrossRef] [PubMed]

33. Owen, S.M.; Peñuelas, J. Volatile isoprenoid emission potentials are correlated with essential isoprenoid concentrations in five plant species. *Acta Physiol. Plant.* **2013**, *35*, 3109–3125. [CrossRef]

34. Zuo, Z.; Weraduwage, S.M.; Lantz, A.T.; Sanchez, L.M.; Weise, S.E.; Wang, J.; Childs, K.L.; Sharkey, T.D. Isoprene acts as a signaling molecule in gene networks important for stress responses and plant growth. *Plant Physiol.* **2019**, *180*, 124–152. [CrossRef] [PubMed]

35. Morfopoulos, C.; Prentice, I.C.; Keenan, T.F.; Friedlingstein, P.; Medlyn, B.E.; Peñuelas, J.; Possell, M. A unifying conceptual model for the environmental responses of isoprene emissions from plants. *Ann. Bot.* **2013**, *112*, 1223–1238. [CrossRef] [PubMed]

36. Sinclair, T.R.; Hammer, G.L.; Van Oosterom, E.J. Potential yield and water-use efficiency benefits in sorghum from limited maximum transpiration rate. *Funct. Plant Boil.* **2005**, *32*, 945–952. [CrossRef]

37. Querejeta, J.I.; Prieto, I.; Torres, M.P.; Campoy, M.; Alguacil, M.D.M.; Roldan, A. Water-spender strategy is linked to higher leaf nutrient concentrations across plant species colonizing a dry and nutrient-poor epiphytic habitat. *Environ. Exp. Bot.* **2018**, *153*, 302–310. [CrossRef]

38. Chesson, P.; Gebauer, R.L.E.; Schwinning, S.; Huntly, N.; Wiegand, K.; Ernest, S.K.M.; Sher, A.; Novoplansky, A.; Weltzin, J.F. Resource pulses, species interactions, and diversity maintenance in arid and semi-arid environments. *Oecologia* **2004**, *141*, 236–253. [CrossRef]

39. Tattini, M.; Loreto, F.; Fini, A.; Guidi, L.; Brunetti, C.; Velikova, V.; Gori, A.; Ferrini, F. Isoprenoids and phenylpropanoids are part of the antioxidant defense orchestrated daily by drought-stressed Platanus × acerifolia plants during Mediterranean summers. *New Phytol.* **2015**, *207*, 613–626. [CrossRef]

40. Velikova, V.; Brunetti, C.; Tattini, M.; Doneva, D.; Ahrar, M.; Tsonev, T.; Stefanova, M.; Ganeva, T.; Gori, A.; Ferini, F.; et al. Physiological significance of isoprenoids and phenylpropanoids in drought response of Arundinoideae species with contrasting habitats and metabolism. *Plant Cell Environ.* **2016**, *39*, 2185–2197. [CrossRef]

41. Harvey, C.; Li, Z.; Tjellström, H.; Blanchard, G.J.; Sharkey, T.D. Concentration of isoprene in artificial and thylakoid membranes. *J. Bioenerg. Biomembr.* **2015**, *47*, 419–429. [CrossRef]

42. Brilli, F.; Loreto, F.; Baccelli, I. Exploiting Plant Volatile Organic Compounds (VOCs) in agriculture to improve sustainable defense strategies and productivity of crops. *Front. Plant Sci.* **2019**, *10*, 264. [CrossRef] [PubMed]

43. Robertson, G.P.; Swinton, S.M. Reconciling agricultural productivity and environmental integrity: A grand challenge for agriculture. *Front. Ecol. Environ.* **2005**, *3*, 38–46. [CrossRef]

44. Porter, J.R.; Semenov, M. Crop responses to climatic variation. *Philos. Trans. R. Soc. B Boil. Sci.* **2005**, *360*, 2021–2035. [CrossRef] [PubMed]

45. Cossani, C.M.; Reynolds, M. Physiological traits for improving heat tolerance in wheat. *Plant Physiol.* **2012**, *160*, 1710–1718. [CrossRef] [PubMed]

46. Lantz, A.T.; Allman, J.; Weraduwage, S.M.; Sharkey, T.D. Isoprene: New insights into the control of emission and mediation of stress tolerance by gene expression. *Plant Cell Environ.* **2019**, *42*, 2808–2826. [CrossRef] [PubMed]

47. Rosenthal, J.P.; Dirzo, R. Effects of life history, domestication and agronomic selection on plant defence against insects: Evidence from maizes and wild relatives. *Evol. Ecol.* **1997**, *11*, 337–355. [CrossRef]

48. Rodriguez-Saona, C.; Vorsa, N.; Singh, A.P.; Johnson-Cicalese, J.; Szendrei, Z.; Mescher, M.C.; Frost, C.J. Tracing the history of plant traits under domestication in cranberries: Potential consequences on anti-herbivore defences. *J. Exp. Bot.* **2011**, *62*, 2633–2644. [CrossRef]

49. Nakhforoosh, A.; Bodewein, T.; Fiorani, F.; Bodner, G. Identification of water use strategies at early growth stages in durum wheat from shoot phenotyping and physiological measurements. *Front. Plant Sci.* **2016**, *7*, 925. [CrossRef]

50. Turner, N.C. Adaptation to water deficits: A changing perspective. *Funct. Plant Boil.* **1986**, *13*, 175–190. [CrossRef]

Article

Genome-Wide Identification of Barley ABC Genes and Their Expression in Response to Abiotic Stress Treatment

Ziling Zhang, Tao Tong, Yunxia Fang, Junjun Zheng, Xian Zhang, Chunyu Niu, Jia Li, Xiaoqin Zhang * and Dawei Xue *

College of Life and Environmental Sciences, Hangzhou Normal University, Hangzhou 311121, China; zhangziling@stu.hznu.edu.cn (Z.Z.); tongtao@stu.hznu.edu.cn (T.T.); yxfang12@163.com (Y.F.); zhengjunjun0415@163.com (J.Z.); zhangxian@hznu.edu.cn (X.Z.); niuchunyu@stu.hznu.edu.cn (C.N.); lijia@stu.hznu.edu.cn (J.L.)
* Correspondence: xiaoqinzhang@163.com (X.Z.); dwxue@hznu.edu.cn (D.X.)

Received: 22 August 2020; Accepted: 27 September 2020; Published: 28 September 2020

Abstract: Adenosine triphosphate-binding cassette transporters (ABC transporters) participate in various plant growth and abiotic stress responses. In the present study, 131 ABC genes in barley were systematically identified using bioinformatics. Based on the classification method of the family in rice, these members were classified into eight subfamilies (ABCA–ABCG, ABCI). The conserved domain, amino acid composition, physicochemical properties, chromosome distribution, and tissue expression of these genes were predicted and analyzed. The results showed that the characteristic motifs of the barley ABC genes were highly conserved and there were great diversities in the homology of the transmembrane domain, the number of exons, amino acid length, and the molecular weight, whereas the span of the isoelectric point was small. Tissue expression profile analysis suggested that ABC genes possess non-tissue specificity. Ultimately, 15 differentially expressed genes exhibited diverse expression responses to stress treatments including drought, cadmium, and salt stress, indicating that the ABCB and ABCG subfamilies function in the response to abiotic stress in barley.

Keywords: barley; ABC gene family; gene expression; abiotic stress

1. Introduction

Named after the binding frame of adenosine triphosphate (ATP), ATP-binding cassette transporters (ABC transporters) are widely found in eukaryotes and prokaryotes [1]. A previous study found that ABC transporters, as one of the most widely functional protein superfamilies, are involved in plant physiological processes [2], such as plant hormone transport, nutrient uptake by organisms, stomatal regulation [3], environmental stress responses, and the interaction between plants and microorganisms [4]. Plant ABC transporters possess nucleotide-binding domains (NBDs) and transmembrane domains (TMDs), and the NBD is a hydrophilic domain with several highly conserved motifs, characterized by the Walker A and Walker B sequences, the ABC signature motif (also known as Walker C) [5], and the H loop, and the Q loop [6]. In contrast to the NBD domain, ABC proteins contain low homologous hydrophobic transmembrane domains (TMDs), and they typically consist of at least six transmembrane α-helices. The NBD domain provides energy through combining and hydrolyzing ATP, whereas the TMD domain is able to select substrates for transportation across membranes through the energy channel provided by the former [7]. TMDs function as selectors for substrates to translocate membrane proteins through the NBD energy channel. A typical ABC full-size transporter have a core unit of two pore-forming TMDs and two cytosolic NBDs [2]. Half-size transporters, composed of only one TMD and one NBD, are thought to form homodimerize or heterodimerize that act as functional pump [8]. In many identified

ABC transporters in bacteria, some NBDs and TMDs are present on different polypeptides, called 1/4 molecular transporters [9]. Meanwhile, a few ABC transporters are not directly involved in transport and have been found to participate in other cellular processes such as DNA repair and the transcription and regulation of gene expression [10–14]. In conclusion, although the amino acid sequence of ABC transporters is homologous, the function of ABC transporters is diverse owing to their different structures.

As a consequence of the rapid development of whole-genome sequencing, the ABC gene family had been identified in an increasing number of plants, including 131 in *Arabidopsis thaliana* [15], 125 in rice, 314 in rape [16], and 130 in maize [17]. A large number of plant ABC transporters are plant secondary metabolites that have evolved in response to a particular living environment. Owing to the wide range and large number of ABC transporters, several methods have been proposed to ABC protein nomenclature [18]. According to the homologous relationships, phylogenetic relationships, and domain organization, the new nomenclature system of HUGO system (Human Genome Organization) categorizes ABC transporters into eight subfamilies, ABCA to ABCH subfamily. However, ABCH has not been characterized in plants [19,20]. Afterwards, only ABCI subfamily containing "prokaryotic"-type ABCs has been identified in plants [18]. In total, eight subfamilies (ABCA–ABCG and ABCI) have been identified in plant genomes [21].

It is well universal that abiotic stresses, such as temperature, drought, salt, and heavy metals, seriously restrict plant growth and affect the yield and quality of crops [22]. In severe cases, abiotic stress will directly result in plant death. A growing number of studies have demonstrated that ABC transporters play a pivotal role in crop yield [23], quality formation [24], and the resistance response [25]. Given the importance of ABC transporters in plant life activities, increasingly, plant ABC transporters have been identified, cloned, and functionally analyzed [26]. To date, the ABC gene family has not been identified and analyzed in barley, and its association with abiotic stress has not been explored. Barley (*Hordeum vulgare* L.) is one of the oldest cultivated crops in the world. It is widely adaptable and exhibits strong drought, cold, and salt tolerance characteristics [27]. In the research, we used bioinformatics method to conduct a whole-genome study of the ABC gene family in barley and identified 131 ABC transporters. We performed sequence characteristics, physicochemical properties, gene phylogeny, and expression profile analysis of ABC proteins at the genomic level in barley. We also investigated the expression patterns of barley under cadmium (Cd), drought, and salt stresses by quantitative real-time (qRT)-PCR. Our findings improve understanding of the function of the ABC gene family and will facilitate further studies on detailed molecular and biological functions in barley.

2. Materials and Methods

2.1. Identification of ABC Gene Family Members in Barley

The ABCs domain-containing protein and genome sequence were retrieved from Pfam database (http://pfam.xfam.org/) [28]. The identified ABC sequences of barley and rice were confirmed for the presence of PFAM domain PF00005 (ABC transport domain) and PF00664 (ABC transmembrane domain) in barley and rice using HMMER program. In order to ensure accuracy analysis, we uploaded conserved sequences into NCBI-CDD [29] and SMART database [30] (http://smart.embl -heidelberg.de/) for protein prediction and unannotated sequences were removed. At the same time, the final protein-coding sequences were verified by searching NCBI non-redundant protein sequence database with BLASTP. Protein features including molecular weight and isoelectric point (pI) of the HvABC proteins were predicted and analyzed by using tools from ExPASy website (https://web.expasy.org/protparam/) [31,32].

2.2. Multiple Sequence Alignment and Phylogenetic Analysis of the ABC Gene Family in Barley

For the sake of understanding the phylogenetic relationship of ABC proteins between barley and rice, the phylogenetic tree was constructed using all the identified ABC amino acid sequences of barley and rice. Multiple alignments of sequences were conducted using MUSCLE [33] of the EMBL-EBI [34] software with the default options. Then, MUSCLE website was utilized to construct the phylogenetic tree by the neighbor-joining (NJ) method with a bootstrap test of 1000-fold (https://www.ebi.ac.uk/Tools/msa/muscle/) [35]. The results were displayed using iTOL visualization.

2.3. Analysis of ABC Gene Structure and Chromosome Location in Barley

The information of barley ABC gene family, including intron, exon, physical location on chromosome and gene annotation file information, was retrieved in the Ensemble Plants database.

The exon–intron organizations of all the *HvABC* genes were exhibited using the online program Gene Structure Display Server (http://gsds.cbi.pku.edu.cn/) [36]. The MEME online program for protein sequence analysis was used to identify conserved motifs of ABC proteins (http://meme-suite.org/tools/meme) [37]. The MG2C was used to draw the location images of HvABCs on chromosomes. The protein sequence was used to predict subcellular localization of barley ABC gene family using WoLF PSORT (https://wolfpsort.hgc.jp/) [38].

2.4. Construction of ABC Gene Expression Profiles in Barley

To create the expression profile of *HvABC* genes among different organs and development stages, the RNA-seq data from various tissues in barley were retrieved from IPK (https://webblast.ipk-gatersleben.de/barley_ibsc/index.php). The dataset, 14 stages, included the, root (ROO1, ROO2), leaves (LEA, SEN), inflorescences (INF2, LOD, PAL, LEM, RAC), grain (CAR5, CAR15), etiolated seeding (ETI), tillers (NOD), and epidermal strips (EPI). The transcript abundance of *HvABC* genes was calculated as fragments per kilobase of exon model per million mapped reads (FPKM) and log2 (FPKM+ 1) values of invertase genes in these tissues were used to depict heatmaps. The cluster results were shown using the Multiple Experiment Viewer (MeV) (J. Craig Venter Institute, La Jolla, CA, USA).

2.5. Stress Treatment, Total RNA Extraction, and qRT-PCR Analysis

The barley cultivar, Morex, was selected for stress treatments. Seedlings were grown on nutrient solution [39] in growth chambers at 26 °C under a 14/10 h light/dark photoperiod and photosynthetically activated radiation at 18,000 lx. Two-leaf-stage plants were treated with different abiotic stress. For drought, salt, and cadmium treatments, the seedlings were treated with 20% PEG6000, 200 mmol·L^{-1} NaCl and 50 µmol·L^{-1} CdCl$_2$ nutrient solution for 24 h, respectively. All these leaf samples were snap-frozen in liquid nitrogen and the total RNAs were isolated from young leaves using an RNA kit (AxyPrep, USA). Then, the RNA was reverse-transcribed using the HifairTM II 1stStrand cDNA Synthesis Kit following the manufacturer's instructions. RNA extraction and cDNA synthesis from all samples were stored at −80 °C for RNA extraction.

Based on NCBI, ABC transporters related to abiotic stress, such as rice, wheat and maize, were retrieved. Then, the phylogenetic tree with barley ABC transporter (the method is the same as above) was constructed, combined with the expression information of barley ABC gene, expression site, and gene intron, 15 *HvABC* genes were selected to conduct qRT-PCR. cDNA obtained was used for quantitative RT-PCR using SYBR Green Master Mix and a BioRad CFX96 real-time system. The qRT-PCR experiments were performed with three biological and technical replicates. The relative expression levels were calculated using the formula $2^{-\Delta\Delta CT}$ [40]. The result was analyzed using SigmaPlot v10.0. Primers for qRT-PCR were designed using Primer Premier v5.0., using the website of Ensemble Plants to verify the specificity of primers. The barley actin gene *HvActin (HORVU1Hr1G002840)* was used as an internal control. The primers sequence used are listed in Table 1.

Table 1. Primer sequence used for qRT-PCR amplification of ABC gene family in barley.

Gene Name	Primer	Forward Primer Sequence (5′-3′)	Reverse Primer Sequence (5′-3′)
HvABCG45	ABC1	GGCGGAACTGCTGTCATCT	AGTCGGCAACACCCTTTCT
HvABCG48	ABC2	CAGCCTGGGTTCGTTTGAG	TCGGAGTGATCGCCGTTGT
HvABCG38	ABC3	GGTTTGGATGCTCGTGCTG	TGATTTGACCGCCTCTTTT
HvABCA3	ABC4	TGGGCTCATTCCACCTACA	CCGTCAATGTTTCCCAGAG
HvABCF5	ABC5	TGGCTGGAAGAAACACTGAA	TCGGGTCTGCACATACTGGTC
HvABCF4	ABC6	CACATGCAGAACAAGACCCTC	GCTTCGCAGATCCATGACC
HvABCC16	ABC7	GCCATTCGGCGACCATACA	CACGAGCAAGCTGAACACG
HvABCC11	ABC8	GTCCTTGACGCTGATACTGG	TAGCACTGGTGCCTCCTCC
HvABCB24	ABC9	TGATACTGGGATTTGGTTAGG	CGAATGGCACTGAGAATGAG
HvABCB13	ABC10	CGTTCAACTCGGAGGACAAGA	CCATGCAGCGTACCACAGG
HvABCG27	ABC11	GAGGGAGGCAGCGTCAAGCA	GCAGGATGGCGAACTGGTTG
HvABCG29	ABC12	AGGGCTTCCCGTTGTAGGTG	TCGCATCCGTCATCACCATG
HvABCG25	ABC13	GTTCTGGATCGAGATGGGTGT	GAAGATGGTCGCCAGGATGA
HvABCG21	ABC14	ATACCGGCATACTGGCTGTGG	CCAGCACTCGCTCCTTCACC
HvABCG18	ABC15	TGCTCACCGCCAACTCATTC	TCCTTGCTCGCCACGAAGT
HvActin	Actin	TGGATCGGAGGGTCCATCCT	GCACTTCCTGTGGACGATCGCTG

3. Results

3.1. Identification and Physicochemical Properties of the ABC Gene Family in Barley

Through multiple bioinformatics analyses, a total of 131 ABC transporter genes were identified in barley (Supplementary Table S1). The barley *ABC* genes were classified according to their sequence similarity with rice *ABC* genes and were further named *HvABCA–HvABCG, HvABCI*. Among the 131 HvABC proteins, all of the proteins contained one or more NBDs domains based on the domain composition analysis of the ABC proteins.

Comprehensive information on the HvABCs, including the domain structure, predicted protein length, exon number, molecular weight (MW), isoelectric point (PI), and subcellular localization, is provided in Supplementary Table S1. The amino acid numbers scoped from 171 aa (HvABCG1) to 1628 aa (HvABCC9), and the corresponding molecular weight changed from 19391.19 to 182783.15 kD. The protein lengths of ABCA, ABCE, ABCF, and ABCI had few differences, while the protein lengths of ABCB, ABCC, and ABCG varied greatly. In contrast, the variation in PI was small, with the majority constituting basic proteins. Based on the subcellular localization prediction of 131 barley *ABC* genes, 99 *ABC* genes were detected to be localized on the plasma membrane, which confirmed that most of the ABC transporters are bound to the plasma membrane and are responsible for the efflux of intracellular substances, while only a few are present on the vacuoles, chloroplasts, and mitochondria in these organelles. ABC transporters mainly regulate the division of exogenic substances into the organelles, which also reflect the endosymbiotic origin of the plastids and mitochondria [41]. Previous studies on the ABCC subfamily have shown that ABCCs are involved in cellular detoxification, which may contribute to the complexation of toxins and heavy metal ions with glutathione or organic acids for storage in the vacuoles or transportation out of cells [42,43]. All 22 genes of the ABCC subfamily in barley reside in the plasma membrane.

3.2. Phylogenetic Analysis of the ABC Gene Family in Barley

Previous studies have consistently demonstrated that the ABC gene family underwent species-specific amplification after the divergence between dicotyledonous and monocotyledonous plants [44]. Exploring the relationship between HvABC proteins and OsABC proteins could help classify and assess the potential functions of HvABCs. For further classification, ABC sequences from two different plant species, including 131 HvABC proteins and 100 OsABC proteins, were subjected to phylogenetic analysis. Based on the phylogenetic relationships with OsABCs, the HvABCs were divided into eight subfamilies (ABCA–ABCG, ABCI) (Figure 1). In the light of phylogenetic tree, except ABCI were dispersed, the HvABC proteins of each subfamily were clustered together. Among those subfamilies, the ABCB and ABCG subfamily had the greatest number of members with 32 genes and 49 genes, accounting for 21.97% and 37.12%, respectively. Only three members were identified in ABCE (Supplementary Table S1). Further investigation revealed that the HvABCE and HvABCF proteins contained two NBDs but, as expected, no TMD. Although their NBDs domains share sequence homology with other members of the ABC gene family in barley, they are not transposons in the conventional sense and have no obvious transport function. Analysis of the phylogenetic tree terminal branches indicated that there were 67 pairs of orthologous proteins between species, among which the ABCG and ABCB subfamily were the greatest with 23 pairs and 18 pairs, respectively, indicating that the ABC gene family retained very high homology in the evolution of barley and rice. In addition, there were 20 pairs of paralogs, among which the ABCA, ABCB, ABCC, and ABCG subfamilies of barley contained one, four, three, and five pairs, respectively, and there were seven pairs of paralogs in rice; that is, the ABCB, ABCC, and ABCG subfamilies contained two, one, and four pairs, respectively. The above findings can infer that the members of the ABC gene family of barley may have evolved independently and expanded in a species-specific way.

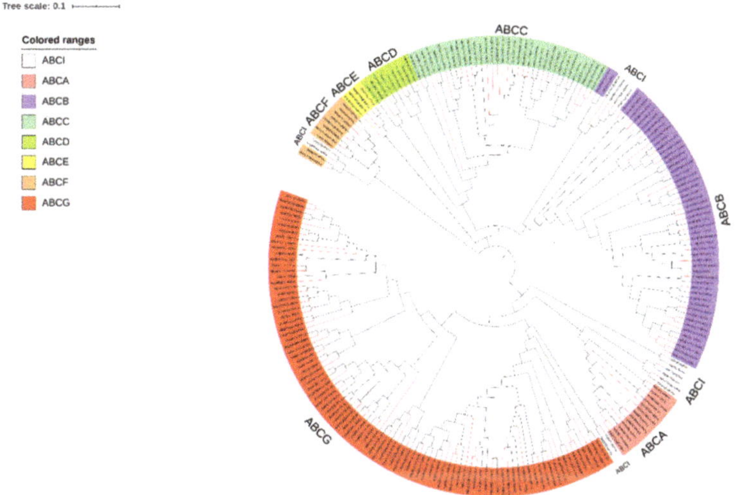

Figure 1. Unrooted Neighbor-Joining tree constructed with ABC proteins of *Hordeum vulgare* L. (HORVV) and *Oryza sativa* L. (ORYSJ). The domains clustered into eight subgroups (ABCA-ABCG, ABCI). Different colored shadings indicated eight ABC transporter subfamilies. The red branch is barley, and the black branch is rice.

3.3. Chromosome Mapping of the ABC Gene Family in Barley

A total of 131 *HvABCs* were mapped on the seven chromosomes, and the remaining three genes were distributed on unanchored scaffolds (Figure 2). Chromosome mapping revealed that the *HvABC* genes were mostly concentrated on or near the end of the chromosomes where they exhibited a high variation in their distribution. In addition, 31 genes, the maximum number, were located on chromosome 3H. On the contrary, chromosome 6H contained only nine genes. As shown in Figure 2, eight *HvABCs* clusters (*HvABCA4/HvABCA7*, *HvABCB11/HvABCB16/HvABCB22*, *HvABCB4/HvABCB27*, *HvABCG29/HvABCB23*, *HvABCC22/HvABCC2*, *HvABCC6/HvABCC8*, *HvABCC16/HvABCC15/HvABCI1*, and *HvABCC13/HvABCG1*) containing 18 genes were identified on chromosomes 1H, 3H, 4H, and 7H. Tandem duplication was an important recent gene duplication pattern in the expansion of HvABCs gene family. These may be caused by tandem repeat genes. Therefore, these gene clusters may be tandem repeat arrays.

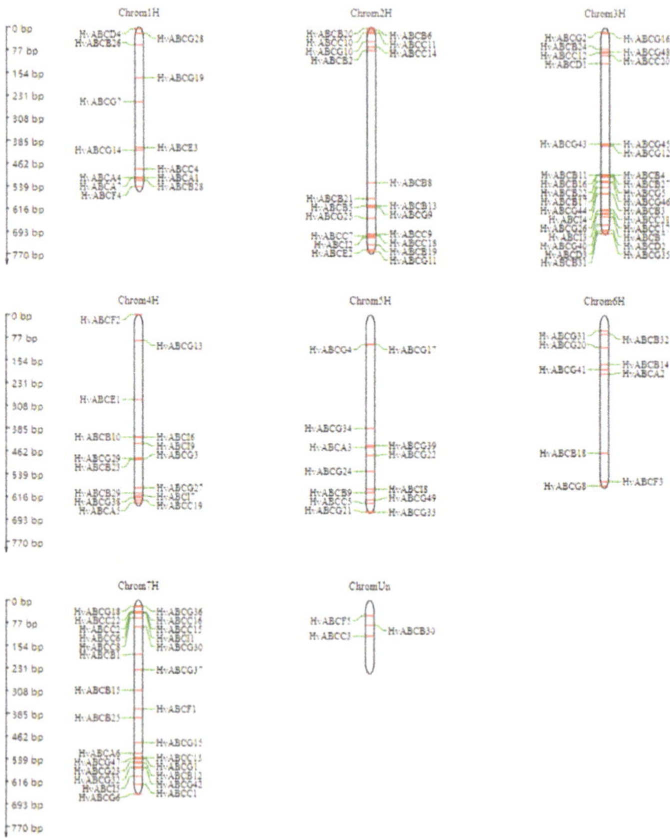

Figure 2. Mapping of the *HvABC* genes on *Hordeum vulgare* L. chromosomes. The chromosome number is indicated at the top of each chromosome. Three putative *HvABC* genes could not be localized on a specific chromosome.

3.4. Analysis of Exon-Intron Structure and Conserved Domain of the ABC Gene Family in Barley

To further explore the conservation and diversity of protein structures and compositions of HvABCs, the corresponding analysis is carried out by MEME and GSDS website. The result (Supplementary Figure S1) showed that most of the HvABC proteins (122 of 131, 93.1%) presented multiple introns, while nine *HvABC* genes (*HvABCG14*, *HvABCG15*, *HvABCG12*, *HvABCG26*, *HvABCG16*, *HvABCG28*, *HvABCG17*, *HvABCG3*, *HvABCG1*, *HvABCI7*, and *HvABCI2*) had no introns. These nine genes may have originated from the transposable events of reverse transcription. The number and size of the introns varied in the remaining 122 genes, indicating significant differences in gene structure. It can also be speculated that intron deletion and insertion events occurred in the evolutionary process, which may be caused by the differentiation of ABC gene family members after replication. The exons of all identified *HvABC* genes ranged from 1 to 30, among which the number of *ABC* genes with 10 exons was the highest, accounting for 9.92%.

MEME motif analysis identified seven conserved motifs in the HvABC proteins. Combining the domain characteristics of each subfamily, we analyzed the conservative motifs (Supplementary Figure S2). The number of conserved motifs in each HvABC protein varies from one to five. The results (Figure 3) indicate that all seven highly conserved motifs belong to the NBD domain, which also suggests that the NBD sequence identity was higher than that of the TMDs. Among them, motif 2 is the Walker A in the nucleotide binding domains, and motif 6 is the Walker B in the nucleotide binding domains, and the motif between Walker A and Walker B is the ABC characteristic motif; that is, motif 1, motif 3, motif 4, motif 5, and motif 7. However, the [LIVMFY] subunit in motif 1 contains other residues (Table 2), and motif 6 in Walker B is interrupted by a hydrophilic residue. This phenomenon is currently only observed in plant ABC transporters [45]. The result also shows that several motifs are widely distributed in the HvABC proteins, such as motifs 1 and 2. In contrast, other motifs are specific to only one or two subfamilies. For instance, only ABCB and ABCC subfamilies contain motif 3, and motif 7 exists only in ABCC and ABCI. ABCG and ABCF contain the specific motif 4, and these motifs are probably required for specific protein functions. The functional differentiation in HvABCs during the evolutionary process may be due to the diversity of motif components in the different subfamilies.

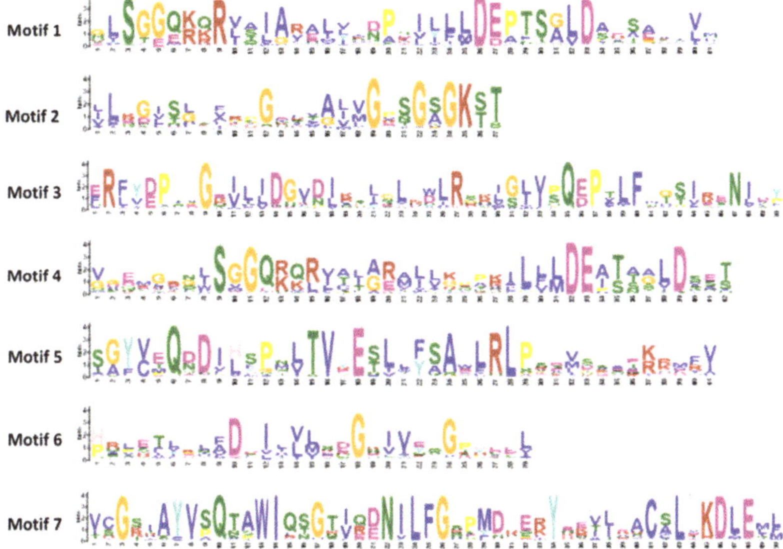

Figure 3. Conserve amino acid in seven motifs of ABC gene family in barley. Motif analysis and the sequence logos was performed using MEME website.

Table 2. Seven conservative motif protein sequences given by the MEME online tool.

Motif	Protein Sequence
Motif1	GLSGGQKQRVAIARALLABPSILLLDEPTSGLDAESAAIVM
Motif2	LLKGISLSFRPGELVALVGPSGSGKST
Motif3	ERFYDPTAGEILJDGVDIKSJGLHWLRSKJGJVPQEPTLFMGSIRENIDY
Motif4	VGEMGRNLSGGQRQRVALARAJLKPPKILLLDEATAALDSET
Motif5	SGYVEQBDIHSPNLTVYESLLFSAWLRLPSDVSSAEKRMFV
Motif6	HRLETLRLFDDILVLSDGKIVEQGPHEEL
Motif7	VCGSIAYVSQTAWIQSGTIQDNILFGSPMDRERYEEVJEACSLVKDLEML

3.5. Tissue-Specific Expression of ABC Genes in Barley

Tissue-specific expression patterns of genes can help elucidate their function in plant species and predict their role in growth and development. To further elucidate the expression profiles of *HvABCs* in different tissues and developmental stages, we used expression data of different tissues from the IPK website as a resource. A heatmap displaying the expression data of *HvABCs* in different periods as well as in the tissues and organs was generated, and the *HvABCs* were clustered by their expression patterns. Tissue specific expression profiles showed that the *HvABC* genes could be categorized into eight types (Figure 4), which indicated that the expression of *ABC* genes had undergone significant differentiation. Furthermore, the expression data clustering results did not clearly correspond to the subfamilies differentiated by the phylogenetic analysis, indicating that sequence similarity does not completely determine expression pattern and function similarity.

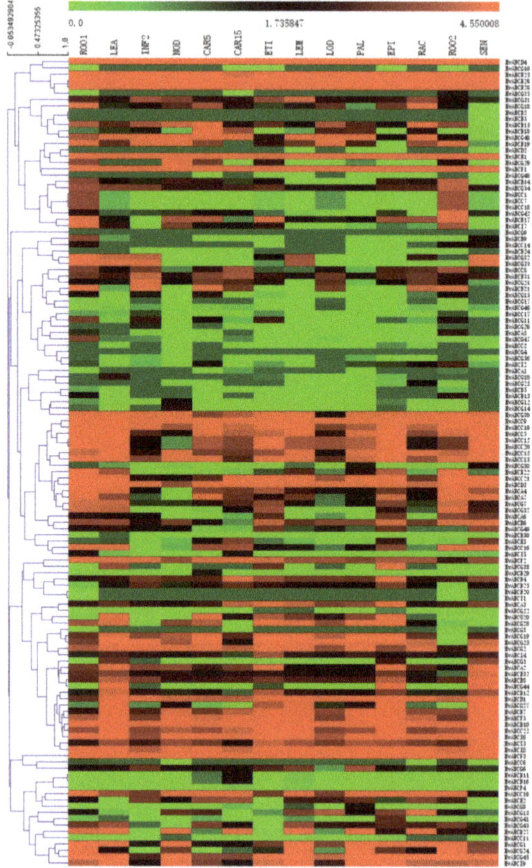

Figure 4. Heatmap showing the expression pattern of *HvABC* genes in developmental stages and tissues, including ROO1 (Roots from seedings), ROO2 (Roots), LEA (Shoots from seedings), ETI (Etiolated seeding, dark cond), INF2 (Developing inflorescences), PAL (Dissected inflorescences), LEM/LOD/RAC (inflorescences, lemma/lodicule/rachis), NOD (Developing tillers), CAR5/CAR15 (Developing grain, 5 DAP/15 DAP), EPI (Epidermal strips), SEN (Senescing leaves). The combined phylogenetic trees of *HvABCs* genes on the left panel. The scale bar at the top represents relative expression value. Red denotes high expression levels, and green denotes low expression levels.

In this study, a total of 127 *HvABCs* were determined as being expressed in at least one tissue (the FPKM values of *HvABCB1*, *HvABCC3*, *HvABCC4*, and *HvABCG9* in the 15 tissues were all 0), and 92 *HvABCs* were expressed in all tissues. The tissue-specific expression profile showed that some HvABC gene family members did not have tissue specificity and highlighted an essential role in almost all growth and developmental stages. As the Figure 4 shows, the majority of *HvABCs* presented different expression patterns, whereas a few exhibited similar expression patterns. Some ABCC subfamily genes were ubiquitously and highly expressed in all the tissues, especially the root, such as *HvABCC1*, *HvABCC7*, and *HvABCC18*. Some *HvABCs* also exhibited tissue-specific expression; for instance, *HvABCG10* only specifically expressed in the leaf tissues, *HvABCB11* and *HvABCB16* specifically expressed in the developing grain, and *HvABCG12*, *HvABCG14*, and *HvABCG29* were high during tiller development, implying that these genes may play specific roles in the relevant tissues.

3.6. Expression Analysis of ABC Genes in Barley in Response to Abiotic Stress

Research has found that plant ABC transporters play a major role in auxin, heavy metal transport, and abiotic stress, especially the ABCG subfamily and ABCB subfamily [46]. In this study, a total of 15 *HvABC* genes were identified using bioinformatics methods, including one *HvABCA*, two *HvABCBs*, two *HvABCCs*, two *HvABCFs*, and eight *HvABCGs*. The expression responses of selected *HvABC* genes under NaCl, PEG, and Cd treatment conditions were examined using qRT-PCR in our study.

A range of expression levels were observed in the selected *HvABCs* at 24 h after exposure to Cd, drought, and salt stress. The analyses revealed that different members within each HvABC subfamily responded differently to the same set of abiotic stresses. Compared with the control levels, the results showed that the expression levels of 13 of the 15 identified *HvABC* genes increased in response to Cd stress. The most pronounced increases were observed in *HvABCA3* and *HvABCG29*, whereas *HvABCG38* and *HvABCG21* were dramatically repressed by Cd stress (Figure 5A). Under salt stress conditions, most *HvABC* genes exhibited the opposite pattern to Cd stress, with *HvABCG48*, *HvABCF4*, *HvABCB24*, and *HvABCG25* being repressed (Figure 5B). Expression analysis following PEG treatment indicated that *ABC* genes were upregulated by 1.0 times, including *HvABCG45*, *HvABCG48*, *HvABCA3*, *HvABCF5*, *HvABCF4*, *HvABCC16*, *HvABCC11*, and *HvABCB13*. Under different abiotic stresses, *HvABC* genes showed different response patterns and different response degrees. Among the 15 *HvABC* genes, *HvABCG38* was repressed after all the stress treatments, whereas the expression of other genes, including *HvABCG45*, *HvABCA3*, *HvABCF5*, *HvABCC16*, *HvABCC11*, *HvABCB13*, *HvABCG27*, *HvABCG29*, and *HvABCG18*, was increased by the stress treatments (Figure 5C). Conversely, the expressions of *HvABCG4*, *HvABCG25*, *HvABCG21*, *HvABCF4*, and *HvABCB24* were enhanced or induced by the stress treatments.

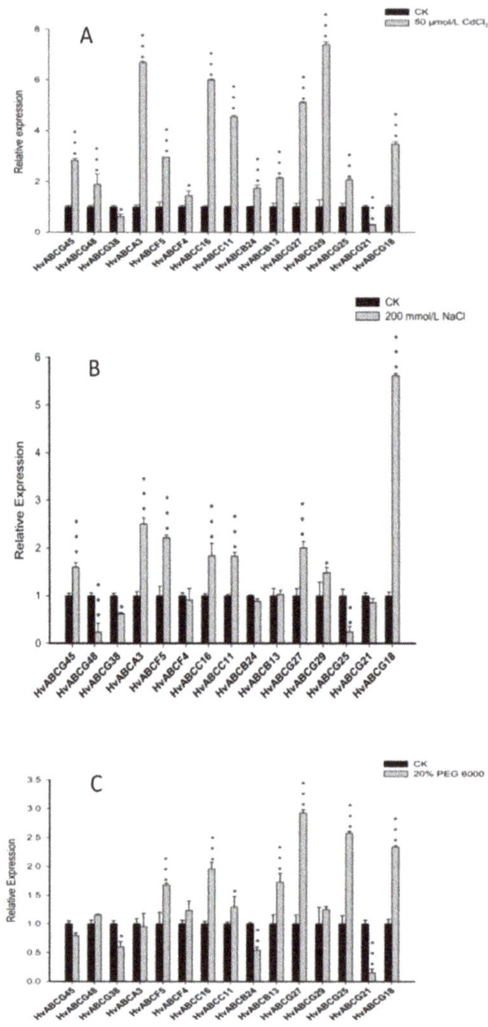

Figure 5. qRT-PCR analysis of 15 *HvABC* genes in response to (**A**) 50 μmol·L^{-1} CdCl$_2$ (**B**) 200 mmol·L^{-1} NaCl (**C**) 20% PEG6000. columns in black represent CK, columns in gray represent abiotic stress. ANOVA and LSD was used to test significance. Asterisks indicate the corresponding gene significantly up- or down-regulated compared with the untreated control (* $p < 0.05$, ** $p < 0.01$, *** $p < 0.001$) Data are the means of three replicates with standard errors represented by bars.

4. Discussion

The ATP-binding cassette (ABC) transporters belong to a large superfamily of proteins, which are ubiquitous and important in all kind of life events for all life organism [9,47]. In plants, ABC transporters participate in the transport of exogenous substances and secondary metabolites and in the abiotic stress response, as well as in many other important physiological and developmental processes [6,21].

Because of repeated genome replication, the number of ABC proteins in plants is much higher than in animals. Given the important regulatory role of ABC transporters in plant growth and development, bioinformatics analysis of ABC family genes has been conducted in *Arabidopsis* [15], rice [48], maize [17],

Lotus corniculatus [49], grape [50], *Brassica napus* [16], and other important plants. Thus far, most research into the functions of ABC transporters has been concentrated on *Arabidopsis*. In this study, 131 HvABC proteins were identified genome-wide in barley (Supplementary Table S1), which is similar to the number of genes identified in rice and *Arabidopsis* [15,48]. Additionally, phylogenetic analysis divided the protein sequences of barley and rice into eight subfamilies, including ABCA–ABCG and ABCI (Figure 1). In each subfamily, the *ABC* orthologous genes reflect the sequence similarity of barley and rice, and relatively, the emergence of paralogous genes indicates differentiation It is presumed that the barley genome experienced whole-genome replication events before the separation of gramineous species. All *HvABC* genes have at least one NBD domain, which indicates that NBD is a unique domain of ABC transporters. The HvABC transporter domain has various organizational forms that can be split into full-size members, half-sized transporters, and 1/4 molecular transporters [8,9]. *HvABC* include multi-domains, whereas few genes generally contain only one NBD domain. This phenomenon has been confirmed in *SLABCB27* of the tomato ABC gene family [51]. Conserved sequence analysis also reflects the uniqueness of the ABC domain and the high conservation of NBDs. On the contrary, the sequences and structures of TMDs differ, which reflects the chemical diversity of the substrates transported by ABC transporters. The members of the HvABC gene family have large functional differences, with large span amino acid, as well as large differences in isoelectric point and pH value (Supplementary Table S1). In short, the structural differences also reflect the different functions.

Gene expression pattern is an important starting point for further evolution and function research. Thus, we conducted expression analysis for all HvABCs in 15 tissues and organs. Except for a few genes, most HvABCs did not exhibit clear tissue specificity (Figure 4), indicating that HvABCs play a role in all aspects of plant growth. Studies have shown that *Arabidopsis AtABCB19* participates in the regulation of the separation of post-embryonic organs and cytoplasmic flow in the inflorescence axis of *Arabidopsis* [52,53], while *AtABCB1* plays a major role in another development, and *AtABCB19* plays a synergistic role [54]. The *AtABCG11* protein is distributed in the stems, leaves, and floral organs of plants and is involved in the transportation of paraffin wax and grease substances on the surface of plants.

Previous studies have found that the ABCB subfamily may be closely associated with abiotic stress in plants [17,55,56]. For example, *AtABCB25* is related to heavy metal resistance in *Arabidopsis*, and the overexpression of *AtABCB25* can improve the resistance of *Arabidopsis* to Cd and lead [57]. *OsABCB23* and *OsABCB24* are induced by drought stress, while *OsABCB6*, *OsABCB9*, and *OsABCB8* are induced by salt stress [48,58]. *OsABCB27*, located on the vacuolar membrane, participates in the response of rice to aluminum stress [59]. Under drought stress, the expression of the *ZmABCB7* and *ZmABCB8* genes increases significantly, while the expression of *ZmABCB18* is significantly inhibited by salt stress [17]. In this study, it was found that the expression of *HvABCB13* was significantly increased under Cd, drought, and salt stress, while *HvABCB24* was inhibited by salt stress and drought, but could respond significantly to Cd stress. ABCG transporters excrete wax and keratinous monomers, thus reducing water loss [60]. *AtABCG12* participates in the transmembrane transport of cuticular wax in the stem epidermis, *AtABCG32* is involved in the formation of the cell wall cuticle, and the *AtABCG32* transporter substrate is a keratinous monomer. When plants suffer from drought stress, *AtABCG40* increases abscisic acid production to timeously close the stomata [61]. At the same time, as the largest subfamily, ABCG also plays an important role in abiotic stress. *SpTUR2* in duckweed is expressed upon exposure to salicylic acid, cold, and high-salt environments [62]. The overexpression of the *AtABCG36* gene can improve the resistance of *Arabidopsis* to salt and drought [63]. The ABCG transporter *OsABCG9* in rice roots responds significantly to PEG, zinc, and Cd stress, and its expression is closely related to redox mechanism and heavy metal stress [64]. Our study shows that the HvABCG subfamily responded to salt, drought, and Cd stress to different degrees. *HvABCG45*, *HvABCG27*, *HvABCG29*, and *HvABCG18* were concurrently involved in three types of stress induction. In addition, *HvABCG38* and *HvABCG21* could significantly respond to salt stress, and their expression levels decreased significantly after drought and Cd treatments. *HvABCG48* and *HvABCG25* were significantly upregulated under drought and Cd stress but were inhibited by salt stress (Figure 5). This also confirms that the HvABCB and HvABCG subfamily are involved in the abiotic stress response, and most HvABC genes are regulated by at least one abiotic stress factor.

A genome-wide analysis of ABC gene family in barley was carried out in the present study. The phylogenetic relationships, gene structures, chromosome locations, and expression profiles of *HvABCs* were studied in detail. Furthermore, 15 genes responding to abiotic stress were preliminarily screened, and the expression levels of the genes following stress treatment were analyzed. Taken together, our study revealed the functional diversity of HvABCs proteins and provided candidate *ABC* genes for future breeding to various stresses.

Supplementary Materials: The following are available online at http://www.mdpi.com/2223-7747/9/10/1281/s1, Figure S1. Intron-exon structures of the ABC genes in barley. Analysis of HvABC genes structure using GSDS online tool. Yellow boxes indicate the coding sequences (CDS), blue rectangles indicate UTR, while black lines indicate introns. Figure S2. The motif analysis of ABC family in barley. Seven conservative motifs were identified and were represented by different colors. The black solid line represents the corresponding HvABC protein and its length. The position of the motif on the sequence was labeled. Table S1: Physicochemical properties, domain and subcellular localization of ABC proteins in barley.

Author Contributions: D.X., X.Z. (Xiaoqin Zhang), and Y.F. managed this project; Z.Z., T.T., and X.Z. (Xian Zhang) analyzed the sequencing data; Z.Z., J.Z., C.N., and J.L. performed the experiments; Z.Z., T.T., and D.X. wrote the manuscript. All authors have read and agreed to the published version of the manuscript.

Funding: This research was supported by the National Natural Science Foundation of China (31401316) and Hangzhou Scientific and Technological Program (20140432B03).

Acknowledgments: We thank LetPub (www.letpub.com) for its linguistic assistance during the preparation of this manuscript.

Conflicts of Interest: The authors declare that they have no competing interests.

References

1. Higgins, C.F.; Linton, K.J. The ATP switch model for ABC transporters. *Nat. Struct. Mol. Biol.* **2004**, *11*, 918–926. [CrossRef] [PubMed]

2. Davidson, A.L.; Dassa, E.; Orelle, C.; Chen, J. Structure, function, and evolution of bacterial ATP-binding cassette systems. *Microbiol. Mol. Biol. Rev.* **2008**, *72*, 317–364. [CrossRef] [PubMed]

3. Lane, T.S.; Rempe, C.S.; Davitt, J.; Staton, M.E.; Peng, Y.; Soltis, D.E.; Melkonian, M.; Deyholos, M.; Leebens-Mack, J.H.; Chase, M. Diversity of ABC transporter genes across the plant kingdom and their potential utility in biotechnology. *BMC Biotechnol.* **2017**, *16*, 47. [CrossRef] [PubMed]

4. Liu, Y.Q.; Zhao, Y.F. Structure and mechanism of ABC transporter. *Chin. Bull. Life Sci.* **2017**, *29*, 223–229.

5. Jasinski, M.; Ducos, E.; Martinoia, E.; Boutry, M. The ATP-binding cassette transporters: Structure, function and gene family comparison between rice and *Arabidopsis*. *Plant Physiol.* **2003**, *131*, 1169. [CrossRef]

6. Rea, P.A. Plant ATP-binding cassette transporters. *Annu. Rev. Plant Biol.* **2007**, *58*, 347–375. [CrossRef]

7. Schneider, E.; Hunke, S. ATP-binding-cassette (ABC) transport systems: Functional and structural aspects of the ATP-hydrolyzing subunits/domains. *FEMS Microbiol. Rev.* **1998**, *22*, 1–20. [CrossRef]

8. Mcfarlane, H.E.; Shin, J.J.H.; Bird, D.A.; Samuels, A.L. Arabidopsis ABCG transporters, which are required for export of diverse cuticular lipids, dimerize in different combinations. *Plant Cell* **2010**, *22*, 3066–3075. [CrossRef]

9. Higgins, C.F. ABC transporters: From microorganisms to man. *Annu. Rev. Cell Biol.* **1992**, *8*, 67–113. [CrossRef]

10. Dong, J.; Lai, R.; Jennings, J.L.; Link, A.J.; Hinnebusch, A.G. The novel ATP-binding cassette protein ARB1 is a shuttling factor that stimulates 40s and 60s ribosome biogenesis. *Mol. Cell. Biol.* **2005**, *25*, 9859–9873. [CrossRef]

11. Dong, J.; Lai, R.; Nielsen, K.; Fekete, C.A.; Qiu, H.; Hinnebusch, A.G. The essential ATP-binding cassette protein RLI1 functions in translation by promoting preinitiation complex assembly. *J. Biol. Chem.* **2004**, *279*, 42157. [CrossRef] [PubMed]

12. Garcia, O.; Bouige, P.; Forestier, C.; Dassa, E. Inventory and comparative analysis of rice and *Arabidopsis* ATP-binding cassette (ABC) systems. *Mol. Biol.* **2004**, *343*, 249–265. [CrossRef] [PubMed]

13. Kärblane, K.; Gerassimenko, J.; Nigul, L.; Piirsoo, A.; Smialowska, A.; Vinkel, K.; Kylsten, P.; Ekwall, K.; Swoboda, P.; Truve, E.; et al. ABCE1 is a highly conserved RNA silencing suppressor. *PLoS ONE* **2015**, *10*, e0116702. [CrossRef] [PubMed]

14. Kovalchuk, A.; Driessen, A.J. Phylogenetic analysis of fungal ABC transporters. *BMC Genomics* **2010**, *11*, 177. [CrossRef] [PubMed]

15. Sanchez-Fernandez, R.; Davies, T.G.; Coleman, J.O.; Rea, P.A. The Arabidopsis thaliana ABC protein superfamily, a complete inventory. *J. Biol. Chem.* **2001**, *276*, 30231–30244. [CrossRef]

16. Zhang, X.D.; Zhao, K.X.; Yang, Z.M. Identification of genomic ATP binding cassette (ABC) transporter genes and Cd-responsive ABCs in *Brassica napus*. *Gene* **2018**, *664*, 139–151. [CrossRef]

17. Pang, K.; Li, Y.; Liu, M.; Meng, Z.; Yu, Y. Inventory and general analysis of the ATP-binding cassette (ABC) gene superfamily in maize (*Zea mays* L.). *Gene* **2013**, *526*, 411–428. [CrossRef]

18. Verrier, P.J.; Bird, D.; Burla, B.; Dassa, E.; Forestier, C.; Geisler, M.; Klein, M.; Kolukisaoglu, U.; Lee, Y.; Martinoia, E.; et al. Plant ABC proteins–a unified nomenclature and updated inventory. *Trends Plant Sci.* **2008**, *13*, 151–159. [CrossRef]

19. Annilo, T.; Chen, Z.Q.; Shulenin, S.; Costantino, J.; Thomas, L.; Lou, H.; Stefanov, S.; Dean, M. Evolution of the vertebrate ABC gene family: Analysis of gene birth and death. *Genomics* **2006**, *88*, 1–11. [CrossRef]

20. Dermauw, W.; Van Leeuwen, T. The ABC gene family in arthropods: Comparative genomics and role in insecticide transport and resistance. *Insect. Biochem. Mol. Biol.* **2014**, *45*, 89–110. [CrossRef]

21. Kang, J.; Park, J.; Choi, H.; Burla, B.; Kretzschmar, T.; Lee, Y.; Martinoia, E. Plant ABC Transporters. *Arabidopsis Book* **2011**, *9*, e0153. [CrossRef] [PubMed]

22. Roy, S.J.; Tucker, E.J.; Tester, M.A. Genetic analysis of abiotic stress tolerance in crops. *Curr. Opin. Plant Biol.* **2011**, *14*, 232–239. [CrossRef] [PubMed]

23. Orsi, C.H.; Tanksley, S.D. Natural variation in an ABC transporter gene associated with seed size evolution in tomato species. *PLoS Genet.* **2009**, *5*, e1000347. [CrossRef] [PubMed]

24. Xu, X.H.; Zhao, H.J.; Liu, Q.L.; Frank, T.; Engel, K.H.; An, G.; Shu, Q.Y. Mutations of the multi-drug resistance-associated protein ABC transporter gene 5 result in reduction of phytic acid in rice seeds. *Theor. Appl. Genet.* **2009**, *119*, 75–83. [CrossRef]

25. Crouzet, J.; Trombik, T.; Fraysse, A.S.; Boutry, M. Organization and function of the plant pleiotropic drug resistance ABC transporter family. *FEBS Lett.* **2006**, *580*, 1123–1130. [CrossRef]

26. Hwang, J.U.; Song, W.Y.; Hong, D.; Ko, D.; Yamaoka, Y.; Jang, S.; Yim, S. Plant ABC transporters enable many unique aspects of a terrestrial plant's lifestyle. *Mol. Plant* **2016**, *9*, 338–355. [CrossRef]

27. Tong, T.; Fang, Y.X.; Zhang, Z.L.; Zheng, J.J.; Xue, D.W. Genome-wide identification, phylogenetic and expression analysis of SBP-box gene family in barley (*Hordeum vulgare* L.). *Plant Growth Regul.* **2020**, *90*, 1–13. [CrossRef]

28. Finn, R.D.; Tate, J.; Mistry, J.; Coggill, P.C.; Sammut, S.J.; Hotz, H.R.; Ceric, G.; Forslund, K.; Eddy, S.R.; Sonnhammer, E.L.; et al. The Pfam protein families database. *Nucleic Acids Res.* **2008**, *36*, D281–D288. [CrossRef]

29. Marchler-Bauer, A.; Derbyshire, M.K.; Gonzales, N.R.; Lu, S.; Chitsaz, F.; Geer, L.Y.; Geer, R.C.; He, J.; Gwadz, M.; Hurwitz, D.I.; et al. CDD: NCBI's conserved domain database. *Nucleic Acids Res.* **2015**, *43*, D222–D226. [CrossRef]

30. Letunic, I.; Doerks, T.; Bork, P. SMART 7: Recent updates to the protein domain annotation resource. *Nucleic Acids Res.* **2012**, *40*, D302–D305. [CrossRef]

31. Artimo, P.; Jonnalagedda, M.; Arnold, K.; Baratin, D.; Csardi, G.; de Castro, E.; Duvaud, S.; Flegel, V.; Fortier, A.; Gasteiger, E.; et al. ExPASy: SIB bioinformatics resource portal. *Nucleic Acids Res.* **2012**, *40*, W597–W603. [CrossRef] [PubMed]

32. Liu, M.Y.; Sun, W.J.; Ma, Z.T. Genome-wide investigation of the AP2/ERF gene family in tartary buckwheat (*Fagopyum Tataricum*). *BMC Plant Biol.* **2019**, *19*, 84. [CrossRef] [PubMed]

33. Edgar, R.C. MUSCLE: Multiple sequence alignment with high accuracy and high throughput. *Nucleic Acids Res.* **2004**, *5*, 1792–1797. [CrossRef]

34. Mcwilliam, H.; Li, W.; Uludag, M. Analysis Tool Web Services from the EMBL-EBI. *Nucleic Acids Res.* **2013**, *41*, W597–W600. [CrossRef] [PubMed]

35. Edgar, R.C. MUSCLE: A multiple sequence alignment method with reduced time and space complexity. *BMC Bioinform.* **2004**, *5*, 113. [CrossRef] [PubMed]

36. Hu, B.; Jin, J.; Guo, A.Y.; Zhang, H.; Gao, G. GSDS 2.0: An upgraded gene feature visualization server. *Bioinformatics* **2014**, *31*, 1296–1297. [CrossRef]

37. Bailey, T.L.; Boden, M.; Buske, F.A.; Frith, M.; Grant, C.E.; Clementi, L.; Ren, J.; Li, W.W.; Noble, W.S. MEME SUITE: Tools for motif discovery and searching. *Nucleic Acids Res.* **2009**, *37*, W202–W208. [CrossRef]

38. Horton, P.; Park, K.J.; Obayashi, T.; Nakai, K. Protein subcellular localization prediction with WOLF PSORT. In Proceedings of the 4th Asia-Pacific Bioinformatics Conference, Taipei, Taiwan, 13–16 February 2006.

39. Zhang, X.; Tong, T.; Tian, B.; Fang, Y.; Xue, D. Physiological, biochemical and molecular responses of barley seedlings to Aluminum stress. *Phyton* **2019**, *88*, 253–260. [CrossRef]

40. Tong, C.; Wang, X.; Yu, J.; Wu, J.; Liu, S. Comprehensive analysis of RNA-seq data reveals the complexity of the transcriptome in *Brassica rapa*. *BMC Genomics* **2013**, *14*, 689. [CrossRef]

41. Theodoulou, F.L. Plant ABC Transporters. *Biochim. Biophys. Acta* **2000**, *1465*, 79–103. [CrossRef]

42. Song, W.Y.; Mendoza-Cózatl, D.G.; Lee, Y.; Schroeder, J.I.; Ahn, S.N.; Lee, H.S.; Wicker, T.; Martinoia, E. Phytochelatin-metal(loid) transport into vacuoles shows different substrate preferences in barley and Arabidopsis. *Plant Cell Environ.* **2013**, *37*, 1192–1201. [CrossRef] [PubMed]

43. Sousa, C.A.; Hanselaer, S.; Soares, E.V. ABCC subfamily vacuolar transporters are involved in Pb (Lead) detoxification in *Saccharomyces cerevisiae*. *Appl. Biochem. Biotechnol.* **2014**, *175*, 65–74. [CrossRef] [PubMed]

44. Theodoulou, F.L.; Kerr, I.D. ABC transporter research: Going strong 40 years on. *Biochem. Soc. Trans.* **2015**, *43*, 1033–1040. [CrossRef] [PubMed]

45. Martinoia, E.; Klein, M.; Geisler, M.; Bovet, L.; Forestier, C.; Kolukisaoglu, U.; Müller-Röber, B.; Schulz, B. Multifunctionality of plant ABC transporters—More than just detoxifiers. *Planta* **2002**, *214*, 345–355. [CrossRef]

46. Martin, C.; Berridge, G.; Mistry, P.; Higgins, C.; Charlton, P.; Callaghan, R. Drug binding sites on P-glycoprotein are altered by ATP binding prior to nucleotide hydrolysis. *Biochemistry* **2000**, *39*, 11901–11906. [CrossRef]

47. Broehan, G.; Kroeger, T.; Lorenzen, M.; Merzendorfer, H. Functional analysis of the ATP-binding cassette (ABC) transporter gene family of *Tribolium castaneum*. *BMC Genomics* **2013**, *14*, 1–18. [CrossRef]

48. Xu, X.; Qiu, J.; Xu, Y.; Xu, C.W. Molecular evolution and expression analysis of subfamily ABCB transporter genes in rice. *Chin. J. Rice Sci.* **2012**, *26*, 127–136.

49. Akifumi, S.; Nobukazu, S.; Shusei, S.; Yasukazu, N.; Satoshi, T.; Kazufumi, Y. Genome-wide analysis of ATP-binding cassette (ABC) proteins in a model legume plant, Lotus japonicus: Comparison with Arabidopsis ABC protein family. *DNA Res.* **2006**, *13*, 205–228.

50. Çakır, B.; Kılıçkaya, O. Whole-genome survey of the putative ATP-binding cassette transporter family genes in *Vitis vinifera*. *PLoS ONE* **2013**, *8*, e78860. [CrossRef]

51. Amoako, O.P.; Ayaka, M.; Mami, S.; Enrico, M.; Stefan, R.; Koh, A.; Daisuke, S.; Shungo, O.; Shogo, M.; Katsuhiro, S. Genome-wide analysis of ATP binding cassette (ABC) transporters in tomato. *PLoS ONE* **2018**, *13*, e0200854.

52. Okamoto, K.; Ueda, H.; Shimada, T.; Tamura, K.; Hara-Nishimura, I. An ABC transporter B family protein, ABCB19, is required for cytoplasmic streaming and gravitropism of the inflorescence stems. *Plant Signal Behav.* **2016**, *11*, e1010947. [CrossRef] [PubMed]

53. Zhao, H.; Liu, L.; Mo, H.; Qian, L.; Cao, Y.; Cui, S.; Li, X.; Ma, L.; Merks, R.M.H.J.P.O. The ATP-Binding cassette transporter ABCB19 regulates postembryonic organ separation in *Arabidopsis*. *PLoS ONE* **2013**, *8*, e60809. [CrossRef] [PubMed]

54. Cecchetti, V.; Brunetti, P.; Napoli, N.; Fattorini, L.; Altamura, M.M.; Costantino, P.; Cardarelli, M. ABCB1 and ABCB19 auxin transporters have synergistic effects on early and late Arabidopsis anther development. *J. Integr. Plant Biol.* **2015**, *57*, 1089–1098. [CrossRef] [PubMed]

55. Chai, C.; Subudhi, P.K. Comprehensive analysis and expression profiling of the OsLAX and OsABCB auxin transporter gene families in Rice (*Oryza sativa*) under phytohormone stimuli and abiotic stresses. *Front. Plant Sci.* **2016**, *3*, 593. [CrossRef]

56. Shen, C.J.; Bai, Y.H.; Wang, B.J.; Zhang, S.N.; Wu, Y.R.; Chen, M.; Jiang, D.A.; Qi, Y.H. Expression profile of PIN, AUX/LAX and PGP auxin transporter gene families in *Sorghum bicolor* under phytohormone and abiotic stress. *FEBS J.* **2010**, *277*, 2954–2969. [CrossRef]

57. Kim, D.Y.; Bovet, L.; Kushnir, S.; Noh, E.W.; Martinoia, E.; Lee, Y. AtATM3 is involved in heavy metal resistance in *Arabidopsis*. *Physiol. Plant.* **2006**, *140*, 922–932. [CrossRef]

58. Jayita, S.; Atreyee, S.; Kamala, G.; Bhaskar, G. Molecular phylogenetic study and expression analysis of ATP-binding cassette transporter gene family in *Oryza sativa* in response to salt stress. *Comput. Biol. Chem.* **2015**, *54*, 18–32.
59. Huang, C.F.; Yamaji, N.; Mitani, N.; Yano, M.; Nagamura, Y.; Ma, J.F. A bacterial-type ABC transporter is involved in aluminum tolerance in rice. *Plant Cell* **2009**, *21*, 655–667. [CrossRef]
60. Bird, D.A. The role of ABC transporters in cuticular lipid secretion. *Plant Sci.* **2008**, *174*, 563–569. [CrossRef]
61. Pighin, J.A.; Zheng, H.; Balakshin, L.J.; Goodman, I.P.; Western, T.L.; Jetter, R.; Kunst, L.; Samuels, A.L. Plant cuticular lipid export requires an ABC transporter. *Science* **2004**, *306*, 702–704. [CrossRef]
62. Von, R.L.E.; Degenkolb, T.; Zerjeski, M. Profiling of Arabidopsis secondary metabolites by capillary liquid chromatography coupled to electrospray ionization quadrupole time-of-flight mass spectrometry. *Plant Physiol.* **2004**, *134*, 548–559.
63. Kim, D.Y.; Jin, J.Y.; Alejandro, S.; Martinoia, E.; Lee, Y. Overexpression of AtABCG36 improves drought and salt stress resistance in *Arabidopsis*. *Physiol Plant.* **2010**, *139*, 170–180. [CrossRef] [PubMed]
64. Moons, A. Ospdr9, which encodes a PDR-type ABC transporter, is induced by heavy metals, hypoxic stress and redox perturbations in rice roots. *FEBS Lett.* **2003**, *553*, 370–376. [CrossRef]

Article

Insights into the Superoxide Dismutase Gene Family and Its Roles in *Dendrobium catenatum* under Abiotic Stresses

Hui Huang [1], Hui Wang [1,2], Yan Tong [1] and Yuhua Wang [1,*]

[1] Yunnan Key Laboratory for Wild Plant Resources, Kunming Institute of Botany, Chinese Academy of Sciences, Kunming 650201, China; huanghui@mail.kib.ac.cn (H.H.); wanghui1@mail.kib.ac.cn (H.W.); tongyan@mail.kib.ac.cn (Y.T.)
[2] University of Chinese Academy of Sciences, Beijing 100049, China
* Correspondence: wangyuhua@mail.kib.ac.cn

Received: 27 September 2020; Accepted: 20 October 2020; Published: 28 October 2020

Abstract: *Dendrobium catenatum* is a member of epiphytic orchids with extensive range of pharmacological properties and ornamental values. Superoxide dismutase (SOD), a key member of antioxidant system, plays a vital role in protecting plants against oxidative damage caused by various biotic and abiotic stresses. So far, little is known about the SOD gene family in *D. catenatum*. In this study, eight SOD genes, including four Cu/ZnSODs, three FeSODs and one MnSOD, were identified in *D. catenatum* genome. Phylogenetic analyses of SOD proteins in *D. catenatum* and several other species revealed that these SOD proteins can be assigned to three subfamilies based on their metal co-factors. Moreover, the similarities in conserved motifs and gene structures in the same subfamily corroborated their classification and inferred evolutionary relationships. There were many hormone and stress response elements in *DcaSODs*, of which light responsiveness elements was the largest group. All *DcaSODs* displayed tissue-specific expression patterns and exhibited abundant expression levels in flower and leaf. According to public RNA-seq data and qRT-PCR analysis showed that the almost *DcaSODs*, except for *DcaFSD2*, were highly expressed under cold and drought treatments. Under heat, light, and salt stresses, *DcaCSD1*, *DcaCSD2*, *DcaCSD3* were always significantly up-regulated, which may play a vital role in coping with various stresses. The expression levels of *DcaFSD1* and *DcaFSD2* were promoted by high light, suggesting their important roles in light response. These findings provided valuable information for further research on *DcaSODs* in *D. catenatum*.

Keywords: *Dendrobium catenatum*; superoxide dismutase (SOD); gene family; gene expression; stresses

---

1. Introduction

Dendrobium catenatum, belonging to *Dendrobium* genus (Orchidaceae), is a member of epiphytic orchids which takes root on the surface of tree bark or rocks [1]. Due to the special living environment, *D. catenatum* evolved novel features and sophisticated defense mechanisms that allow it to exploit its environment and against serious abiotic stresses, including thick leaves, abundant polysaccharides and facultative crassulaceaen acid (CAM) metabolism that is a photosynthetic pathway with high water-use efficiency [2–6]. *D. catenatum* is considered to be drought-resistant material useful for elucidating mechanisms of mitigating drought stress [2,5,6]. Additionally, *D. catenatum* is a well-known traditional Chinese medicinal herb, and has both an extensive range of pharmacological properties and ornamental values. Stem of *D. catenatum* contains a large number of polysaccharides that exhibits anti-inflammatory, immune-enhancing, antioxidant, and anti-glycation activities [7,8]. Light and water affected significantly the accumulation of polysaccharides [2,6]. Consequently, study on the *D. catenatum* not only has important scientific value, but also has important economic value.

The increased production of toxic reactive oxygen species (ROS) is considered to be a universal or common feature of stress conditions. ROS includes hydrogen peroxide (H_2O_2), superoxide anion radicals ($O_2^{\cdot-}$), peroxyl radicals (HOO-), hydroxyl radicals (OH-), and singlet oxygens ($^1O_2$) [9,10]. The antioxidant defense system constitutes the first line of defense against ROS produced in response to abiotic stresses. The superoxide dismutase (SOD) can catalyze the dismutation of supeoxide to H_2O_2 and $O_2^{\cdot}$ and is one of the most effective components of the antioxidant defense system in plant cells against ROS toxicity. SODs, which are localized at different cellular compartments, could be categorized into three subgroups based on metal co-factors: Cu/ZnSOD, FeSOD and MnSOD [11]. Because of their crucial roles in the antioxidant system, SODs have been reported to be involved in protection against abiotic stresses such as heat, cold, drought and salinity [12–15]. For example, Overexpression of Cu/ZnSODs in potato indicated that transgenic plants exhibited increased tolerance to oxidative stress [12]. Asensio et al. [14] indicated that stress conditions, such as nitrate excess or drought markedly increased anti-cytosolic FeSOD (cyt-FeSOD) contents in soybean tissues. Currently, the SOD gene family had been identified in many species, such as banana [15], *Arabidopsis thaliana* [16], tomato [17], cucumber [18], cotton [19], wheat [20], and *Salvia miltiorrhiza* [21]. To date, the characters of *SOD* genes and their roles in stress resistance of *D. catenatum* are still largely unknown. In this study, we perform a genome-wide analysis of *SOD* genes in *D. catenatum* genome, and investigated their characteristics, including physicochemical properties, structural characteristics and evolutionary relationships, and responses to abiotic stress. Together, our study provides a foundation for further investigation into its function of the SOD family in *D. catenatum*.

2. Results

2.1. Identification of SOD Gene Family Members in Dendrobium catenatum

After strict screening, a total of eight *DcaSOD* genes were identified in the *D. catenatum* genome (Table 1, Table S1). These *DcaSOD* genes were termed as *DcaCSD1*, *DcaCSD2*, *DcaCSD3*, *DcaCSD4*, *DcaFSD1*, *DcaFSD2*, *DcaFSD3* and *DcaMSD1*, and were assigned to three subfamilies according to their functional annotations. The number of amino acids of DcaSOD proteins ranged from 269 amino acids (aa) (DcaFSD2) to 76 aa (DcaFSD3) with an average of 196 aa. The molecular weights (MW) changed from 15.31 (DcaCSD2) to 30.78 kDa (DcaFSD2) with isoelectric points of 4.91 (DcaFSD2)—8.61 (DcaMSD1). The subcellular localization of eight DcaSOD proteins were predicted by ProComp. The results showed that DcaCSD1, DcaFSD1, and DcaFSD2 may be located in the chloroplast. DcaCSD2, DcaCSD3 and DcaCSD4 were predictably located in the cytoplasmic. In addition, the DcaMSD1 was predictably located in the mitochondria. The grand average of hydropathicity (GRAVY) of these DcaSOD proteins implied that DcaCSD1 and DcaFSD3 were hydrophobic protein, and other DcaSOD proteins were hydrophilic protein.

2.2. Phylogenetic Analysis of DcaSOD Proteins

To further explore the classification and evolutionary characteristics of these DcaSODs, multiple sequence alignment of DcaSOD protein sequences with their homologs from *Arabidopsis*, *Oryza sativa*, *Phalaenopsis equestris* and *Apostasia shenzhenica* was carried out. An un-rooted phylogenetic tree showed that all *SOD* genes were divided into three groups, named CSD, FSD and MSD subfamily (Figure 1, Table S1). The CSD subfamily had the most members (20), followed by the FSD (15) and MSD subfamily (6). We found that all DcaSOD proteins were much closer to PeqSOD proteins than AshSOD proteins.

2.3. Conserved Motifs, Gene Structures, Distribution, and cis-Elements Analysis

A total of seven motifs ranging from 21 to 50 aa were searched by MEME analysis. As shown in Figure 2, almost all members in the same subfamily shared common motif compositions with each other, suggesting functional similarities among these SOD proteins within the same subfamily. Motif 2 was widely distributed across almost all SOD proteins, except for PeqCSD5 and DcaFSD3. The members of CSD subfamily contained the motifs 1, 2, 4, and 7. The members of MSD subfamily contained the motifs 2, 3, and 6. Motif 5 was presented in the members of FSD subfamily, except for DcaFSD3. The predicted exon-intron structures were analyzed to gain an insight into the variation of *SOD* genes in *D. catenatum*. The results showed that the exons numbers of *DcaSOD* genes ranged from 2 (*DcaFSD3*) to 8 (*DcaCSD1*) (Figure 3a). In addition, the identified eight *DcaSOD* genes were mapped onto scaffolds, which showed that they distributed in eight different scaffolds (Figure 3b). Multiple sequence alignment of the eight DcaSOD amino acid sequences was performed and the results showed that all DcaCSDs contained the Cu/ZnSOD signatures (G[FL]H[VLI]H[DEGS][FY]GD[TI]) and (GNAG[GA]R[LI][AG]CG) (Figure 3c). The conserved metal-binding domain (D[MV]WE[VH][TA][IY][LY]) were found in DcaFSDs and DcaMSD. In addition, the signature (A[EQ][VT]WNHDFFW[EQ]S) responsible for the recognition of iron ion by FeSODs were identified in DcaFSDs, except for DcaFSD3.

To further explore the potential functions of *DcaSOD* genes during plant growth and stress responses, the sequences of the 2.0 kb region upstream of the translation initiation site of each of *DcaSOD* genes were analysis using the PlantCARE. In total, 142 *cis*-elements related to hormones and stresses responses were identified in all identified *DcaSOD* promoters (Figure 4). Among these predicted *cis*-elements, the light responsiveness element was the largest group including Box 4 (25), G-Box (7), GT1-motif (10), I-box (8), TCT-motif (8), TCCC-motif (2) and others (17) (Table S2). The Box 4 (25), the most light-responsive element was present in 6 *DcaSOD* promoters. Fourteen CGTCA-motif and 14 TGACG-motif, both of which are involved in MeJA-responsiveness, were identified in five *DcaSOD* promoters, respectively. ABRE was abscisic acid (ABA) responsiveness element and was present exclusively in *DcaFSD1 and DcaFSD2* promoters. *DcaCSD2* and *DcaFSD1* promoters contained low-temperature-responsive element (LTR) that was involved in response to cold stress. TATC-box and P-box that were gibberellin-responsive element were found to be present in three and three *DcaSOD* promoters, respectively. The TCA-element that was involved in salicylic acid responsiveness were found to be present in *DcaCSD2* and *DcaCSD3* promoters. In addition, eight ARE *cis*-elements involved in anaerobic induction were identified in five *DcaSOD* promoters. There was only a *cis*-element was identified in *DcaFSD3* promoter, which is due to many gaps in its promoter region.

Table 1. Summary of physicochemical characteristics and classification of DcaSOD proteins in *D. catenatum*.

No.	Gene Name	Gene ID	Functional Annotations	Position	Protein Length (aa)	MW	pI	Predicted Subcellular Localization	Grand Average of Hydropathicity (GRAVY)
1	*DcaCSD1*	Dca002609	Cu/ZnSOD	KZ503041.1:12,499,497–2,513,682	224	22,761.91	6.43	Chloroplast	0.086
2	*DcaCSD2*	Dca003218	Cu/ZnSOD	KZ501977.1:4,636,681–4,648,211	152	15,312.98	5.34	Cytoplasmic	−0.208
3	*DcaCSD3*	Dca012338	Cu/ZnSOD	KZ503541.1:64,465–66,952	222	22,926.95	6.29	Cytoplasmic	−0.008
4	*DcaCSD4*	Dca016361	Cu/ZnSOD	KZ502953.1:468,317–473,374	164	16,547.33	6.16	Cytoplasmic	−0.151
5	*DcaFSD1*	Dca023874	FeSOD	KZ501928.1:290,962–306,815	228	26,325.25	6.71	Chloroplast	−0.344
6	*DcaFSD2*	Dca004481	FeSOD	KZ502877.1:8,227,560–8,230,822	269	30,783.39	4.91	Chloroplast	−0.488
7	*DcaFSD3*	Dca024864	FeSOD	KZ503331.1:79–387	76	26,554.21	5.66	None	0.603
8	*DcaMSD1*	Dca024548	MnSOD	KZ502716.1:147,498–155,759	238	26,554.21	8.61	Mitochondrial	−0.343

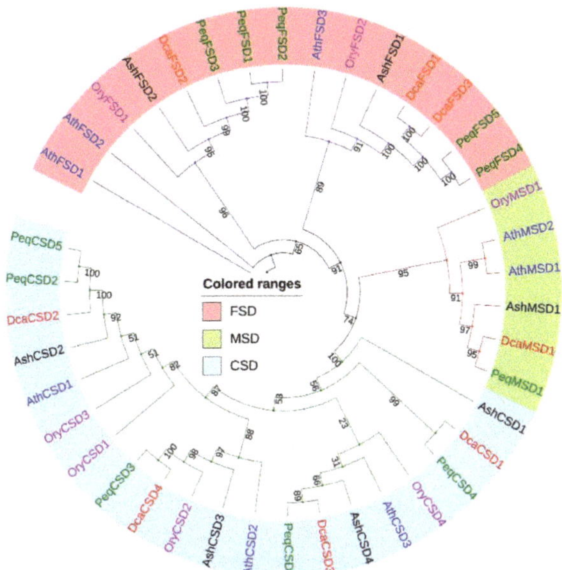

Figure 1. Phylogenetic tree of SOD proteins from *D. catenatum, Arabidopsis, Oryza sativa, Phalaenopsis equestris* and *Apostasia shenzhenica*. The phylogenetic tree was constructed using the Maximum- Likelihood (ML) method with 1000 bootstrap replications. The three subfamilies were distinguished in different colors. The identified DcaSOD proteins were highlighted by red.

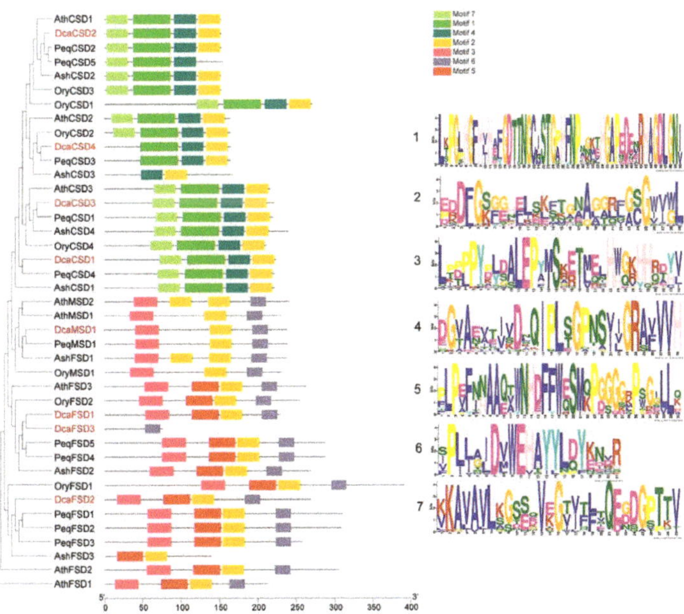

Figure 2. The motif composition and distribution of SOD proteins in *D. catenatum, Arabidopsis, O. sativa* and *P. equestris* and *A. shenzhenica*. The colored boxes with numbers represent seven motif. The identified DcaSOD proteins were highlighted by red.

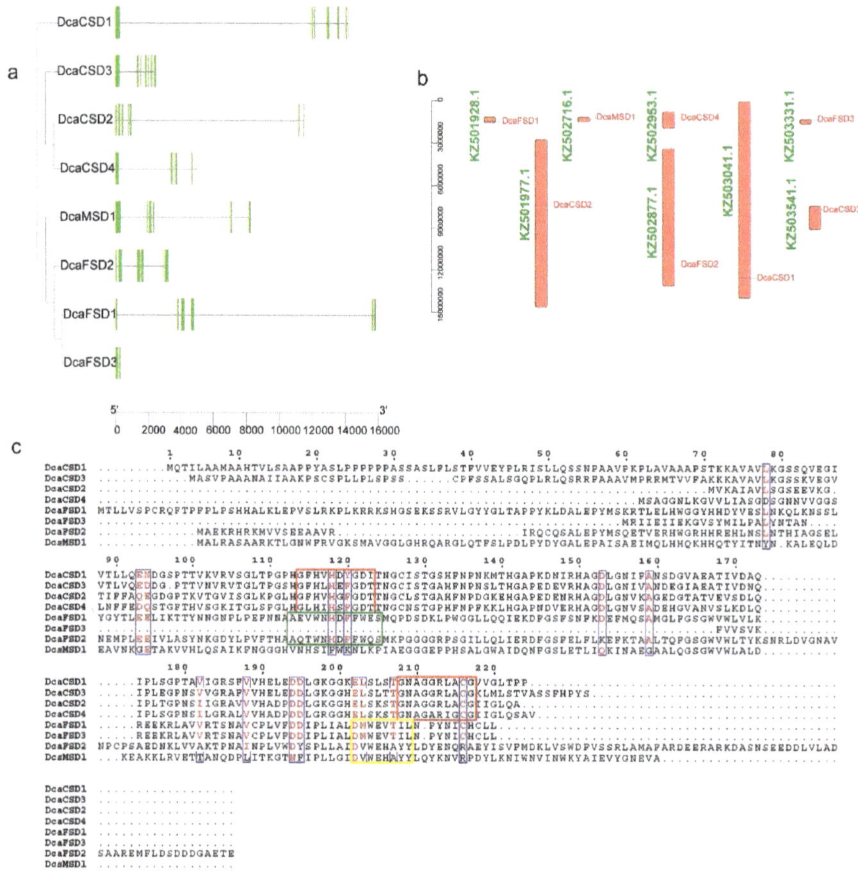

Figure 3. Gene structures, locations, and multiple sequence alignment of *DcaSOD* genes. (**a**) Gene structures of *DcaSOD* genes. Exons and introns were represented by green ellipses and gray lines, respectively. (**b**) Locations of *DcaSOD* genes. The orange bars indicate the different scaffolds of *D. catenatum* genome. (**c**) Multiple sequence alignment of amino acid sequences of DcaSOD proteins. Cu/ZnSOD signatures (G[FL]H[VLI]H[DEGS][FY]GD[TI]) and (GNAG[GA]R[LI][AG]CG) are boxed in red. The conserved metal-binding domain (D[MV]WE[VH][TA][IY][LY]) for Fe-MnSOD is in yellow box. The signature (A[EQ][VT]WNHDFFW[EQ]S) responsible for the recognition of iron ion by FeSODs in boxed in green.

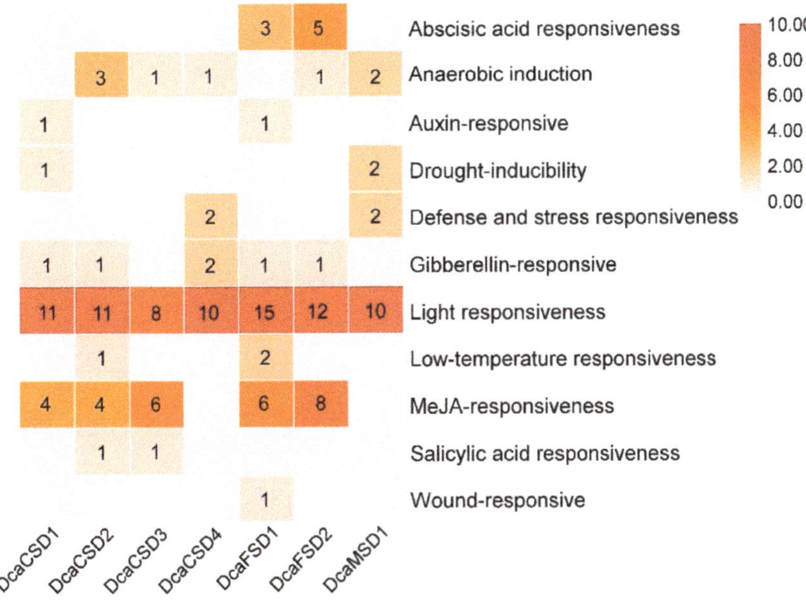

Figure 4. *Cis*-elements in promoters of *DcaSOD* genes that are related to hormone and stresses responses. The bar indicates that the number of *cis*-elements.

2.4. Distinct Expression Profiles of DcaSOD Genes in Different Tissues, Cold and Drought Responses

To analyze the expression profiles of eight *DcaSOD* genes, we investigated their transcripts abundance patterns across multiple tissues including flower, leaf, stem, and root based on public RNA-seq data. The heat map showed that almost all *DcaSOD* genes had tissue-specific expression patterns (Figure 5a). Most of them were highly expression level in flower and leaf, especially *DcaCSD2*, *DcaCSD3*, *DcaCSD4* and *DcaMSD1*, and were lowly expressed in root and stem. Under cold treatment (0 °C for 20 h), the expression levels of *DcaCSD1*, *DcaCSD3*, *DcaCSD4*, *DcaFSD1* and *DcaFSD3* increased by cold treatment (Figure 5b). Of them, the expression level of *DcaFSD3* under cold treatment was more than three times than that of control. *DcaCSD2* and *DcaMSD1* showed constitutive expression with high expression levels in leaves under control and cold treatment.

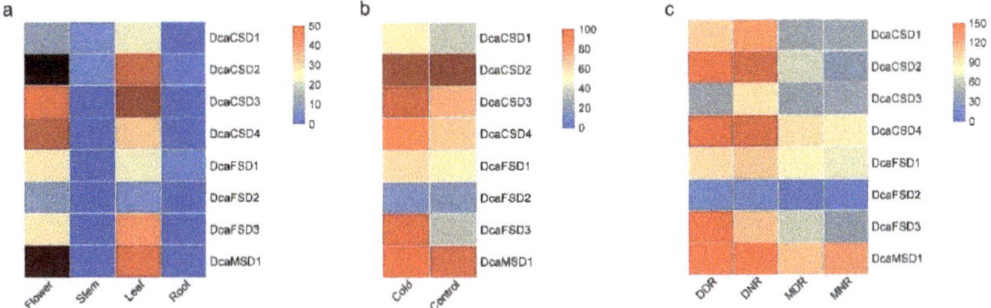

Figure 5. Expression profiles of *DcaSOD* genes in different tissues, cold and drought treatments. (**a**) Tissues. (**b**) Cold treatment. (**c**) Drought treatment. The Fragments Per Kilobase Million (FPKM) values of genes in samples were showed by different colored rectangles. Red indicates high expression level. Blue indicates low expression level.

To achieve a better understanding of the roles of SODs under drought in *D. catenatum*, we used public RNA-seq data to analyze the transcriptomes of leaves under control (30–35% volumetric water content) and serious drought (0% volumetric water content), were collected at 09:00 (designed as moist-day MDR and dried-day DDR) and 21:00 (designed as moist-night MNR and dried-night DNR) (Figure 5c). Under drought treatment, the expression of all *DcaSOD* genes were up-regulated at night and in the daytime. The expression levels of *DcaCSD2* were six and two times under drought than that of control at night and in the daytime, respectively. The expression levels of *DcaCSD1* were three and two times under drought than that of control at night and in the daytime, respectively. It is notable that the four *DcaCSDs* were highly expressed in DNR than in DDR. Among the FSD subfamily, the expression levels of *DcaFSD3* were three and two times under drought than that of control at night and in the daytime, respectively. After differentially expressed genes (DEGs) identification between DDR, DNR, MDR and MNR, a total of 5478 DEGs were identified (fold change $\geq$ 2 and FDR $\leq$ 0.01) (Table S3). *DcaCSD1* and *DcaCSD2* genes were annotated as DEGs. The Pearson correlation analysis was carried out to predict the relationship between the two *DcaSODs* and other DEGs, considering positive ($\geq$0.99) and negative ($\leq$−0.99) relationship with *p*-value < 0.005 (Table 2). In our result, two genes showed negative correlation with the *DcaCSD1*, while 13 genes were detected as positive. There were 12 negative correlation genes and three positive correlation genes with the *DcaCSD2*.

2.5. qRT-PCR Verified the Expression of DcaSOD Genes in Response to Heat, Light and NaCl Treatments

To gain insight into potential functions, qRT-PCR was used to assess the heat (35 °C), high light (HL), dark (LL) and NaCl treatments on the expression of eight *DcaSOD* genes in *D. catenatum* (Figure 6). Under heat stress, *DcaCSD1* and *DcaCSD3* were markedly up-regulated with the extension of time, especially in 48 h. The expression levels *DcaCSD2* significantly increased in 4 h and 24 h, and followed with a decrease expression level in 48 h. The *DcaFSD1*, *DcaFSD2*, *DcaFSD3* and *DcaMSD1* were down-regulated under heat stresses, except for the *DcaFSD1* in 48 h, *DcaFSD3* in 24 h and *DcaMSD1* in 4 h (Figure 6). Expression levels of seven *DcaSOD* genes including *DcaCSD1*, *DcaCSD2*, *DcaCSD3*, *DcaCSD4*, *DcaFDS1*, *DcaFDS2* and *DcaMSD1*, were markedly up-regulated in leaves after HL treatments. The expression level of *DcaCSD3* under HL treatment for 24 h was two times than that of control. It is notable that LL suppressed the expression of almost *DcaSOD* genes (Figure 7). Under salt stress, six *DcaSOD* genes including *DcaCSD1*, *DcaCSD2*, *DcaCSD3*, *DcaCSD4*, *DcaFSD1*, and *DcaMSD1* were up-regulated in 24 h, especially the members of CSD subfamily. With the extension of salt stress time to 48 h, the expression levels of these *DcaSOD* genes were down-regulated compared with 24 h, except for the *DcaFSD1* (Figure 8).

Table 2. The Pearson correlation coefficients between two *DcaCSD* genes and other DEGs. The FPKM values of these genes are listed in Table S3. The genes with a correlation coefficients > 0.99 with *DcaCSD1* have been indicated by red. The genes with a correlation coefficients > 0.99 with *DcaCSD2* have been indicated by blue.

Gene ID	Correlation Coefficient	*p*-Value	Description
DcaCSD1			
DN40231_c3_g1_i3	0.999	0.0004	MADS box protein DOMADS2
DN32316_c5_g1_i3	0.999	0.0005	HVA22-like protein e
DN28998_c0_g1_i2	−0.999	0.0007	Cytochrome P450 90B2
DN33200_c0_g6_i4	0.999	0.0008	/
DN37544_c1_g1_i1	0.999	0.0012	CDT1-like protein a, chloroplastic
DN31187_c4_g1_i2	0.999	0.001	Sucrose synthase
DN36361_c1_g5_i2	−0.997	0.0026	Trimethyltridecatetraene synthase
DN33357_c2_g2_i1	0.996	0.0038	Putative glutathione peroxidase 7, chloroplastic
DN24425_c0_g1_i2	0.996	0.0039	Dermcidin
DN32586_c2_g4_i1	0.996	0.0041	Aldehyde dehydrogenase family 2 member C4
DN36550_c0_g2_i4	0.996	0.0041	DEAD-box ATP-dependent RNA helicase 50
DN29711_c0_g1_i2	0.996	0.0044	Keratin, type I cytoskeletal 14
DN78496_c0_g1_i1	0.996	0.0044	Apolipoprotein E
DN39306_c2_g1_i5	0.995	0.0048	Keratin, type I cytoskeletal 10
DN29035_c0_g1_i1	0.995	0.0049	Probable histone H2A.2
DcaCSD2			
DN29768_c0_g1_i3	−0.995	0.0046	Proline-rich receptor-like protein kinase PERK13
DN33141_c1_g5_i1	−0.995	0.0049	/
DN33417_c1_g1_i2	−0.999	0.0013	Zinc-finger homeodomain protein 2
DN34829_c2_g1_i11	−0.997	0.0027	Receptor-like serine/threonine-protein kinase SD1-7
DN36012_c3_g1_i1	−0.999	0.0008	WAT1-related protein At5g64700
DN36361_c1_g1_i2	−0.996	0.0037	Trimethyltridecatetraene synthase
DN36986_c4_g1_i2	0.997	0.0033	Starch branching enzyme I
DN37712_c8_g3_i2	−0.999	0.0003	3-O-acetylpapaveroxine carboxylesterase CXE1
DN38190_c0_g2_i6	−0.997	0.0027	Ricin B-like lectin R40C1
DN38200_c5_g1_i1	−0.995	0.0049	/
DN38969_c0_g3_i6	−0.998	0.0020	RNA-directed DNA polymerase homolog
DN40032_c2_g2_i2	−0.999	0.0008	Transposon Ty3-I Gag-Pol polyprotein
DN40530_c1_g4_i2	0.998	0.0025	Probable alpha-mannosidase At5g13980
DN40586_c9_g5_i1	0.997	0.0030	Homeobox-leucine zipper protein HOX32
DN40622_c1_g3_i1	−0.998	0.0018	Protein ASPARTIC PROTEASE IN GUARD CELL 1

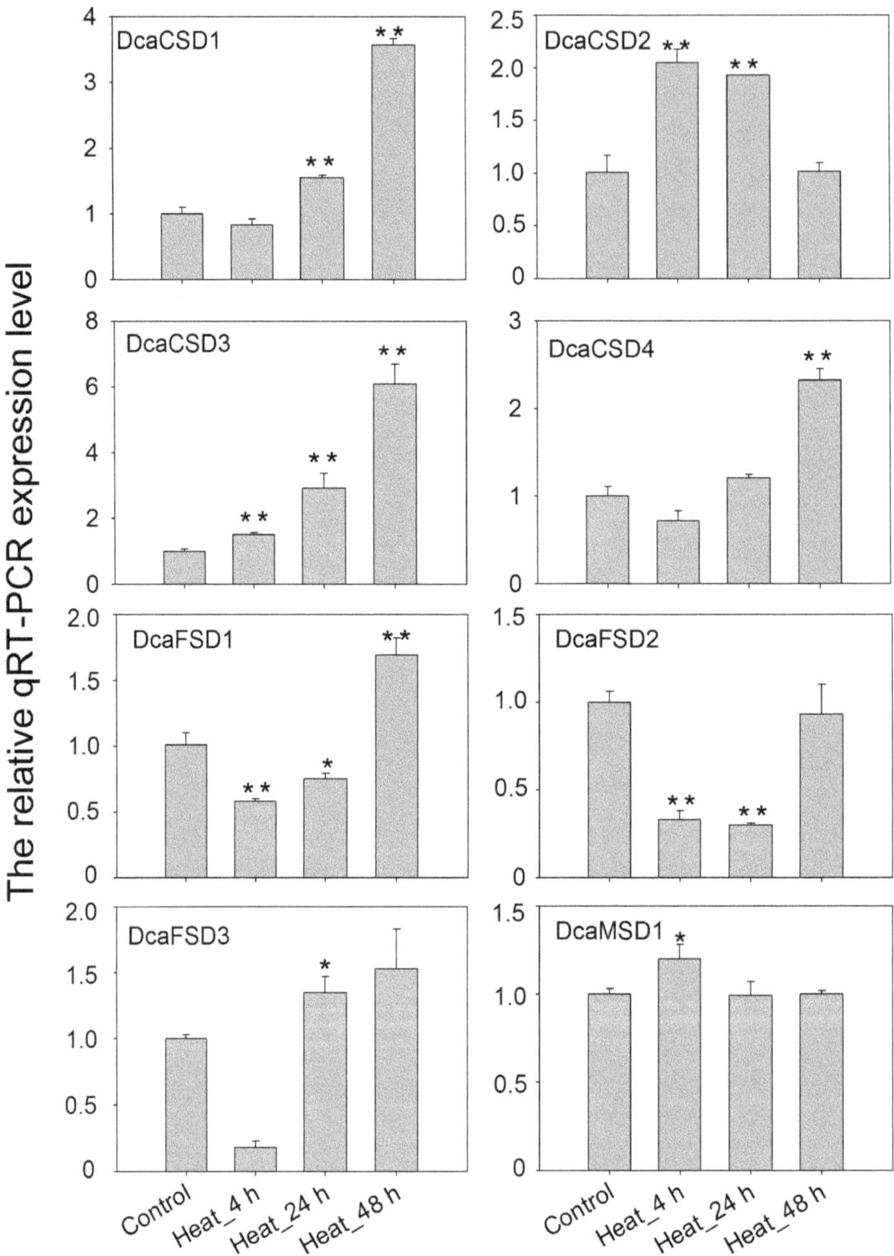

Figure 6. qTR-PCR analysis of the expression patterns of *DcaSOD* genes under heat treatments. The relative qRT-PCR expression levels were calculated with $2^{-\Delta\Delta CT}$ and the *DcaActin* gene was used as endogenous reference gene. Error bars represent the standard deviation of three replications. Bars marked with asterisks indicate significant differences (Student's *t*-test) to corresponding control samples for the time points under treatments (* $p < 0.05$, ** $p < 0.01$).

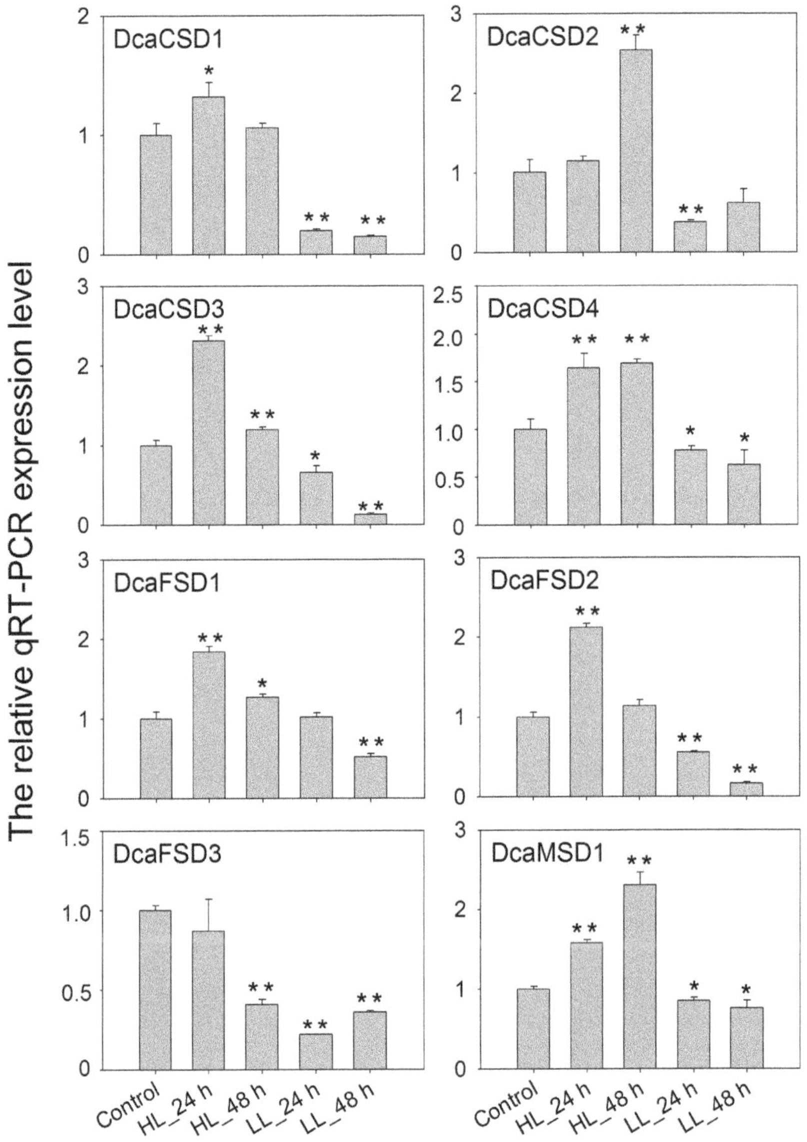

Figure 7. qTR-PCR analysis of the expression patterns of *DcaSOD* genes under high light and dark treatments. The relative qRT-PCR expression levels were calculated with $2^{-\Delta\Delta CT}$ and the *DcaActin* gene was used as endogenous reference gene. Error bars represent the standard deviation of three replications. Bars marked with asterisks indicate significant differences (Student's *t*-test) to corresponding control samples for the time points under treatments (* $p < 0.05$, ** $p < 0.01$).

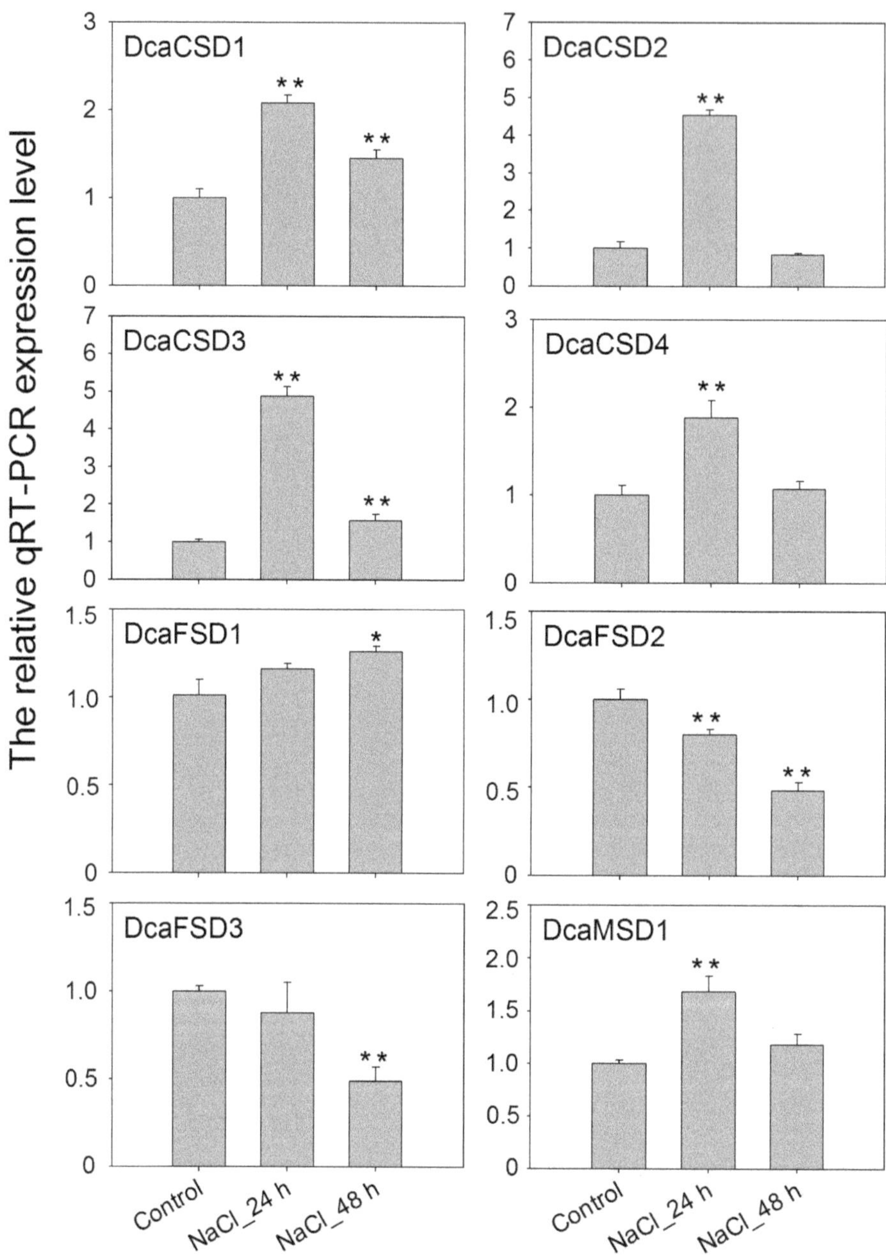

Figure 8. qTR-PCR analysis of the expression patterns of *DcaSOD* genes under NaCl treatment. The relative qRT-PCR expression levels were calculated with $2^{-\Delta\Delta CT}$ and the *DcaActin* gene was used as endogenous reference gene. Error bars represent the standard deviation of three replications. Bars marked with asterisks indicate significant differences (Student's *t*-test) to corresponding control samples for the time points under treatments (* $p < 0.05$, ** $p < 0.01$).

3. Discussion

Environmental stresses pose considerable challenges for plant growth and development. SODs are the core of antioxidant enzymes, and can effectively reduce oxidative damage via scavenging the active oxygen produced by organisms under stress. The crucial roles of SOD genes in the acclimation of plants to abiotic stresses have been demonstrated in many previous studies. However, detailed information concerning *DcaSODs* characters and functions, particularly their role in stresses responses of *D. catenatum*, remained unclear. Therefore, a systematic analysis of the SOD gene family was performed in *D. catenatum*.

In this study, a total of eight *DcaSOD* genes were identified in *D. catenatum* genome. Compared with other plant species, the number of *SOD* genes in *D. catenatum* is close to *A. thaliana* (8), tomato (8), cucumber (9), and *S. miltiorrhiza* (8), which is less than that in banana (12), cotton (*Gossypium hirsutum*, 18), Wheat (26). Intragenome syntenic relationship analysis indicated that *MaCSD2A* and *2B*, and *MaMSD1A* and *1C* or *1D* in banana genome were derived from whole genome duplication [15]. Differences in the number of *SOD* genes between plant species may be attributed to gene duplication, which comprises tandem and segmental duplication, and plays a crucial role in the expansion of *SOD* genes for diversification. Analysis of the evolutionary relationships of SOD proteins among *D. catenatum*, *Arabidopsis*, *O. sativa*, *P. equestris* and *A. shenzhenica* showed that SOD proteins could be divided into three subfamily based on their metal co-factors, namely CSD, FSD, and MSD subfamily (Figure 1). Previous studies showed that there are only FeSOD and MnSOD in algae and bryophytes. Cu/ZnSOD only exist in higher plants, implying that FeSOD and MnSOD evolved first, and then Cu/ZnSOD appeared later to cope with the complex external environment that affects plant growth and development [22]. We found that all DcaSOD proteins were much closer to PeqSOD proteins than AshSOD proteins. There is a much closer evolutionary relationship between *D. catenatum* and *P. equestris* than *D. catenatum* and *A. shenzhenica* based on the whole genome sequences analysis [23]. The conserved motifs and gene structures provided further support for the classification of DcaSOD proteins in *D. catenatum*. The DcaFSD3 has a shorter length compared to other DcaSODs and FSDs in other species. We infer that the gene has been truncated, and key domains are missing. The failure of subcellular location prediction, and few identified motifs and exons in DcaFSD3 might attribute to its truncated sequence. Most DcaSOD proteins of the same group apparently had similar motifs constituents (Figure 2), which is in accordance with the results in other plant species [15,17,18]. Taken together, the similarities in conserved motifs and gene structures in the same subfamily corroborate their classification and inferred evolutionary relationships.

In *D. catenatum*, most *DcaSOD* genes displayed distinct tissue-specific expression patterns (Figure 5a). Most of them showed higher expression level in flower, which is similar to previous studies on foxtail millet [24] and *Zostera marina* [25]. There is a high production of ROS during organogenesis and reproductive metabolism, and high expression levels of *SOD* gene in flowers [18,19]. Moreover, most *DcaSOD* genes were highly expressed in leaf. *MaCSD1D* and *MaFSD1A* in banana exhibited the highest expression levels in leaves. However, other *MaSODs* were expressed moderately in leaves [15]. In cucumber, the expression of *CsCSD1*, *CsCSD2*, *CsFSD1*, *CsFSD2* were also abundantly expressed in leaves [18]. The preferential expression patterns of *SOD* genes imply their specific roles in the development and biological function of different tissues. In addition, there might be functional divergence of *SODs* in different species.

A lot of evidence demonstrated that *SOD* genes participated in abiotic stresses responses [12–15]. A total of 142 *cis*-elements related to hormones and stresses responses were identified in all identified *DcaSOD* promoters (Figure 4). Among these predicted *cis*-elements, the light responsiveness elements (77) were the largest group element and appeared in six *DcaSOD* promoters (Figure 4). In cucumber, the light-responsive elements were also the largest group of elements [18]. A relatively large number of light-responsive *cis*-elements was also observed in tomato *SiSOD* promoters [17]. These results might suggest that the *SOD* genes participate in light response, which was confirmed by our qRT-PCR analysis (Figure 6). Under high and low light conditions, almost *DcaSOD* genes showed different expression levels. The high light promoted the expression of all *DcaSOD* genes, except for *DcaFSD3*. Transgenci pea with an overexpressing chloroplastic Cu/ZnSOD showed increased resistant to high light [26]. Although only three low-temperature responsiveness elements were found in *DcaCSD2* and *DcaFSD1* genes (Figure 4), almost *DcaSODs* were up-regulated under cold treatment (Figure 5b), demonstrating that those *DcaSODs* might play a predominant antioxidant role under cold stress. The introduction of *MnSOD* into the mitochondria and chloroplast of alfalfa resulted in an improvement of freeze tolerance [27]. Several *cis*-elements involved in drought response including ABA responsiveness, drought-inducibility and defense and stress responsiveness were identified in *DcaCSD4*, *DcaFSD2*, *DcaFSD3* and *DcaMSD1* (Figure 4). RNA-Seq data analysis results showed that all *DcaSODs* were up-regulated under drought treatment (Figure 5). Drought tolerance of sugarcane was attributed to the elevated activity of SOD [28]. Han et al. [21] demonstrated that *SmCSD2* and *SmCSD3* were highly expressed under drought stress in *S. miltiorrhiza*. Furthermore, the Pearson correlation analysis revealed that several genes function as stress resistance had close relationship with *DcaCSD1* and *DcaCSD2*. For example, a *glutathione peroxidase* (DN33357_c2_g2_i1) that constitutes a glutathione peroxidase-like protective system against oxidative stresses [29], has a close relationship with *DcaCSD1*. The *proline-rich receptor-like protein kinase* (PERK) (DN29768_c0_g1_i3) has a close relationship with *DcaCSD2*. PERK suppressed the accumulation of ROS in *Arabidopsis* root [30]. The correlation analysis could provide useful information to reveal the regulation network of important pathways [31,32]. Under heat stress, the expression levels of four *DcaCSDs* increased at three time points, except for *DcaCSD1* and *DcaCSD4* at 4 h and *DcaCSD2* at 48 h. However, the expression patterns of three *DcaFSDs* were opposite to that of *DcaCSDs*, and deceased under heat stress at 4 h. Heat stress can cause photoinhibition of PSII and promotes the accumulation of ROS, which accelerates oxidative stress [33]. Heat stress represses the expression of chloroplastic *MaCSD2A* but strongly induces chloroplastic *MaFSD1A* in banana [15]. It must be aware that there are different mechanisms of antioxidants to response the short-time (minutes and hours) reaction and the long-term adaptation in different light and temperature treatments [34]. Under NaCl stress, four *DcaCSDs* gene and *DcaMSD1* were significantly up-regulated in 24 h, especially *DcaCSD2* and *DcaCSD3*, and then decreased in 48 h. Contrast to *DcaCSDs* and *DcaMSD1*, three *DcaFSDs* were down-regulated or unchanged. We infer that *DcaCSDs* and *DcaMSD1* genes play important role in coping with salt stress. Wang et al. [35] reported that transgenic *Arabidopsis* overexpressing *MnSOD* enhanced salt-tolerance. Further studies are needed to clarify the role of *DcaSODs* in the future.

4. Materials and Methods

4.1. Identification and Sequence Analysis of DcaSOD Genes in D. catenatum

The SOD proteins sequences of *A. thaliana* and *O. sativa* were retrieved from The Arabidopsis Information Resource (TAIR) (https://www.arabidopsis.org/) and Phytozome (http://www.phytozome.net/). The genome sequences of the *D. catenatum*, *P. equestris* and *A. shenzhenica* were downloaded from NCBI under the accession codes JSDN00000000 [6], PRJNA389183 and PRJNA310678 [23]. Firstly, SOD proteins sequences from *Arabidopsis* and *O. sativa* as query sequences to search the *D. catenatum*, *P. equestris* and *A. shenzhenica* protein database for candidate sequences by BLASTP (E-value $< 1 \times 10^{-5}$). In addition, then, the hmmsearch program of the HMMER software (version 3.2.1) (http://hmmer.org/download.html) was also applied to the identification of Fe/MnSOD (PF02777 and PF00081) and Cu/ZnSOD (PF00080) in Pfam 32.0 database (http://pfam.xfam.org/). For further screening, the above obtained protein sequences were analyzed by SMART (http://smart.embl-heidelberg.de/) and NCBI Conserved Domain-search (https://www.ncbi.nlm.nih.gov/cdd). Physicochemical characteristics of the DcaSOD proteins were computed using the online ExPASy-ProtParam tool (http://web.expasy.org/protparam/), including the number of amino acids, molecular weight (MW), isoelectric point (pI) and grand average of hydropathicity (GRAVY). The subcellular localization of DcaSOD proteins were predicted by ProComp 9.0 (http://linux1.softberry.com).

4.2. Phylogenetic Analysis

All predicted DcaSODs together with SODs of *Arabidopsis*, *O. sativa*, *P. equestris* and *A. shenzhenica* were aligned with CLUSTAL. A Maximum-Likelihood (ML) phylogenetic tree was constructed by MEGA X (version 10.1.7) [36], with the bootstrap values of 1000 replicates. The phylogenetic tree was visualized by iTOL (https://itol.embl.de/) [37].

4.3. Conserved Motifs, Gene Structures, Locations, Multiple Sequence Alignments and Cis-elements Analyses of DcaSOD Genes

The distribution of the conserved motifs based on amino acid sequence was conducted with the MEME (http://meme.nbcr.net/meme/) [38] and was visualized by using TBtools [39]. The exon-intron structures of *DcaSOD* genes were analyzed by the Gene Structure Display Server (GSDS) (http://gsds.cbi.pku.edu.cn/). According to the *D. catenatum* genome annotation file, the start and end location information of *DcaSOD* genes were extracted and visualized by TBtools. Multiple sequence alignments of the amino acid sequences of DcaSOD proteins were performed by ClustalW software (https://www.genome.jp/tools-bin/clustalw) and then redrawn with ESPript 3.0 (http://espript.ibcp.fr/ESPript/ESPript/index.php). For *cis*-acting regulatory elements predication, the DNA sequences (2000 bp) upstream of the initiation codon for each candidate gene were extracted, and the *cis*-elements were predicted with PlantCARE (http://bioinformatics.psb.ugent.be/webtools/plantcare/html/) [40].

4.4. Transcriptome Analysis

The raw RNA-seq data of different tissues under cold and dry treatments of *D. catenatum* were downloaded from the NCBI Sequence Read Archive (SRA) database (http://www.ncbi.nlm.nih.gov/sra) under the BioProject number PRJNA283237, PRJNA314400, and PRJNA432825 [41]. Hisat [42] was used for mapping reads to the *D. catenatum* reference genome with default parameters. The Stringtie [43] was used to analyze gene expression level, and then the Fragments Per Kilobase Million (FPKM) value was used to normalize gene expression level. The DEseq2 [44] was employed to identify the differentially expressed genes (DEGs) with a threshold of fold change ≥ 2 and false discovery rate (FDR) ≤ 0.01. The heatmaps of gene expression and the correlation coefficient calculation were performed by R software.

4.5. Plant Material, Growth Conditions and Treatments

D. catenatum three-year old plantlets were used in this study. For light treatments, the plantlets were treated under dark (designed as LL), high light (250 µmol photons $m^{-2}s^{-1}$, designed as HL) and 12 h light (80 µmol photons $m^{-2}s^{-1}$)/12 h dark photoperiod (control) for 24 and 48 h in greenhouse. For heat treatment, the plantlets were treated at 35 °C for 4, 24 and 48 h, and 25 °C was used as control in a growth chamber. Under salt stress, plantlets were treated with 0.5 M NaCl and sterile water (control) for 24 and 48 h in greenhouse. After treatments, mature leaves were collected and frozen immediately in liquid nitrogen, and stored at −80 °C for RNA extraction. Three biological replicate samples were contained in each treatment.

4.6. Real-Time PCR Experiment

Total RNAs were extracted using RNAprep Pure kit (DP441, Tiangen, Beijing, China), and were used as template to synthesize the first-strand cDNA by using the FastKing RT kit (Tiangen, Beijing China). The primers were designed based on *DcaSOD* genes sequences using Premier 5.0 software (Table S4). The *DcaActin* gene was selected as an internal standard. qRT-PCR was performed on ABI PRISM® 7500 Sequence Detection System (Applied Biosystems, Foster City,

CA, USA). qRT-PCR assay was performed as described Yao et al. [45]. The $2^{-\Delta\Delta CT}$ method was used to analyze relative transcript abundances. Student's *t*-test was employed using SPSS software (version 18.0) to calculated levels of significance (* $p < 0.05$, ** $p < 0.01$).

Supplementary Materials: Supplementary Materials can be found at http://www.mdpi.com/2223-7747/9/11/1452/s1. Table S1: The SOD protein sequences of D. catenatum, Arabidopsis, O. sativa, P. equestris and A. shenzhenica. Table S2: The predicted cis-elements of DcaSOD genes promoters. Table S3: The annotation and expression level of DEGs under drought. Table S4: qRT-PCR primers used in this study.

Author Contributions: H.H. and Y.W. conceived and designed the experiments; H.H., H.W. and Y.T. performed the experiments and analyzed the data; H.H. wrote and revised the paper. All authors read and approved the final manuscript.

Funding: This work was supported by the Strategic Priority Research Program of Chinese Academy of Sciences (No. XDA20050204), the Second Tibetan Plateau Scientific Expedition and Research (STEP) program (2019QZKK0502), and the grant from the Beijing DR PLANT Biotechnology Co., Ltd.

Conflicts of Interest: The authors declare no conflict of interest.

References

1. Atwood, J. The size of Orchidaceae and the systematic distribution of epiphytic orchids. *Selbyana* **1986**, *9*, 171–186.

2. Su, W.H.; Zhang, G.F. The photosynthesis pathway in leaves of *Dendrobium catenatum*. *Chin. J. Plant Ecol.* **2003**, *27*, 631–637.

3. Zhang, Z.; Dongxian, H.; Rongfu, G. Concomitant CAM and C3 photosynthetic pathways in *Dendrobium catenatum* plants. *J. Am. Soc. Hortic. Sci.* **2014**, *139*, 290–298. [CrossRef]

4. Zhang, S.B.; Dai, Y.; Hao, G.Y.; Li, J.W.; Fu, X.W.; Zhang, J.L. Differentiation of water-related traits in terrestrial and epiphytic *Cymbidium* species. *Front. Plant Sci.* **2015**, *6*, 1–10. [CrossRef]

5. Yan, L.; Wang, X.; Liu, H.; Tian, Y.; Lian, J.; Yang, R.; Hao, S.; Wang, X.; Yang, S.; Li, Q.; et al. The genome of *Dendrobium officinale* illuminates the biology of the important traditional chinese orchid herb. *Mol. Plant* **2015**, *8*, 922–934. [CrossRef]

6. Zhang, G.-Q.; Xu, Q.; Bian, C.; Tsai, W.-C.; Yeh, C.-M.; Liu, K.-W.; Yoshida, K.; Zhang, L.-S.; Chang, S.-B.; Chen, F.; et al. The *Dendrobium catenatum* Lindl. genome sequence provides insights into polysaccharide synthase, floral development and adaptive evolution. *Sci. Rep.* **2016**, *6*, 19029. [CrossRef]

7. Hsieh, Y.S.; Chien, C.; Liao, S.K.; Liao, S.F.; Hung, W.T.; Yang, W.B.; Lin, C.C.; Cheng, T.J.; Chang, C.C.; Fang, J.M. Structure and bioactivity of the polysaccharides in medicinal plant *Dendrobium huoshanense*. *Bioorganic Med. Chem.* **2008**, *16*, 6054–6068. [CrossRef]

8. Ng, T.B.; Liu, J.; Wong, J.H.; Ye, X.; Sze, S.C.W.; Tong, Y.; Zhang, K.Y. Review of research on Dendrobium, a prized folk medicine. *Appl. Microbiol. Biotechnol.* **2012**, *93*, 1795–1803. [CrossRef]

9. Karuppanapandian, T.; Moon, J.C.; Kim, C.; Manoharan, K.; Kim, W. Reactive oxygen species in plants: Their generation, signal transduction, and scavenging mechanisms. *Aust. J. Crop Sci.* **2011**, *5*, 709–725.

10. Gupta, S.; Dong, Y.; Dijkwel, P.P.; Mueller-Roeber, B.; Gechev, T.S. Genome-wide analysis of ROS antioxidant genes in resurrection species suggest an involvement of distinct ROS detoxification systems during desiccation. *Int. J. Mol. Sci.* **2019**, *20*, 3101. [CrossRef]

11. Wang, W.; Xia, M.; Chen, J.; Yuan, R.; Deng, F.; Shen, F. Gene expression characteristics and regulation mechanisms of superoxide dismutase and its physiological roles in plants under stress. *Biochemistry* **2016**, *81*, 465–480. [CrossRef] [PubMed]

12. Perl, A.; Perl-Treves, R.; Galili, S.; Aviv, D.; Shalgi, E.; Malkin, S.; Galun, E. Enhanced oxidative-stress defense in transgenic potato expressing tomato Cu, Zn superoxide dismutases. *Theor. Appl. Genet.* **1993**, *85*, 568–576. [CrossRef] [PubMed]

13. Pilon, M.; Ravet, K.; Tapken, W. The biogenesis and physiological function of chloroplast superoxide dismutases. *BBA Bioenerg.* **2011**, *1807*, 989–998. [CrossRef] [PubMed]

14. Asensio, A.C.; Gil-Monreal, M.; Pires, L.; Gogorcena, Y.; Aparicio-Tejo, P.M.; Moran, J.F. Two Fe-superoxide dismutase families respond differently to stress and senescence in legumes. *J. Plant Physiol.* **2012**, *169*, 1253–1260. [CrossRef] [PubMed]

15. Feng, X.; Lai, Z.; Lin, Y.; Lai, G.; Lian, C. Genome-wide identification and characterization of the superoxide dismutase gene family in *Musa acuminata* cv. Tianbaojiao (AAA group). *BMC Genom.* **2015**, *16*, 823. [CrossRef] [PubMed]

16. Kliebenstein, D.J.; Monde, R.A.; Last, R.L. Superoxide dismutase in Arabidopsis: An eclectic enzyme family with disparate regulation and protein localization. *Plant Physiol.* **1998**, *118*, 637–650. [CrossRef] [PubMed]

17. Feng, K.; Yu, J.; Cheng, Y.; Ruan, M.; Wang, R.; Ye, Q.; Zhou, G.; Li, Z.; Yao, Z.; Yang, Y.; et al. The SOD gene family in tomato: Identification, phylogenetic relationships, and expression patterns. *Front. Plant Sci.* **2016**, *7*, 1279. [CrossRef]

18. Zhou, Y.; Hu, L.; Wu, H.; Jiang, L.; Liu, S. Genome-wide identification and transcriptional expression analysis of cucumber superoxide dismutase (SOD) family in response to various abiotic stress. *Int. J. Genom.* **2017**, *2017*, 7243973.

19. Wang, W.; Zhang, X.; Deng, F.; Yuan, R.; Shen, F. Genome-wide characterization and expression analyses of superoxide dismutase (SOD) genes in *Gossypium hirsutum*. *BMC Genom.* **2017**, *18*, 376. [CrossRef]

20. Jiang, W.Q.; Yang, L.; He, Y.Q.; Zhang, H.T.; Li, W.; Chen, H.G.; Ma, D.F.; Yin, J.L. Genome-wide identification and transcriptional expression analysis of superoxide dismutase (SOD) family in wheat (*Triticum aestivum*). *PeerJ* **2019**, *7*, e8062. [CrossRef]

21. Han, L.M.; Hua, W.P.; Cao, X.Y.; Yan, J.A.; Chen, C.; Wang, Z.Z. Genome-wide identification and expression analysis of the *superoxide dismutase* (SOD) gene family in *Salvia miltiorrhiza*. *Gene* **2020**, *742*, 144603. [CrossRef] [PubMed]

22. Lin, Y.L.; Lai, Z.X. Superoxide dismutase multigene family in longan somatic embryos: A comparison of CuZn-SOD, Fe-SOD, and Mn-SOD gene structure, splicing, phylogeny, and expression. *Mol. Breed.* **2013**, *32*, 595–615. [CrossRef]

23. Zhang, G.Q.; Liu, K.W.; Li, Z.; Lohaus, R.; Hsiao, Y.Y.; Niu, S.-C.; Wang, J.-Y.; Lin, Y.-C.; Xu, Q.; Chen, L.-J.; et al. The *Apostasia* genome and the evolution of orchids. *Nature* **2017**, *549*, 379–386. [CrossRef] [PubMed]

24. Wang, T.; Song, H.; Zhang, B.; Lu, Q.; Liu, Z.; Zhang, S.; Guo, R.; Wang, C.; Zhao, Z.; Liu, J. Genome-wide identification, characterization, and expression analysis of superoxide dismutase (SOD) genes in foxtail millet (*Setaria italica* L.). *3 Biotech* **2018**, *8*, 486. [CrossRef] [PubMed]

25. Zang, Y.; Chen, J.; Li, R.; Shang, S.; Tang, X. Genome-wide analysis of the superoxide dismutase (SOD) gene family in *Zostera marina* and expression profile analysis under temperature stress. *PeerJ* **2020**, *8*, 9063. [CrossRef] [PubMed]

26. Gupta, A.S.; Heinen, J.L.; Holaday, A.S.; Burke, J.J.; Allen, R.D. Increased resistance to oxidative stress in transgenic plants that overexpress chloroplastic Cu/Zn superoxide dismutase. *Proc. Natl. Acad. Sci. USA* **1993**, *90*, 1629–1633. [CrossRef] [PubMed]

27. McKersie, B.D.; Chen, Y.; de Beus, M.; Bowley, S.R.; Bowler, C.; Inze, D.; D'Halluin, K.; Botterman, J. Superoxide dismutase enhances tolerance of freezing stress in transgenic alfalfa (*Medicago sativa* L.). *Plant Physiol.* **1993**, *103*, 1155–1163. [CrossRef]

28. Jain, R.; Chandra, A.; Venugopalan, V.K.; Solomon, S. Physiological changes and expression of SOD and P5CS genes in response to water deficit in Sugarcane. *Sugar Tech* **2015**, *17*, 276–282. [CrossRef]

29. Milla, M.A.R.; Maurer, A.; Huete, A.R.; Gustafson, J.P. Glutathione peroxidase genes in *Arabidopsis* are ubiquitous and regulated by abiotic stresses through diverse signaling pathways. *Plant J.* **2003**, *36*, 602–615. [CrossRef]

30. Hwang, Y.; Lee, H.; Lee, Y.S.; Cho, H.T. Cell wall-associated ROOT HAIR SPECIFIC 10, a proline-rich receptor-like kinase, is a negative modulator of Arabidopsis root hair growth. *J. Exp. Bot.* **2016**, *67*, 2007–2022. [CrossRef]

31. Pillet, J.; Yu, H.W.; Chambers, A.H.; Whitaker, V.M.; Folta, K.M. Identification of candidate flavonoid pathway genes using transcriptome correlation network analysis in ripe strawberry (*Fragaria* × *ananassa*) fruits. *J. Exp. Bot.* **2015**, *66*, 4455–4467. [CrossRef] [PubMed]

32. Ding, Z.; Fu, L.; Tan, D.; Sun, X.; Zhang, J. An integrative transcriptomic and genomic analysis reveals novel insights into the hub genes and regulatory networks associated with rubber synthesis in *H. brasiliensis*. *Ind. Crop. Prod.* **2020**, *153*, 112562. [CrossRef]

33. Lu, T.; Meng, Z.; Zhang, G.; Qi, M.; Sun, Z.; Liu, Y.; Li, T. Sub-high temperature and high light intensity induced irreversible inhibition on photosynthesis system of tomato plant (*Solanum lycopersicum* L.). *Front. Plant Sci.* **2017**, *8*, 365. [CrossRef] [PubMed]

34. Szymańska, R.; Ślesak, I.; Orzechowska, A.; Kruk, J. Physiological and biochemical responses to high light and temperature stress in plants. *Environ. Exp. Bot.* **2017**, *139*, 165–177. [CrossRef]

35. Wang, Y.; Ying, Y.; Chen, J.; Wang, X. Transgenic *Arabidopsis* overexpressing MnSOD enhanced salt-tolerance. *Plant Sci.* **2004**, *4*, 671–677. [CrossRef]

36. Kumar, S.; Stecher, G.; Li, M.; Knyaz, C.; Tamura, K. MEGA X: Molecular evolutionary genetics analysis across computing platforms. *Mol. Biol. Evol.* **2018**, *35*, 1547–1549. [CrossRef]

37. Letunic, I.; Bork, P. Interactive tree of life (iTOL) v3: An online tool for the display and annotation of phylogenetic and other trees. *Nucleic Acids Res.* **2016**, *44*, 242–245. [CrossRef]

38. Bailey, T.L.; Williams, N.; Misleh, C.; Li, W.W. MEME: Discovering and analyzing DNA and protein sequence motifs. *Nucleic Acids Res.* **2006**, *34*, 369–373. [CrossRef]

39. Chen, C.; Chen, H.; Zhang, Y.; Thomas, H.R.; Frank, M.H.; He, Y.; Xia, R. TBtools-an integrative toolkit developed for interactive analyses of big biological data. *Mol. Plant* **2020**, *13*, 1194–1202. [CrossRef]

40. Lescot, M.; Déhais, P.; Thijs, G.; Marchal, K.; Moreau, Y.; Van de Peer, Y.; Rouzé, P.; Rombauts, S. PlantCARE, a database of plant cis-acting regulatory elements and a portal to tools for in silico analysis of promoter sequences. *Nucleic Acids Res.* **2002**, *30*, 325–327. [CrossRef]

41. Wan, X.; Zou, L.H.; Zheng, B.Q.; Tian, Y.Q.; Wang, Y. Transcriptomic profiling for prolonged drought in *Dendrobium catenatum*. *Sci. Data* **2018**, *5*, 180233. [CrossRef] [PubMed]

42. Kim, D.; Langmead, B.; Salzberg, S.L. HISAT: A fast spliced aligner with low memory requirements. *Nat. Methods* **2015**, *12*, 357–360. [CrossRef] [PubMed]

43. Pertea, M.; Pertea, G.M.; Antonescu, C.M.; Chang, T.C.; Mendell, J.T.; Salzberg, S.L. StringTie enables improved reconstruction of a transcriptome from RNA-seq reads. *Nat. Biotechnol.* **2015**, *33*, 290–295. [CrossRef]

44. Love, M.I.; Huber, W.; Anders, S. Moderated estimation of fold change and dispersion for RNA-seq data with DESeq2. *Genome Biol.* **2014**, *15*, 550. [CrossRef] [PubMed]

45. Yao, Q.Y.; Huang, H.; Tong, Y.; Xia, E.H.; Gao, L.Z. Transcriptome analysis identifies candidate genes related to triacylglycerol biosynthesis, flower coloration and flowering time control in *Camellia reticulata* (Theaceae), a well-known ornamental and oil-producing plant. *Front. Plant Sci.* **2016**, *7*, 163. [CrossRef] [PubMed]

Brief Report

An Abiotic Stress Responsive U-Box E3 Ubiquitin Ligase Is Involved in OsGI-Mediating Diurnal Rhythm Regulating Mechanism

Yo-Han Yoo, Xu Jiang and Ki-Hong Jung *

Graduate School of Biotechnology & Crop Biotech Institute, Kyung Hee University, Yongin 17104, Korea; directorhan@khu.ac.kr (Y.-H.Y.); kangwuk97@khu.ac.kr (X.J.)
* Correspondence: khjung2010@khu.ac.kr

Received: 21 July 2020; Accepted: 17 August 2020; Published: 20 August 2020

Abstract: The plant U-box (PUB) protein is the E3 ligase that plays roles in the degradation or post-translational modification of target proteins. In rice, 77 U-box proteins were identified and divided into eight classes according to the domain configuration. We performed a phylogenomic analysis by integrating microarray expression data under abiotic stress to the phylogenetic tree context. Real-time quantitative reverse transcription polymerase chain reaction (qRT-PCR) expression analyses identified that eight, twelve, and eight *PUB* family genes are associated with responses to drought, salinity, and cold stress, respectively. In total, 16 genes showed increased expression in response to three abiotic stresses. Among them, the expression of *OsPUB2* in class II and *OsPUB33*, *OsPUB39*, and *OsPUB41* in class III increased in all three abiotic stresses, indicating their involvement in multiple abiotic stress regulation. In addition, we identified the circadian rhythmic expression for three out of 16 genes responding to abiotic stress through meta-microarray expression data analysis. Among them, OsPUB4 is predicted to be involved in the rice GIGANTEA (OsGI)-mediating diurnal rhythm regulating mechanism. In the last, we constructed predicted protein-protein interaction networks associated with OsPUB4 and OsGI. Our analysis provides essential information to improve environmental stress tolerance mediated by the PUB family members in rice.

Keywords: abiotic stress; diurnal regulation; OsGI; rice; U-box E3 ligase

---

1. Introduction

Ubiquitination is a protein degradation system that regulates the amount of intracellular accumulation of signaling substances by selectively degrading certain proteins [1]. Especially in plants, various hormone signaling mechanisms, development, biotic, and abiotic stress signaling mechanisms have been reported to be closely related to ubiquitination [2,3]. Ubiquitin is a small 8-kDa protein in all eukaryotes. It is attached to specific proteins by the three enzymes Ub-activating enzyme (E1), Ub-conjugating enzyme (E2) and Ub-ligase (E3). Specific proteins that are subsequently polyubiquitinated are degraded by the 26S proteasome [4,5]. Among these, E3 ligase plays a major role in determining the specificity of the substrate and is largely divided into single-subunit E3 ligase and multi-subunit E3 ligase depending on the structure.

The single-subunit E3 ligase acts as an E3 ligase by itself, without any additional protein [6]. Single-subunit E3 ligase is subdivided into three proteins according to the domain: RING (for Really Interesting New Gene), U-box, and HECT (for Homology to E6-AP carboxyl terminus). Interestingly, other single-subunit E3 ligases exist in similar numbers among eukaryotes, but the number of U-box E3 ligases in plants is higher than in other eukaryotes. For example, 21 and 2 U-box E3 ligases were found in human and yeast [7,8], while Arabidopsis and rice were known to be 64 and 77, respectively [9,10].

Thus, the diversity of plant U-box genes suggests that the U-box domain may play an important role in performing plant-specific intracellular processes.

The circadian clock is an evolutionary system adapted to fluctuating environmental changes in the earth [11]. In particular, circadian clocks of immovable plants have more authority than animals and participate in various developmental processes [12]. The circadian clock is divided into an input that accepts the external environment signal, an oscillator that generates the rhythm according to the change cycle of the external environment, and an output that is controlled by this oscillator [13]. The most representative function of the output in the plant circadian clock is flowering. In other words, the circadian clock of the plant recognizes the photoperiod and determines the timing of the flowering [14]. For example, *Arabidopsis* induces the degradation of CYCLING DOF FACTOR 1 (CDF1), which prevents the expression of *CONSTANS* (*CO*), an important gene for flowering, by inducing the binding between two proteins of GIGANTEA (GI) and FLAVIN-BINDING, KELCH REPEAT, F-BOX 1 (FKF1) in the long-day condition. However, in a short-day condition, the binding between the two proteins GI and FKF1 is decreased, so that the expression of *CO* is kept low and the flowering time is delayed [15].

In rice, 77 U-box E3 ligase genes are known, of which only six genes have been reported [16–20]. In other words, many rice U-box E3 ligase genes have not yet been studied, and phenotypic interference due to functional redundancy may be one of the reasons [21]. Therefore, we analyzed transcriptome data using the phylogenomics tool and tried to obtain information on the environmental response characteristics of individual genes. This study comprehensively analyzed the expression characteristics in response to abiotic stress (drought, salinity, and cold) and the circadian clock which have not been studied in the previous genome-wide PUB family. Based on this, we will provide important fundamental data for studying the functions of individual *PUB* gene family.

2. Materials and Methods

2.1. Multiple Alignment and Phylogenetic Analysis

To perform our phylogenetic analysis of the PUB family, we used the protein sequences of 77 *PUB* genes identified in a previous global analysis of the rice PUB family [10]. The protein sequences for our phylogenomic analysis were downloaded from the Rice Genome Annotation Project Website [22]. After multiple-alignment of those sequences with ClustalX [23], we generated a phylogenetic tree using the Neighbor-Joining method, as incorporated in the MEGA5 tool kit for phylogenetic analysis [24].

2.2. Meta-Analysis of Gene Expression Data and Heatmap Development

Affymetric- and agilent-microarray data (GSE6901, GSE36040 and GSE38023), and RNA-seq data (GSE92989) were downloaded from the NCBI Gene Expression Omnibus (GEO, http://www.ncbi.nlm.nih.gov/geo/). We then uploaded the normalized expression data to the Multi Experiment Viewer and visualized the data via heatmaps (http://www.tm4.org/mev.html).

2.3. Plant Materials and Abiotic Stress Treatment

Rice (*O. sativa* L.cv. Dongjin) seeds were germinated on the Murashige Skoog medium for 14 days at 28 °C. Subsequently, the seedlings were washed with sterilized water to completely remove the agar and were air-dried for 0, 2, 6, and 12 h at 28 °C for drought stress treatment [25]. To simulate salinity stress, we exposed 14-day-old plants to 200 mM NaCl for 0, 2, 6, and 12 h at 28 °C [26]. In the last, we exposed 14-day-old plants to 4 °C ± 1 °C for 0, 2, 6, and 12 h for cold stress treatment. The control plants remained at 28 °C. Leaves and roots of three plants were pooled for one biological replicate and each treatment had three biological repeats.

2.4. Plant Materials for Diurnal Rhythm

To investigate the functional associations of PUB family members with the diurnal rhythm and *OsGI (GIGANTEA)*-mediating regulatory pathway, Wild-type (WT) plant and *osgi* mutant seeds (*LOC_Os01g08700*) were germinated on the Murashige Skoog medium for 7 days at 28 °C [27]. They were then transferred to individual pots and grown in an incubator (12-h light/12-h dark, 28 °C/22 °C) for 30 days. After that, their leaves were sampled at 2-h intervals for 24 h.

2.5. RNA Extraction and Real-Time Quantitative PCR

Samples were frozen in liquid nitrogen and total RNAs were extracted using RNAiso Plus (Takara Bio, Kyoto, Japan). Using MMLV Reverse Transcriptase (Promega, Madison, WI, USA) and the oligo(dT) 15 primer, first-strand cDNA was synthesized as we recently reported [28,29]. For normalizing the amplified transcripts, we used a primer pair for rice *ubiquitin 5* (*OsUbi5/Os01g22490*) [30]. All primers for these analyses are summarized in Table S1.

2.6. Analysis of a Predicted Protein–Protein Interaction Network

Using the STRING tool (https://version-10-5.string-db.org/) [31], we generated a hypothetical protein–protein interaction network involving E3 ubiquitin ligase, transcription factors (TFs), and flowering regulatory genes. The network was edited with the Cytoscape tool (https://cytoscape.org/; version 3.6.0) (The Cytoscape Consortium, New York, NY, USA) [32,33].

3. Results and Discussion

3.1. Integration of Abiotic Stress Expression Patterns with a Phylogenetic Tree Context of the Rice PUB Family Reveals the Key PUB Family Members for the Stress Responses

According to the recent report on the PUB family in rice, 77 estimated U-box proteins were identified through a whole-genome analysis algorithm and divided into eight classes according to the domain configuration (Figure 1 and Figure S1) [10]. We have constructed a phylogenetic tree using protein sequences for each of the five classes except I, VI, and VIII, which have only one or two genes in the eight classes. The expression of 77 genes was visualized using drought- and salinity-treated RNA-seq data (GSE92989) using seedling roots [26], and cold-treated microarray data (GSE6901 and GSE38023) from using seedling leaves. When the resultant microarray data were examined according to criteria where *t*-test *p*-values were <0.01 and upregulation showed a greater than 1 (log$_2$ scale)-fold change for control versus abiotic stress, we were able to identify 16 genes (Figure 1). The remaining genes in the heat-map were visualized in gray color. As a result, we identified eight (*OsPUB2, OsPUB4, OsPUB5, OsPUB8, OsPUB33, OsPUB39, OsPUB41* and *OsPUB67*), twelve (*OsPUB2, OsPUB3, OsPUB5, OsPUB6, OsPUB33, OsPUB39, OsPUB41, OsPUB46, OsPUB51, OsPUB63, OsPUB64* and *OsPUB67*), and eight (*OsPUB2, OsPUB10, OsPUB33, OsPUB39, OsPUB41, OsPUB43, OsPUB46* and *OsPUB64*) upregulated genes under drought, salinity, and cold stress conditions [25,26]. Interestingly, expression of *OsPUB2* in class II and *OsPUB33, OsPUB39*, and *OsPUB41* in class III increased in all three abiotic stresses. In addition, most genes with increased expression in drought, salinity, and cold stress were included in classes II, III, and VII. These results indicate that among the 77 PUB families, there are pivotal classes and genes that respond to environmental stress.

3.2. Real-Time Quantitative PCR Analysis Confirmed Expression Patterns in Response to Drought, Salt and Cold Stresses of 16 PUB Family Genes in Rice

To confirm the global transcriptome data, we carried out quantitative reverse transcription polymerase chain reaction (qRT-PCR) analysis of PUB family genes in the drought, salinity, and cold stress conditions. Fourteen-day-old seedlings were treated with drought, salinity, and cold stress for 0, 2, 6, and 12 h, respectively. Root samples under drought and salinity stress and leaf samples under cold stress are collected for cDNA synthesis.

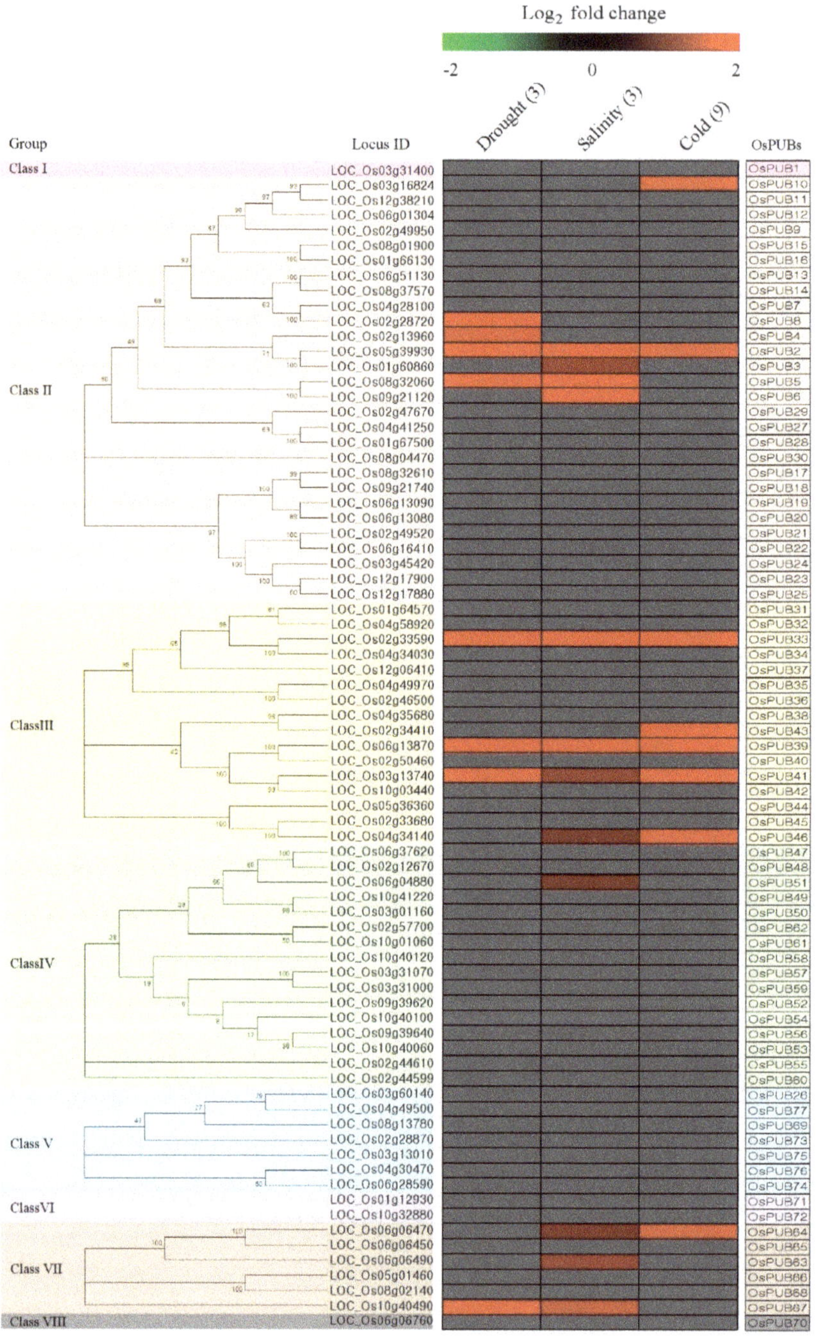

Figure 1. Meta-analysis of *OsPUB* genes expression patterns using drought- salinity- and cold-treated microarray data. On the left side of the heat-map is a phylogenetic tree for each class. Green, low level of log$_2$ intensity; red, high level; gray, *t*-test *p*-value > 0.01 or enrichment values of <1 (log$_2$)-fold.

As the first step, we tested the expression of *OsDREB1A*, a drought and cold stress marker gene, for drought (0, 2, 6, and 12 h) and cold (0, 2, 6, and 12 h) stress samples, and of expression of *OsbZIP23*, a salt stress marker gene, for salt (0, 2, 6, and 12 h) stress samples [34,35]. As expected, expressions of *OsDREB1A* and *OsbZIP23* were significantly stimulated compared to the control at all tested time points of stress treatment (Figure 2), indicating that samples under drought, salt, and cold stress treatments are well qualified for the further differential expression analyses. To validate expression patterns of selected PUB genes, we chose a 2 h sample (salinity) and 12 h samples (drought and cold) out of the time series stress treatments showing stronger and more stable upregulation of marker genes. Subsequently, we confirmed that expressions of 16 PUB genes were significantly upregulated under drought, salinity and cold stresses (Figure 2). These results are in agreement with expression patterns analyzed using transcriptome data associated with abiotic stress.

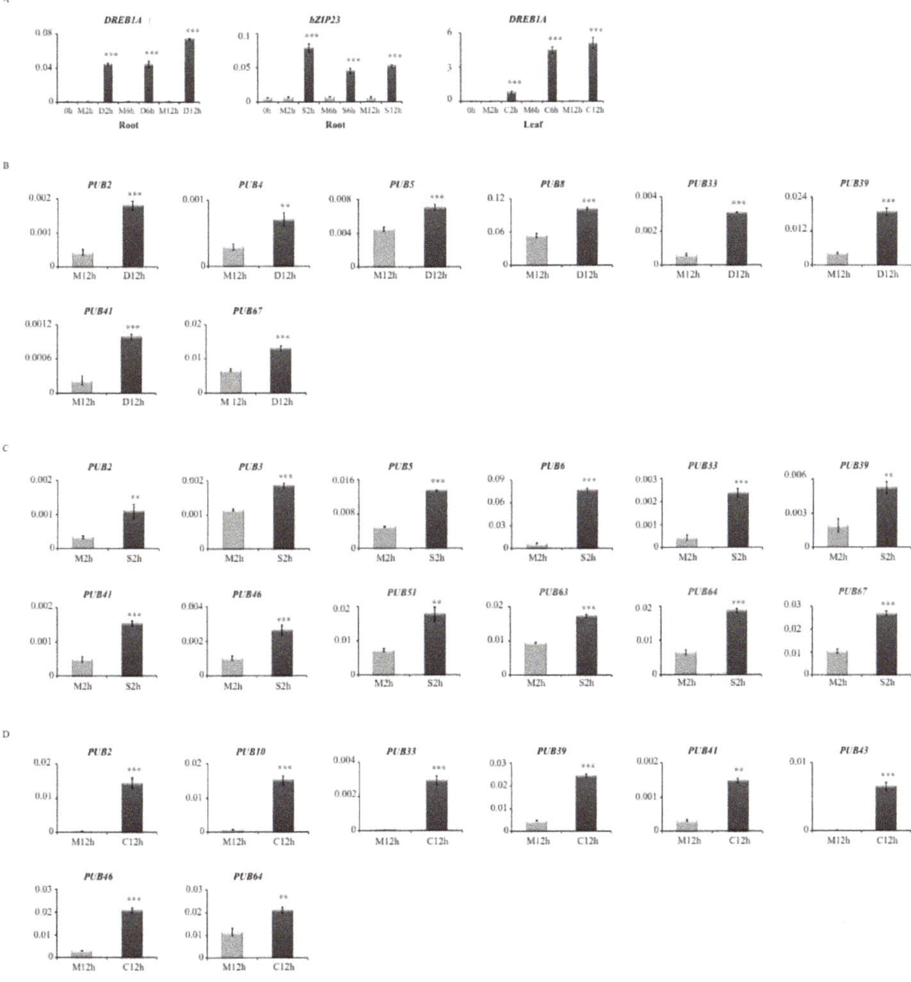

Figure 2. qRT expression profiles for 16 *OsPUB* genes selected from global transcriptome data analysis. OsDREB1A and OsbZIP23 were used as marker genes for abiotic stress (**A**). Abiotic stress samples were prepared from drought (**B**), salinity (**C**), and cold (**D**) (0, 2, 6, and 12 h) in root or leaf. Rice ubiquitin (*OsUbi5*) was served as internal control. ** *p* value < 0.01; *** *p* value < 0.001. N = 3.

3.3. PUB Family Genes Are Involved in OsGI Mediating Diurnal Regulation Pathway

Diurnal rhythm in plants is regulated by light and the circadian clock, and metabolism, physiology, and behavior change between day and night [36]. In addition, recent studies have shown that circadian rhythm correlates with abiotic stress [37–39]. To identify the *PUB* genes in rice associated with diurnal rhythm, we analyzed expression patterns using publicly available Agilent 44k array data (GSE36040) obtained in rice leaves harvested under diurnal rhythm in nine different developmental stages [40]. Of the 77 *PUB* genes, nine (*OsPUB2, OsPUB4, OsPUB16, OsPUB20, OsPUB34, OsPUB47, OsPUB52, OsPUB63,* and *OsPUB77*) genes were observed to show diurnal rhythm in the leaves (Figure S1). Among the 16 genes with increased expression in Abiotic stress, three (*OsPUB2, OsPUB4,* and *OsPUB63*) genes were associated with diurnal rhythm (Figure 1). To confirm expression associated with diurnal rhythm, 37-day-old leaves were sampled at 2 h intervals for 24 h, and we confirm that diurnal rhythms of *OsPUB4* and *OsPUB63* were observed through Real-Time qPCR analyses. There was no difference in expression in the dark state of *OsPUB4*, but expression increased when the plant first recognized the light (Figure 3). In contrast, *OsPUB63* showed no expression difference in the light state but increased expression in the dark state (Figure 3). Unfortunately, *OsPUB2* was associated with diurnal rhythms in the Agilent 44k array data, but no significant change in expression was observed in Real-Time qPCR analysis (Figure S2).

To obtain the insight into the mechanism on the regulation of the diurnal rhythm of these PUB genes, we used rice *gi* mutants with defects in the diurnal rhythm [27]. In *osgi*, diurnal expression of a well-known marker gene for diurnal rhythm, *LATE ELONGATED HYPOCOTYL (LHY)* [41], was dramatically down-regulated across all time points (Figure 3B). Interestingly, *OsPUB4*, like *LHY*, disappeared from diurnal rhythm expression patterns in *osgi* mutants (Figure 3B). In contrast, *OsPUB63* was able to observe the same diurnal rhythm expression patterns in both the control (dongjin) and *osgi* mutants. These results indicate that the *OsPUB4* gene is involved to the *OsGI*-mediating diurnal rhythm regulating mechanism.

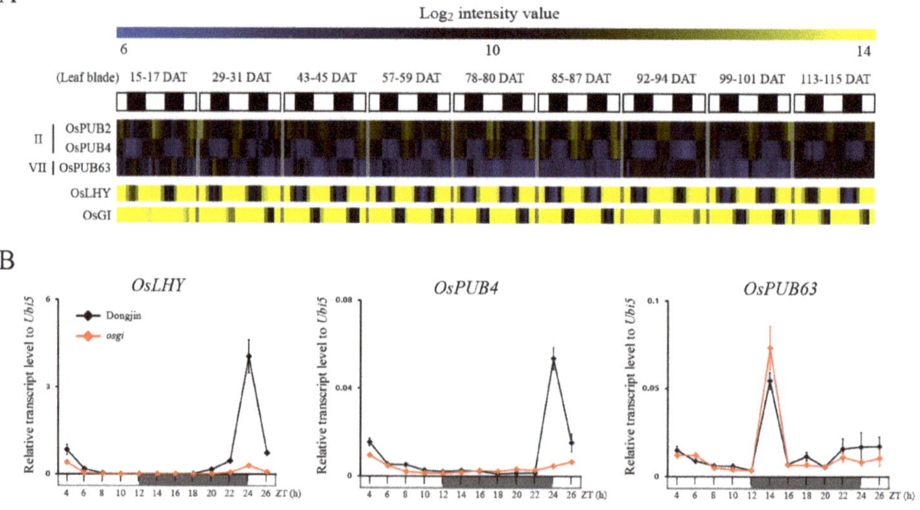

Figure 3. Diurnal expression patterns of two *OsPUB* genes in mature leaves, using available Agilent 44k array data over the entire plant life span (**A**), or evaluated at 12 time points over 24-h period in "Dongjin" rice and *osgi* mutant (**B**). *OsLHY* was standard marker gene for diurnal rhythm. *OsUbi5* was served as the internal control. The continuous white and black bars indicate day and night time, respectively. ZT, zeitgeber time (ZT = 0 at lights-on).

3.4. OsPUB4 Is Under the Control of OsGI, One of Main Regulators of the Circadian Clock

GI is involved in maintaining the circadian clock of downstream genes. It has been reported that the circadian rhythms of flowering regulatory genes such as *Ehd1* (*Early heading date 1*), *Hd3a* (*Heading date 3a*), *RFT1* (*RICE FLOWERING LOCUS T 1*), *Hd1* (*Heading date 1*) and *OsMADS51* are significantly reduced in *osgi* mutants [27,42,43]. Interestingly, like the flowering regulation genes mentioned earlier, the circadian rhythm of *OsPUB4* also decreases in the *osgi* mutant (Figure 3). This result indicates that OsPUB4 might be under the control of OsGI, one of the main regulators of the circadian clock. We created a putative network on the STRING website (https://string-db.org/cgi/input.pl?sessionId=dsUIDFue7qrX&input_page_show_search=on) to check the correlation between OsPUB4 and OsGI. As expected, the network includes photoreceptors such as PHYA (Phytochrome A), PHYB (Phytochrome B), and transcription factors related to flowering time such as Ghd7 (Grain number, plant height, and heading date7), HD3A and HD2 (Figure 4).

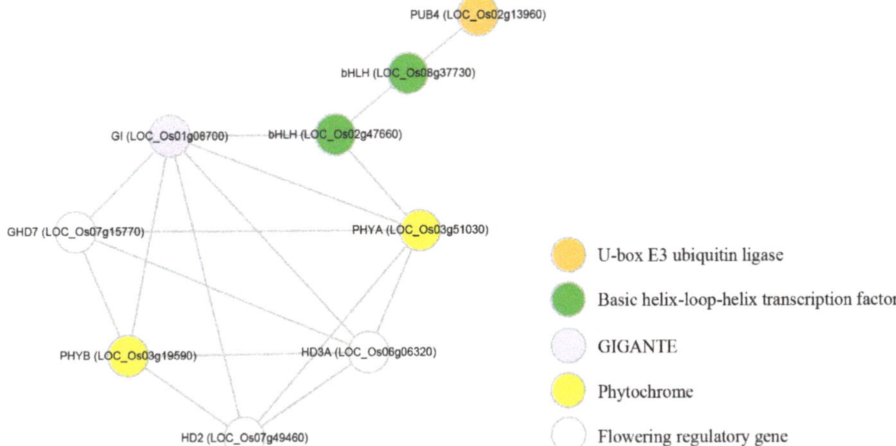

Figure 4. Construction of predicted protein-protein interaction networks associated with OsPUB4 and OsGI. Using the STRING tool, we found seven proteins that are expected to interact with OsPUB4 (orange circle) and OsGI (blue circle). Three of the seven proteins were identified as flowering regulatory genes (White circles), two as phytochrome (yellow circles), and two as bHLH TF (green circles).

4. Conclusions

In this study, we selected genes that respond to abiotic stress in the *OsPUB* family, and further confirmed their circadian clock. Recent studies have shown that genes associated with the circadian clock and flowering time are associated with abiotic stress [38,39]. For example, in Arabidopsis, the *GI*-overexpressed transgenic plants show increased salt sensitivity, while the *osgi* mutants show a salt tolerance phenotype [44]. In addition, *LOV KELCH protein 2* (*LKP2*), which regulates circadian rhythm and flowering time in plants, increases dehydration tolerance when overexpressed [45]. The key clock component (TOC1, the timing of CAB expression 1) that binds to the promoter of the ABA-related gene increases drought tolerance in *toc1-RNAi* plants. Conversely, overexpression of *TOC1* increases water loss in drought conditions, leading to a decrease in survival rate [46]. Interestingly, there is no correlation between the effects of mutations on clock function and abiotic stress resistance. Instead, changes in the expression level of the clock gene in the mutants are presumed to have a direct effect on the regulation of the abiotic stress response [38]. For example, half of the genes responsive to drought, salinity, heat, and osmoticum were found to have diurnal rhythm [47]. This transcriptomic analysis suggests that many genes that respond to abiotic stress are under the control of the circadian clock.

Therefore, we speculate that OsPUB4 will play a role similar to COP1 (Constitutive photomorphogenic 1). COP1 is an E3 ubiquitin ligase containing RING-finger and WD40 domains, and is known to be involved in the control of seedling development, flowering time, and circadian rhythm [48]. In particular, HY5 (Long hypocotyl 5), PHYA, PHYB, PIL1 (Phytochrome interacting factor 3-like 1), CO (CONSTANS), GI (GIGANTEA), etc. were identified as substrates of COP1 [49–54]. Accordingly, we expect OsPUB4 to participate in circadian rhythms and abiotic stress responses by controlling the stability of various proteins.

Supplementary Materials: The following are available online at http://www.mdpi.com/2223-7747/9/9/1071/s1, Figure S1: Heat-map expression data associated with the diurnal rhythm of the *OsPUB* family genes. Figure S2: Diurnal expression patterns of *OsPUB2* gene evaluated at 12 time points over 24-h period in "Dongjin" rice and *osgi* mutant. Table S1: Summary of primer sequences used for qRT-PCR analyses of this study.

Author Contributions: Conceptualization, Y.-H.Y.; Formal analysis, Y.-H.Y.; Investigation, X.J.; Supervision, K.-H.J.; Writing—original draft, Y.-H.Y.; Writing—review & editing, K.-H.J. All authors have read and agreed to the published version of the manuscript.

Funding: This work was supported by grants from the Next-Generation BioGreen 21 Program (PJ01325901 and PJ01366401 to KHJ) and a Program (PJ01492703 to JKH), the Rural Development Administration, Korea.

Acknowledgments: We thank Gynheung An for providing valuable advice at all times.

Conflicts of Interest: There is no conflict of interest to declare.

References

1. Vierstra, R.D. The ubiquitin–26S proteasome system at the nexus of plant biology. *Nat. Rev. Mol. Cell Biol.* **2009**, *10*, 385–397. [CrossRef] [PubMed]

2. Santner, A.; Estelle, M. The ubiquitin-proteasome system regulates plant hormone signaling. *Plant J.* **2010**, *61*, 1029–1040. [CrossRef] [PubMed]

3. Stone, S.L. The role of ubiquitin and the 26S proteasome in plant abiotic stress signaling. *Front. Plant Sci.* **2014**, *5*, 135. [CrossRef] [PubMed]

4. Smalle, J.; Vierstra, R.D. The ubiquitin 26S proteasome proteolytic pathway. *Annu. Rev. Plant Biol.* **2004**, *55*, 555–590. [CrossRef]

5. Pickart, C.M.; Eddins, M.J. Ubiquitin: Structures, functions, mechanisms. *Biochim. Biophys. Acta-Mol. Cell Res.* **2004**, *1695*, 55–72. [CrossRef]

6. Lee, J.H.; Kim, W.T. Regulation of abiotic stress signal transduction by E3 ubiquitin ligases in arabidopsis. *Mol. Cells* **2011**, *31*, 201–208. [CrossRef]

7. Koegl, M.; Hoppe, T.; Schlenker, S.; Ulrich, H.D.; Mayer, T.U.; Jentsch, S. A Novel Ubiquitination Factor, E4, Is Involved in Multiubiquitin Chain Assembly. *Cell* **1999**, *96*, 635–644. [CrossRef]

8. Hatakeyama, S.; Nakayama, K.I. U-box proteins as a new family of ubiquitin ligases. *Biochem. Biophys. Res. Commun.* **2003**, *302*, 635–645. [CrossRef]

9. Wiborg, J.; O'Shea, C.; Skriver, K. Biochemical function of typical and variant *Arabidopsis thaliana* U-box E3 ubiquitin-protein ligases. *Biochem. J.* **2008**, *413*, 447–457. [CrossRef]

10. Zeng, L.R.; Park, C.H.; Venu, R.C.; Gough, J.; Wang, G.L. Classification, expression pattern, and E3 Ligase Activity Assay of Rice U-box-containing proteins. *Mol. Plant* **2008**, *1*, 800–815. [CrossRef]

11. McClung, C.R. Plant Circadian Rhythms. *Plant Cell Online* **2006**. [CrossRef] [PubMed]

12. Harmer, S.L. The Circadian System in Higher Plants. *Annu. Rev. Plant Biol.* **2009**. [CrossRef] [PubMed]

13. Dunlap, J.C. Molecular bases for circadian clocks. *Cell* **1999**, *96*, 271–290. [CrossRef]

14. Suárez-López, P.; Wheatley, K.; Robson, F.; Onouchi, H.; Valverde, F.; Coupland, G. CONSTANS mediates between the circadian clock and the control of flowering in Arabidopsis. *Nature* **2001**, *410*, 1116–1120. [CrossRef] [PubMed]

15. Sawa, M.; Nusinow, D.A.; Kay, S.A.; Imaizumi, T. FKF1 and GIGANTEA complex formation is required for day-length measurement in Arabidopsis. *Science* **2007**, *318*, 261–265. [CrossRef]

16. Byun, M.Y.; Cui, L.H.; Oh, T.K.; Jung, Y.-J.; Lee, A.; Park, K.Y.; Kang, B.G.; Kim, W.T. Homologous U-box E3 Ubiquitin Ligases OsPUB2 and OsPUB3 Are Involved in the Positive Regulation of Low Temperature Stress Response in Rice (*Oryza sativa* L.). *Front. Plant Sci.* **2017**, *8*, 16. [CrossRef]

17. Zeng, L.R.; Qu, S.; Bordeos, A.; Yang, C.; Baraoidan, M.; Yan, H.; Xie, Q.; Nahm, B.H.; Leung, H.; Wang, G. Spotted leaf11, A Negative Regulator of Plant Cell Death and Defense, Encodes a U-Box/Armadillo Repeat Protein Endowed with E3 Ubiquitin Ligase Activity. *Plant Cell* **2004**, *16*, 2795–2808. [CrossRef]

18. Park, J.J.; Yi, J.; Yoon, J.; Cho, L.H.; Ping, J.; Jeong, H.J.; Cho, S.K.; Kim, W.T.; An, G. OsPUB15, an E3 ubiquitin ligase, functions to reduce cellular oxidative stress during seedling establishment. *Plant J.* **2011**, *65*, 194–205. [CrossRef]

19. Ishikawa, K.; Yamaguchi, K.; Sakamoto, K.; Yoshimura, S.; Inoue, K.; Tsuge, S.; Kojima, C.; Kawasaki, T. Bacterial effector modulation of host E3 ligase activity suppresses PAMP-triggered immunity in rice. *Nat. Commun.* **2014**, *5*, 5430. [CrossRef]

20. Hu, X.; Qian, Q.; Xu, T.; Zhang, Y.; Dong, G.; Gao, T.; Xie, Q.; Xue, Y. The U-Box E3 Ubiquitin Ligase TUD1 Functions with a Heterotrimeric G α Subunit to Regulate Brassinosteroid-Mediated Growth in Rice. *PLoS Genet.* **2013**, *9*, e1003391. [CrossRef]

21. Jung, K.-H.; Cao, P.; Sharma, R.; Jain, R.; Ronald, P.C. Phylogenomics databases for facilitating functional genomics in rice. *Rice* **2015**, *8*, 26. [CrossRef] [PubMed]

22. Ohyanagi, H. The Rice Annotation Project Database (RAP-DB): Hub for Oryza sativa ssp. japonica genome information. *Nucleic Acids Res.* **2005**, *34*, D741–D744. [CrossRef] [PubMed]

23. Higgins, D.G.; Thompson, J.D.; Gibson, T.J. Using CLUSTAL for multiple sequence alignments. *Methods Enzymol.* **1996**, *266*, 383–402. [PubMed]

24. Tamura, K.; Peterson, D.; Peterson, N.; Stecher, G.; Nei, M.; Kumar, S. MEGA5: Molecular evolutionary genetics analysis using maximum likelihood, evolutionary distance, and maximum parsimony methods. *Mol. Biol. Evol.* **2011**, *28*, 2731–2739. [CrossRef]

25. Yoo, Y.-H.; Chandran, A.K.N.; Park, J.-C.; Gho, Y.-S.; Lee, S.-W.; An, G.; Jung, K.-H. OsPhyB-Mediating Novel Regulatory Pathway for Drought Tolerance in Rice Root Identified by a Global RNA-Seq Transcriptome Analysis of Rice Genes in Response to Water Deficiencies. *Front. Plant Sci.* **2017**, *8*, 580. [CrossRef] [PubMed]

26. Chandran, A.K.N.; Kim, J.W.; Yoo, Y.H.; Park, H.L.; Kim, Y.J.; Cho, M.H.; Jung, K.H. Transcriptome analysis of rice-seedling roots under soil–salt stress using RNA-Seq method. *Plant Biotechnol. Rep.* **2019**, *13*, 567–578. [CrossRef]

27. Lee, Y.S.; An, G. OsGI controls flowering time by modulating rhythmic flowering time regulators preferentially under short day in rice. *J. Plant Biol.* **2015**, *58*, 137–145. [CrossRef]

28. Yoo, Y.-H.; Choi, H.-K.; Jung, K.-H. Genome-wide identification and analysis of genes associated with lysigenous aerenchyma formation in rice roots. *J. Plant Biol.* **2015**, *58*, 117–127. [CrossRef]

29. Yu, G.H.; Huang, S.H.; He, R.; Li, Y.Z.; Cheng, X.G. Transgenic Rice Overexperessing a Tomato Mitochondrial Phosphate Transporter, SlMPT3;1, Promotes Phosphate Uptake and Increases Grain Yield. *J. Plant Biol.* **2018**, *61*, 383–400. [CrossRef]

30. Jain, M.; Nijhawan, A.; Tyagi, A.K.; Khurana, J.P. Validation of housekeeping genes as internal control for studying gene expression in rice by quantitative real-time PCR. *Biochem. Biophys. Res. Commun.* **2006**, *345*, 646–651. [CrossRef]

31. Szklarczyk, D.; Morris, J.H.; Cook, H.; Kuhn, M.; Wyder, S.; Simonovic, M.; Santos, A.; Doncheva, N.T.; Roth, A.; Bork, P.; et al. The STRING database in 2017: Quality-controlled protein-protein association networks, made broadly accessible. *Nucleic Acids Res.* **2017**. [CrossRef] [PubMed]

32. Su, G.; Morris, J.H.; Demchak, B.; Bader, G.D. Biological Network Exploration with Cytoscape 3. *Curr. Protoc. Bioinform.* **2014**, *47*, 8–13. [CrossRef] [PubMed]

33. Li, Y.; Luo, C.; Chen, Y.; Xiao, X.; Fu, C.; Yang, Y. Transcriptome-based Discovery of AP2/ERF tRanscription Factors and Expression Profiles Under Herbivore Stress Conditions in Bamboo (*Bambusa emeiensis*). *J. Plant Biol.* **2019**, *62*, 297–306. [CrossRef]

34. Dubouzet, J.G.; Sakuma, Y.; Ito, Y.; Kasuga, M.; Dubouzet, E.G.; Miura, S.; Seki, M.; Shinozaki, K.; Yamaguchi-Shinozaki, K. OsDREB genes in rice, *Oryza sativa* L., encode transcription activators that function in drought-, high-salt- and cold-responsive gene expression. *Plant J.* **2003**, *33*, 751–763. [CrossRef] [PubMed]

35. Xiang, Y.; Tang, N.; Du, H.; Ye, H.; Xiong, L. Characterization of OsbZIP23 as a Key Player of the Basic Leucine Zipper Transcription Factor Family for Conferring Abscisic Acid Sensitivity and Salinity and Drought Tolerance in Rice. *Plant Physiol.* **2008**, *148*, 1938–1952. [CrossRef]

36. McClung, R.C. Plant Circadian Rhythms. *Plant Cell* **2006**, *18*, 792–803. [CrossRef]

37. Sanchez, A.; Shin, J.; Davis, S.J. Abiotic stress and the plant circadian clock. *Plant Signal. Behav.* **2011**, *6*, 223–231. [CrossRef]

38. Grundy, J.; Stoker, C.; Carré, I.A. Circadian regulation of abiotic stress tolerance in plants. *Front. Plant Sci.* **2015**, *6*, 648. [CrossRef]

39. Seo, P.J.; Mas, P. STRESSing the role of the plant circadian clock. *Trends Plant Sci.* **2015**, *20*, 230–237. [CrossRef]

40. Sato, Y.; Takehisa, H.; Kamatsuki, K.; Minami, H.; Namiki, N.; Ikawa, H.; Ohyanagi, H.; Sugimoto, K.; Antonio, B.A.; Nagamura, Y. RiceXPro Version 3.0: Expanding the informatics resource for rice transcriptome. *Nucleic Acids Res.* **2013**, *41*. [CrossRef]

41. Ogiso, E.; Takahashi, Y.; Sasaki, T.; Yano, M.; Izawa, T. The Role of Casein Kinase II in Flowering Time Regulation Has Diversified during Evolution. *Plant Physiol.* **2010**, *152*, 808–820. [CrossRef]

42. Izawa, T.; Mihara, M.; Suzuki, Y.; Gupta, M.; Itoh, H.; Nagano, A.J.; Motoyama, R.; Sawada, Y.; Yano, M.; Hirai, M.Y.; et al. Os- *GIGANTEA* Confers Robust Diurnal Rhythms on the Global Transcriptome of Rice in the Field. *Plant Cell* **2011**, *23*, 1741–1755. [CrossRef]

43. Cho, L.H.; Pasriga, R.; Yoon, J.; Jeon, J.S.; An, G. Roles of Sugars in Controlling Flowering Time. *J. Plant Biol.* **2018**, *61*, 121–130. [CrossRef]

44. Kim, W.Y.; Ali, Z.; Park, H.J.; Park, S.J.; Cha, J.Y.; Perez-Hormaeche, J.; Quintero, F.J.; Shin, G.; Kim, M.R.; Qiang, Z.; et al. Release of SOS2 kinase from sequestration with GIGANTEA determines salt tolerance in Arabidopsis. *Nat. Commun.* **2013**, *4*, 1–13. [CrossRef]

45. Miyazaki, Y.; Abe, H.; Takase, T.; Kobayashi, M.; Kiyosue, T. Overexpression of LOV KELCH PROTEIN 2 confers dehydration tolerance and is associated with enhanced expression of dehydration-inducible genes in Arabidopsis thaliana. *Plant Cell Rep.* **2015**, *34*, 843–852. [CrossRef] [PubMed]

46. Legnaioli, T.; Cuevas, J.; Mas, P. TOC1 functions as a molecular switch connecting the circadian clock with plant responses to drought. *EMBO J.* **2009**, *28*, 3745–3757. [CrossRef] [PubMed]

47. Covington, M.F.; Maloof, J.N.; Straume, M.; Kay, S.A.; Harmer, S.L. Global transcriptome analysis reveals circadian regulation of key pathways in plant growth and development. *Genome Biol.* **2008**, *9*, R130. [CrossRef] [PubMed]

48. Kim, J.Y.; Jang, I.C.; Seo, H.S. COP1 controls abiotic stress responses by modulating AtSIZ1 function through its E3 ubiquitin ligase activity. *Front. Plant Sci.* **2016**, *7*, 1182. [CrossRef]

49. Osterlund, M.T.; Hardtke, C.S.; Ning, W.; Deng, X.W. Targeted destabilization of HY5 during light-regulated development of Arabidopsis. *Nature* **2000**, *405*, 462–466. [CrossRef] [PubMed]

50. Soo Seo, H.; Yang, J.Y.; Ishikawa, M.; Bolle, C.; Ballesteros, M.L.; Chua, N.H. LAF1 ubiquitination by COP1 controls photomorphogenesis and is stimulated by SPA1. *Nature* **2003**, *423*, 995–999. [CrossRef]

51. Seo, H.S.; Watanabe, E.; Tokutomi, S.; Nagatani, A.; Chua, N.H. Photoreceptor ubiquitination by COP1 E3 ligase desensitizes phytochrome A signaling. *Genes Dev.* **2004**, *18*, 617–622. [CrossRef] [PubMed]

52. Luo, Q.; Lian, H.L.; He, S.B.; Li, L.; Jia, K.P.; Yang, H.Q. COP1 and phyB physically interact with PIL1 to regulate its stability and photomorphogenic development in Arabidopsis. *Plant Cell* **2014**, *26*, 2441–2456. [CrossRef]

53. Jang, S.; Marchal, V.; Panigrahi, K.C.S.; Wenkel, S.; Soppe, W.; Deng, X.W.; Valverde, F.; Coupland, G. Arabidopsis COP1 shapes the temporal pattern of CO accumulation conferring a photoperiodic flowering response. *EMBO J.* **2008**, *27*, 1277–1288. [CrossRef] [PubMed]

54. Yu, J.W.; Rubio, V.; Lee, N.Y.; Bai, S.; Lee, S.Y.; Kim, S.S.; Liu, L.; Zhang, Y.; Irigoyen, M.L.; Sullivan, J.A.; et al. COP1 and ELF3 Control Circadian Function and Photoperiodic Flowering by Regulating GI Stability. *Mol. Cell* **2008**, *32*, 617–630. [CrossRef] [PubMed]

Article

Responses to Salt Stress in *Portulaca*: Insight into Its Tolerance Mechanisms

Orsolya Borsai [1,2], Mohamad Al Hassan [3], Cornel Negrușier [4], María D. Raigón [5], Monica Boscaiu [6,*], Radu E. Sestraș [7] and Oscar Vicente [5]

[1] Institute of Life Sciences, University of Agricultural Sciences and Veterinary Medicine Cluj-Napoca, Mănăștur St. 3-5, 400372 Cluj-Napoca, Romania; orsolya.borsai@usamvcluj.ro

[2] AgroTransilvania Cluster, Dezmir, Crișeni FN, 407039 Cluj-Napoca, Romania

[3] Wageningen UR Plant Breeding, Wageningen University and Research Centre, 6708 PB Wageningen, The Netherlands; mohamed.alhassan@wur.nl

[4] Department of Soil Sciences, University of Agricultural Sciences and Veterinary Medicine Cluj-Napoca, Mănăștur St. 3-5, 400372 Cluj-Napoca, Romania; cornel.negrusier@usamvcluj.ro

[5] Institute for the Conservation and Improvement of Valencian Agrodiversity (COMAV), Universitat Politècnica de València, Camino de Vera s/n, 46022 Valencia, Spain; mdraigon@qim.upv.es (M.D.R.); ovicente@upvnet.upv.es (O.V.)

[6] Mediterranean Agroforestry Institute (IAM), Universitat Politècnica de València, Camino de Vera s/n, 46022 Valencia, Spain

[7] Faculty of Horticulture, University of Agricultural Sciences and Veterinary Medicine Cluj-Napoca, Mănăștur St. 3-5, 400372 Cluj-Napoca, Romania; rsestras@usamvcluj.ro

* Correspondence: mobosnea@eaf.upv.es

Received: 15 October 2020; Accepted: 25 November 2020; Published: 27 November 2020

Abstract: Climate change and its detrimental effects on agricultural production, freshwater availability and biodiversity accentuated the need for more stress-tolerant varieties of crops. This requires unraveling the underlying pathways that convey tolerance to abiotic stress in wild relatives of food crops, industrial crops and ornamentals, whose tolerance was not eroded by crop cycles. In this work we try to demonstrate the feasibility of such strategy applying and investigating the effects of saline stress in different species and cultivars of *Portulaca*. We attempted to unravel the main mechanisms of stress tolerance in this genus and to identify genotypes with higher tolerance, a procedure that could be used as an early detection method for other ornamental and minor crops. To investigate these mechanisms, six-week-old seedlings were subjected to saline stress for 5 weeks with increasing salt concentrations (up to 400 mM NaCl). Several growth parameters and biochemical stress markers were determined in treated and control plants, such as photosynthetic pigments, monovalent ions (Na^+, K^+ and Cl^-), different osmolytes (proline and soluble sugars), oxidative stress markers (malondialdehyde—a by-product of membrane lipid peroxidation—MDA) and non-enzymatic antioxidants (total phenolic compounds and total flavonoids). The applied salt stress inhibited plant growth, degraded photosynthetic pigments, increased concentrations of specific osmolytes in both leaves and roots, but did not induce significant oxidative stress, as demonstrated by only small fluctuations in MDA levels. All *Portulaca* genotypes analyzed were found to be Na^+ and Cl^- includers, accumulating high amounts of these ions under saline stress conditions, but *P. grandiflora* proved to be more salt tolerant, showing only a small reduction under growth stress, an increased flower production and the lowest reduction in K^+/Na^+ rate in its leaves.

Keywords: abiotic stress; antioxidant activity; growth inhibition; ion homeostasis; proline; salt stress

1. Introduction

Adverse environmental conditions or abiotic stresses such as drought, soil salinity, low or high temperatures share their detrimental effect on the water status of plants, causing a reduction in their photosynthetic capacity and, consequently, limiting their vegetative growth and reproductive success [1,2]. Among the abiotic stresses mentioned, drought and salinity constitute the greatest threat to global food security [3], due to the extent and frequency of their occurrence, as well as their effects on the production of the yields of all major crops [4,5]. With increasing demand for food sources due to the booming world population, the need to combat the effects of drought and soil salinity (reduced crop production) becomes imperative [6]. This is highlighted by the fact that more than 800 million hectares of arable land are currently affected by drought and salinity, and salinity in particular is projected to affect 30% of agricultural land in the next 25 years and about 50% by the end of this century [7]. This increase in the extent of soil salinization (estimated at 1 to 2% per year [8]) is being aggravated by climate change, urbanization and pollution [9].

Soil salinization and drought overlap in inducing osmotic imbalance in affected plants [10]; however, the toxic influx of ions or ionic toxicity is salinity-specific [11]. Increased salinity restricts vegetative growth due to reduced photosynthetic capacity, as affected plants reduce their gas exchange by closing their stomata to reduce water loss [12,13]. As a result, further accumulation of reactive oxygen species (ROS) would begin, disrupting cellular processes and inducing oxidative damage, particularly to the photosynthetic machinery [14,15]. Affected plants respond by activating a series of constitutively expressed defense mechanisms to evade and mitigate the effects of such stresses [16], until optimal conditions are restored or successful propagation of the genetic material is ensured. These include the accumulation of soluble solutes or "osmolytes" for osmotic adjustment [17], the activation and over-expression of ROS sequestering enzymes such as catalase, ascorbate peroxidase and superoxide dismutase [18], and ionic avoidance [19] and compartmentalization [20].

The adaptation and response of plants to adverse environmental conditions, in particular soil salinization, differs greatly among affected species. This ranges from improved growth under stress conditions in the case of obligate halophytes, to complete cessation of growth, and even death for susceptible glycophytes [21]. Almost 98% of terrestrial plant species are considered glycophyte, including all major crops [22], this is partly due to the erosion of the natural resistance inherent to abiotic stresses over millennia of unidirectional breeding for yield, taste and color [23]. A strategy that is recently gaining ground among the scientific community in an attempt to combat the anticipated spread of soil salinization and its underlying effects on global food security is the development of salt-tolerant crops [24]. However, unlike plant resistance to biotic stresses, which depends mostly on monogenic traits, genetically complex responses to abiotic stresses are multigenic and therefore more difficult to control, breed and manipulate [25].

Therefore, a better understanding of the underlying mechanisms that confer tolerance is considered necessary to facilitate the development of more tolerant crops [24], especially because these response strategies differ greatly among plant groups [26]. One area that has been little explored is research on the stress responses of wild relatives of crops, halophytes and minor, ornamental and industrial crops, which could serve as an underexploited source of genetic tools for deciphering the fine tuning of crop tolerance. This would identify targets for molecular improvement and the development of stress markers, which could accelerate ongoing programs towards more stress resistant crops.

Salt tolerance in halophytes is developed through an efficient and well-defined coordination of physiological and metabolic pathways [27] that enable them to complete their life cycles even under conditions of high salinity [28]. The adaptation capacity of halophytes to salt stress is based on several physiological adjustments that take place both at the cellular and molecular level, among them osmotic regulation and ionic homeostasis through extrusion and/or compartmentalization [14]. In previous reports, the inhibitory effects of excessive saline stress on several physiological and biochemical processes in plants were highlighted. Among the former, the function of the photosynthetic apparatus is the most likely to be affected due to stomatal closure and the consequent reduction in carbon dioxide absorption under stress conditions [14]. Therefore, the effects of stress on photosynthesis can be easily evaluated by monitoring changes at the level of photosynthetic pigments [28,29]. It has also been observed that, under stress conditions, halophytes accumulate different compatible solutes, such as sugars (sucrose), betaine (glycinebetaine), amino acids (proline) and sugar alcohols (sorbitol) depending on phylogeny and functional needs [30–32]. In addition, different stressful environmental conditions were reported to induce oxidative stress in plants due to increased "reactive oxygen species" (ROS). However, in both halophytes and glycophytes, ROS production and detoxification occur under optimal conditions, although the balance is greatly altered in stressful situations, particularly high salinity [33]. To that extent, malondialdehyde (MDA), a product of lipid peroxidation that is considered a reliable marker of oxidative stress in plants, was examined in parallel with studies on stress for ROS accumulation. An increase in MDA content in response to abiotic stresses in different plant species is reported in several publications [34]. Since different environmental stresses cause secondary oxidative stress in plants, a general adaptive reaction of plants is the activation of antioxidant compounds and enzymes. A complex group of phenolic compounds and especially the subgroup of flavonoids that include many secondary metabolites are also very important for these antioxidant responses [35].

The genus *Portulaca* comprises approximately 130 species known today, including annuals and perennials (which grow in the tropics and subtropics) and are generally herbaceous, with hairs (conspicuous or inconspicuous) developed at the axils of the succulent, cylindrical leaves [36]. *Portulaca* species have a wide range of distribution throughout the world, from sea level to 2600 m of altitude. *Portulaca oleracea* is often used as a vegetable, but many of its varieties, as well as other species of this genus, are widely known as ornamentals, much appreciated for their flowers of diverse color, type and size, and for their multiple uses in garden design. *Portulaca* can tolerate moderate to high salt concentrations and produce a considerable amount of dry mass even at relatively high salinities compared to standard crops [37,38]. Albeit the relative salt tolerance of other species in this genus has not yet been adequately studied.

The main objective of this research was to investigate the responses to salt stress of three different species of *Portulaca* (*P. oleracea* L. subsp. *oleracea*, *P. grandiflora* Hook. and *P. halimoides* L.) and a purslane cultivar (*P. oleracea* var. "Toucan Scarlet Shades"), under controlled experimental conditions, trying to specify the mechanisms of salinity tolerance in this genus promoting thus their use in saline agriculture and sustainable development, as a source of plant nutrients or ornamental species.

2. Results

2.1. Electrical Conductivity of the Substrate

Electrical conductivity ($EC_{1:5}$) values in control soil samples varied between 0.55–1.08 dS m^{-1} throughout the treatment period. When salt stress was applied, EC increased in parallel to time (as salt accumulated in the substrate) and in a concentration dependent manner (400 mM registering the highest measured EC), with approximately 3, 4.5 and 7.5 dS m^{-1} in 100, 200 and 400 mM soils, respectively at the end of the treatments. It must be noted that differences in between the substrates of stressed plants of the 4 studied genotypes undergoing the same level of salt concentration treatment were statistically insignificant.

2.2. Growth Parameters

Our findings show that plant growth and productivity were adversely affected in all the four *Portulaca* genotypes studied, though in varying degrees as presented in Figure 1.

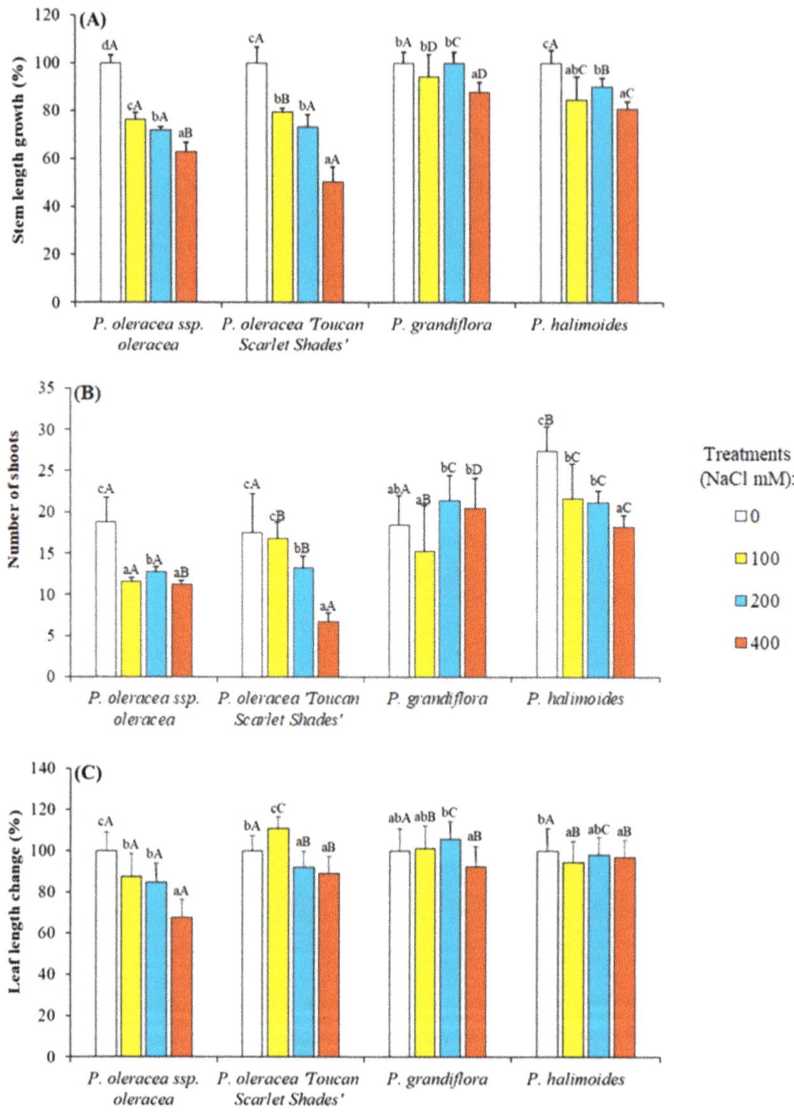

Figure 1. Measured growth parameters in the four studied *Portulaca* genotypes after 5 weeks of applied salt stress. (**A**) Stem length growth (%). For each genotype, values (means ± SE) are shown as percentages of the mean stem length of the control plants, considered 100%, (**B**) number of developed shoots by time of harvest, and (**C**) leaf length change (%). For each genotype, values (means ± SE) are shown as percentages of the mean leaf length of the control plants, considered 100%. Different lowercase letters within each genotype indicate significant differences among the treatments and different capital letters indicate significant differences among the genotypes undergoing the same treatment according to Tukey's HSD test ($P < 0.05$). Absolute values for the controls' stem length are: 31.00 ± 1.04 cm (*P. oleracea* L. subsp. *oleracea*), 32.25 ± 2.20 cm (*P. oleracea* "Toucan Scarlet Shades"), 19.60 ± 0.92 cm (*P. grandiflora*) and 18.20 ± 0.96 cm (*P. halimoides*). Absolute values for the controls' leaf length are: 4.00 ± 0.22 cm (*P. oleracea* L. subsp. *oleracea*), 2.93 ± 0.17 cm (*P. oleracea* "Toucan Scarlet Shades"), 2.6 1 ± 0.12 cm (*P. grandiflora*), 3.36 ± 0.09 cm) and 1.13 ± 0.05 cm (*P. halimoides*).

Stem lengths at time of harvest (5 weeks) for controls ranged from 17.37 cm to 31.00 cm in *P. halimoides* and *P. oleracea* L. subsp. *oleracea*, respectively. Applied soil salinity had a detrimental impact on the main stem growth of all four investigated genotypes, inducing a decrement in lengthening in parallel to increasing applied salt concentration. This varied between the different taxa, being very acute in the two studied *P. oleracea* genotypes, where stem length growth was reduced by nearly 40% and 50% in *P. oleracea* L. spp. *oleracea* and *P. oleracea* "Toucan Scarlet Shades", respectively (Figure 1A). On the other hand, *P. grandiflora*, and *P. halimoides* seemed to be less affected with all applied salt concentrations, reporting a mere decrease in stem growth of 12.24% and 19.23%, respectively, in 400 mM NaCl treated plants.

The number of lateral shoots was significantly lower in three of the studied genotypes under applied salt conditions in comparison to their corresponding controls (all but *P. grandiflora*) (Figure 1B). The highest averaged shoot number was 27.4 in *P. halimoides*, while the lowest 17.5 was reported in *P. oleracea* "Toucan Scarlet Shades" (Figure 1B). The most sensitive to the highest applied salt concentration was *P. oleracea* "Toucan Scarlet Shades" reporting a decrease of 61.5% in shoot number during the salt treatments in comparison to its controls. On the other hand, the least affected accession in shoot number was *P. grandiflora* which showed an increase in shoot formation under 400 mM NaCl. Another monitored growth parameter was leaf length. Here again the effect of applied salt stress for 5 weeks, ranged from severe (causing a reduction by 35% relative to its controls leaf length) in *P. oleracea* L. spp. *oleracea* and mild (reporting a decrease from control leaf length by a mere 3% under 400 mM NaCl) in *P. halimoides* (Figure 1C). The other two genotypes, *P. oleracea* "Toucan Scarlet Shades" and *P. grandiflora* did not show significant changes in their leaf length neither under 100 and 200 mM NaCl nor under 400 mM NaCl treatments (10% reduction in comparison to their controls).

Flower bud formation and flowering differed in unstressed plants among the four studied genotypes. *P. oleracea* L. subsp. *oleracea* produced the highest number of formed flower buds (about 108) (Figure 2A) though none developed into flowers (Figure 2B). On the other hand, only *P. oleracea* "Toucan Scarlet Shades" and *P. grandiflora* had any flower formation, where interestingly the latter had a higher number of flowers under applied salt stress. The number of formed flower buds decreased significantly in all investigated genotypes except in *P. grandiflora* where again it seems that the applied stress induced reproductive development with a 25% increase under 400 mM NaCl in comparison to control.

Upon completion of the applied salt treatments, plants were harvested and roots and shoots were weighted. A fraction of material was used for water content determination. Fresh weight in all studied taxa was dramatically reduced under the applied stress conditions (Figure 3A), displaying a reduction of almost 90% under 400 mM NaCl, compared to their respective controls for both *P. oleracea* "Toucan Scarlet Shades" and *P. halimoides*. On the other hand, fresh weight was reduced by almost 30% for *P. oleracea* L. subsp. *oleracea* under similar conditions, while *P. grandiflora* had the best resistance as it maintained 70% of its control FW when exposed to the highest stress treatment applied.

After oven drying the leaves, water content was calculated from the measured dry weight and afore measured leaf fresh weight for every harvested plant. Interestingly, water content in leaves and roots recorded only a small decrease in comparison to the loss in FW (Figure 3A–C). All studied genotypes showed only a small decrease in their roots' WC under salt stress (Figure 3C). Only *P. halimoides* showed a significant decrease in its root's WC when 400 mM NaCl was applied, showing a decrease to 60% from 80% measured in its unstressed plants.

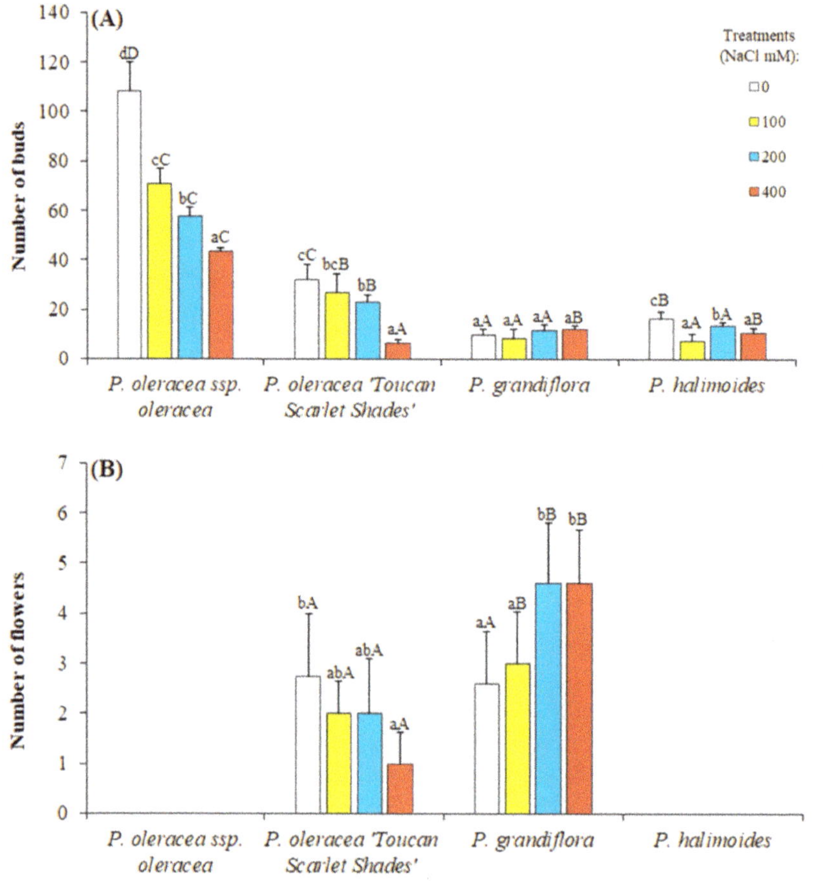

Figure 2. The influence of applied salt stress on the reproductive development of the four studied *Portulaca* genotypes. (**A**) Number of flower bud occurrence and (**B**) number of flowers. Different lowercase letters within each genotype indicate significant differences among the treatments and different capital letters indicate significant differences among the genotypes undergoing the same treatment according to Tukey's HSD test ($P < 0.05$).

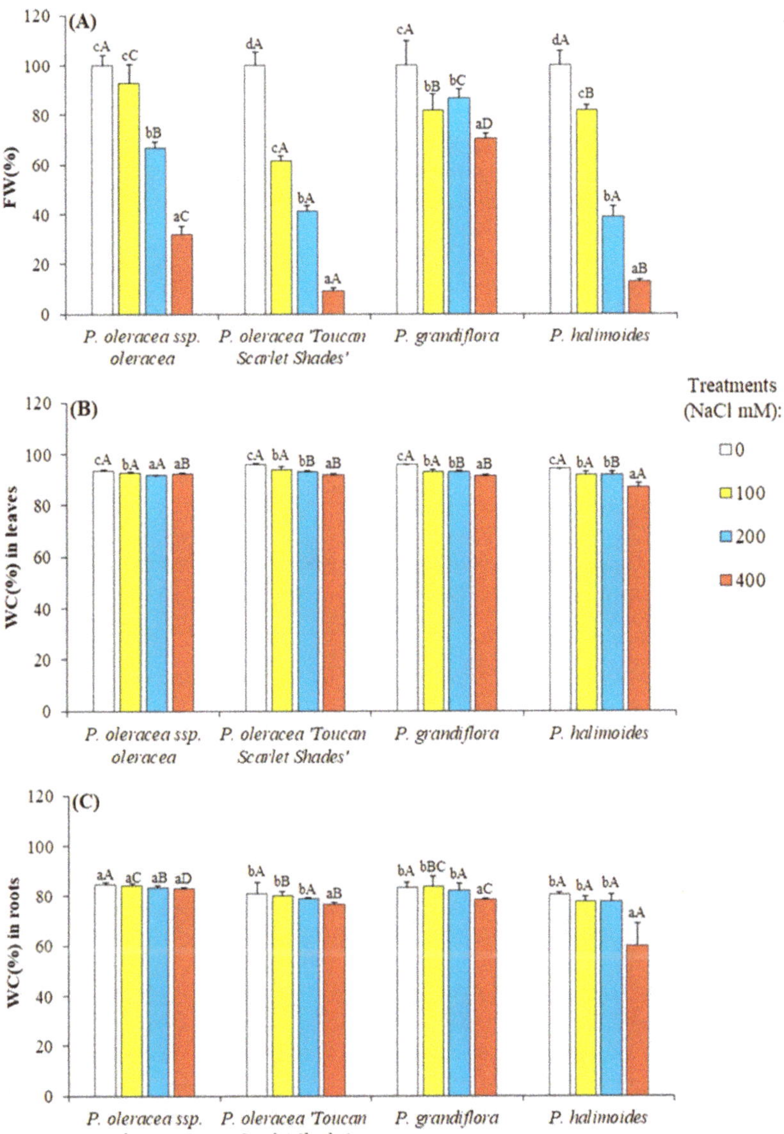

Figure 3. Total fresh yield and water content reduction in leaves and roots after 5 weeks of applied stress on four *Portulaca* genotypes. (**A**) fresh weight percentage are shown as percentages of the mean fresh weight of the control plants, considered as 100% (absolute values FW values in grams are 62.56 ± 5.82 g for *P. oleracea* L. subsp. *oleracea*, 112.00 ± 13.47 g for *P. oleracea* "Toucan Scarlet Shades", 20.53 ± 4.43 g for *P. grandiflora*, and 11.85 ± 1.54 for *P. halimoides*), (**B**) water content percentage in the leaves (WC%), and (**C**) water content percentage in the roots. Different lowercase letters above the bars within each accession indicate significant differences among treatments and different capital letters denote significant differences among the accessions undergoing the same treatment according to Tukey's HSD test (*P* < 0.05).

2.3. Photosynthetic Pigments

Chlorophyll levels decreased in some of the studied genotypes under stress (Table 1), notably *P. oleracea* L. subsp. *oleracea* and to a lesser extent in *P. grandiflora*. Furthermore, an increase in these pigments was registered in *P. oleracea* "Toucan Scarlet Shades" and *P. halimoides*. Carotenoids contents increased under 400 mM NaCl in all studied genotypes except *P. oleracea* L. subsp. *oleracea* where it nearly remained equal to that measured in its control plants (Table 1). In terms of carotenoids' increase, *P. halimoides* reported the biggest increment (2.71-fold change as compared to its control), followed by *P. oleracea* "Toucan Scarlet Shades" (1.35-fold change).

Table 1. Photosynthetic pigments in the leaves of the four studied *Portulaca* genotypes. Chlorophyll a (Chl a), chlorophyll b (Chl b) and carotenoids (Caro) contents after 5 weeks of applied salt stress. Values shown are means ± SE. For each pigment, different lowercase letters in a column indicate significant differences between treatments; different capital letters in a row denote significant differences among the genotypes undergoing the same treatment.

Photosynthetic Pigment (mg g^{-1} DW)	Treatment (mM NaCl)	*P. oleracea* L. subsp. *oleracea*	*P. oleracea* "Toucan Scarlet Shades"	*P. grandiflora*	*P. halimoides*
Chl a	0	7.82 ± 0.83 bC	3.18 ± 0.21 bA	4.08 ± 0.69 cB	3.25 ± 0.17 aA
	100	9.90 ± 0.67 cC	2.57 ± 0.42 aA	3.27 ± 0.34 bAB	3.76 ± 0.79 bB
	200	7.87 ± 1.13 bC	3.60 ± 0.19 cAB	3.07 ± 0.36 aA	4.28 ± 0.51 cB
	400	5.45 ± 0.29 aB	3.21 ± 0.61 bcA	3.12 ± 0.46 aA	5.57 ± 0.31 dB
Chl b	0	5.83 ± 0.97 bC	1.20 ± 0.19 aA	3.13 ± 0.96 bB	1.29 ± 0.24 aA
	100	8.79 ± 1.20 cB	0.95 ± 0.14 aA	1.69 ± 0.41 aA	1.21 ± 0.26 aA
	200	6.09 ± 1.76 bB	1.23 ± 0.08 aA	1.48 ± 0.29 aA	1.23 ± 0.07 aA
	400	4.32 ± 0.42 aB	1.27 ± 0.13 aA	1.41 ± 0.19 aA	1.79 ± 0.14 bA
Caro	0	0.49 ± 0.32 aA	0.62 ± 0.04 aB	0.50 ± 0.06 aA	0.66 ± 0.06 aB
	100	0.23 ± 0.29 aA	0.68 ± 0.12 aB	0.52 ± 0.20 aB	0.97 ± 0.25 bC
	200	0.32 ± 0.32 aA	0.91 ± 0.13 bB	0.45 ± 0.15 aA	1.20 ± 0.14 cC
	400	0.42 ± 0.10 aA	0.84 ± 0.23 bB	0.66 ± 0.03 bA	1.78 ± 0.15 dC

2.4. Osmolytes

The findings of the current research show that proline (Pro) content of the leaves in the control *Portulaca* genotypes were relatively low ranging from 2.19 µmol g^{-1} DW (*P. oleracea* "Toucan Scarlet Shades") to 8.50 µmol g^{-1} DW (*P. oleracea* L. subsp. *oleracea*) (Figure 4A). Similarly, Pro levels were also low in the roots, ranging from 0.35 µmol g^{-1} DW (*P. oleracea* "Toucan Scarlet Shades") to 1.6 µmol g^{-1} DW (*P. halimoides*) in control plants (Figure 4B). Proline accumulation was found both in leaves and roots of the plants subjected to salt stress treatments. After 5 weeks of applied stress, Pro accumulation in the leaves was the highest in *P. halimoides* with a 5.66-fold change increase, and similarly a 12-fold increment from control levels in their roots. In addition, *P. oleracea* "Toucan Scarlet Shades" leaves registered the smallest increase under stress (a mere 9.67 µmol g^{-1} DW in 400 mM NaCl stressed plants). Therefore, a positive correlation was shown between proline level and salt treatment intensity; however, proline accumulation was not also related to the degree of salt tolerance in these species (Figure 4A,B).

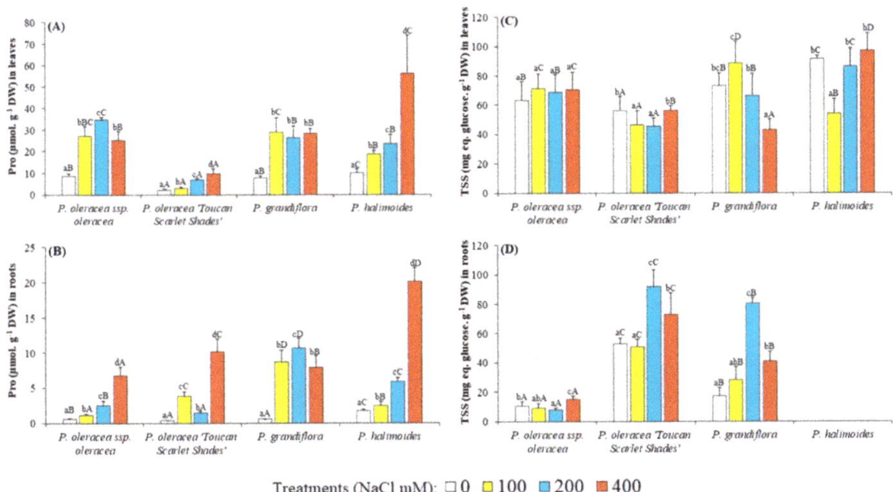

Figure 4. Osmolytes quantified in the four *Portulaca* genotypes subjected to salt stress treatments. Proline (Pro) and total soluble sugars (TSS) accumulation in the leaves (**A,C**, respectively) and roots (**B,D**, respectively). The values shown are means ± SE ($n = 5$). Different lowercase letters above the bars indicate significant differences between the control plants, and different capital letters indicate significant differences between different genotypes.

Among the tested *Portulaca* genotypes, the soluble sugar content of the leaves in the untreated controls varied from 46 (*P. oleracea* "Toucan Scarlet Shades") to 91 (*P. halimoides*) mg eq. glucose g^{-1} DW (Figure 4C). Under salt stress conditions no clear correlation was shown between TSS content and salt stress intensity. In the leaves, TSS followed different accumulation patterns: (i) Soluble sugar level increased at 100 mM NaCl concentration, followed by a decrease at 200 mM NaCl and increasing again at the highest salt concentration (e.g., *P. oleracea* subsp. *oleracea*), (ii) soluble sugar level increased at 100 mM NaCl concentration but decreased at 200 and 400 mM NaCl concentrations (e.g., *P. grandiflora*) and (iii) TSS content decreased at the lowest salt concentration but increased at 200 and 400 mM NaCl concentrations (*P. oleracea* "Toucan Scarlet Shades" and *P. halimoides*). *P. grandiflora* showed a significant decrease (41.20%) in soluble sugar content at the highest salinity level as compared to control (42.93 mg eq. glucose g-1 DW vs. control 73.024 mg eq. glucose g^{-1} DW).

In the roots, a variable pattern of TSS content was found as the salt concentrations increased. It was observed that the TSS content of all genotypes was significantly higher at the highest salt concentrations applied than in controls. Statistically significant differences were recorded at 200 and 400 NaCl concentrations as compared to controls (Figure 4D). No data was obtained in *P. halimoides* due to an accidental loss of material during lab work.

2.5. Malodialdehyde and Non-Enzymatic Antioxidants

After 5 weeks of salt treatment, it was expected that the MDA in the leaves of stressed plants would show an increase consistent with the increased salinity level, but in the *Portulaca* genotypes investigated, this was not the case, (Figure 5A), as MDA levels in the leaves fluctuated under applied stress.

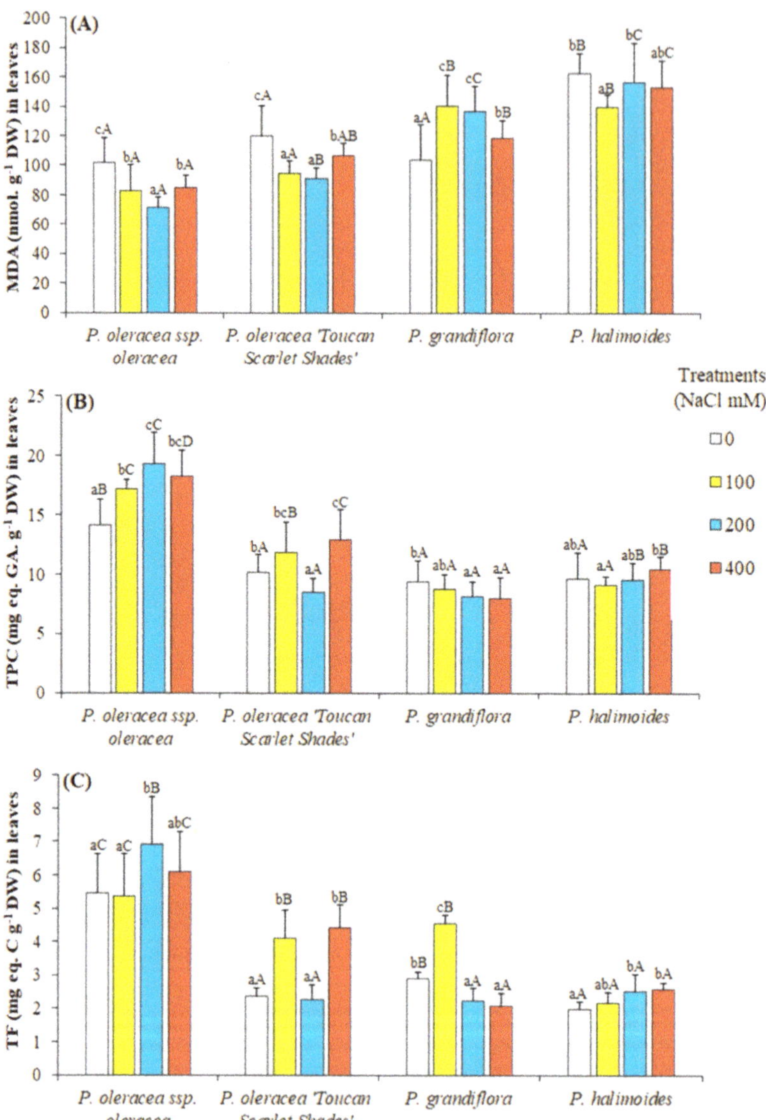

Figure 5. Malondialdehyde (MDA) and chemical antioxidants accumulation in the leaves of *Portulaca* genotypes after 5 weeks of salt stress treatments. (**A**) MDA, (**B**) total phenolic compounds (TPC), and (**C**) total flavonoids (TF). The values shown are means ± SE (*n* = 5). Different lowercase letters above the bars indicate significant differences between the control plants, and different capital letters indicate significant differences between different genotypes.

The pattern of variation of total flavonoid content in response to increasing salt concentrations was very similar to that of TPC (Figure 5B,C). Among the untreated control plants, it was observed that flavonoid content varied, with the highest content in *P. oleracea* L. subsp. *oleracea* (5.45 C. g^{-1} DW), while the lowest was in *P. halimoides* L. (1.97 C. g^{-1} DW). Under applied salt stress, TPC and TF levels fluctuated in the leaves of all four studied genotypes, showing no strong correlation with the increasing concentrations of applied NaCl, except for *P. oleracea* L. subsp. *oleracea*, though *P. oleracea*

"Toucan Scarlet Shades" reported a significant increment in its accumulated TPC and TF when 400 mM NaCl was applied (in comparison to its controls).

2.6. Ionic Content

Salinity induced the accumulation of Na^+ and Cl^- in high amounts in roots and leaves of the four genotypes in response to salt stress acting as Na^+ and Cl^- includers. Sodium levels in NaCl-treated plants of all genotypes showed an applied-salt-concentration-dependent increase (Figure 6A,B). It increased up to 4.5 mmol g^{-1} DW in the leaves of 400 mM NaCl stressed plants of *P. oleracea* L. subsp. *oleracea* and *P. halimoides*, while *P. oleracea* "Toucan Scarlet Shades" and *P. grandiflora* reported merely half those levels under similar conditions (Figure 6A). Control contents of sodium in the leaves of studied plants were similar in all four genotypes, and the increase was comparable when 100 and 200 mM NaCl were applied. Interestingly however, was the trend in chloride ions accumulation in the leaves, which were expected to mimic that of sodium. *P. grandiflora* reported chloride levels 4 folds lower than that of *P. oleracea* L. subsp. *oleracea* and nearly 2.5 times that of *P. halimoides* and *P. oleracea* "Toucan Scarlet Shades" (Table 2). Chloride accumulation in the leaves seemed to reach a plateau after 100 mM NaCl in *P. grandiflora* and 200 mM in *P. halimoides*. As for potassium in the leaves (Figure 6C), all studied genotypes reported an increment under 400 mM NaCl, except for *P. halimoides* which showed a 30% decrement. When correlating the aforementioned data through K^+/Na^+ ratios, *P. oleracea* L. subsp. *oleracea* had the biggest decrease though it had the highest ratio in non-stressed conditions (Figure 6E). Moreover, *P. oleracea* "Toucan Scarlet Shades" and *P. grandiflora* had the smallest decrease in this ratio, with the latter retaining the smallest K^+/Na^+ ratio in control conditions.

Table 2. Chloride ions' concentration in roots and leaves and stems of the four investigated *Portulaca* genotypes undergoing 5 weeks of salt treatments.

Chloride Ions (μmol g^{-1} DW)	Treatment (mM NaCl)	*P. oleracea* L. ssp. *oleracea*	*P. oleracea* "Toucan Scarlet Shades"	*P. grandiflora*	*P. halimoides*
Leaves	0	136.35 ± 25.41aA	244.18 ± 45.78aB	256.34 ± 19.01aB	217.45 ± 46.27aB
	100	1001.88 ± 132.14bC	449.37 ± 99.91bA	542.94 ± 133.24bA	738.21 ± 20.06bB
	200	1977.47 ± 164.27cC	537.64 ± 108.71bA	691.44 ± 116.61bA	1105.85 ± 50.17cB
	400	2436.03 ± 182.24dC	908.43 ± 88.77cB	618.84 ± 26.94bA	1027.84 ± 248.03cB
Roots	0	357.26 ± 105.75aB	125.47 ± 17.96aA	337.74 ± 64.09aB	N.A.
	100	794.84 ± 181.52bA	710.152 ± 44.68bA	901.01 ± 108.31bA	N.A.
	200	846.95 ± 236.21bA	1444.22 ± 230.83cB	1121.48 ± 211.66bcAB	N.A.
	400	1489.29 ± 119.33cA	2328.77 ± 173.74dB	1336.98 ± 179.31cA	N.A.

Values shown are means ± SE. For each tissue type, different lowercase letters in a column indicate significant differences between treatments; different capital letters in a row denote significant differences among the genotypes undergoing the same treatment.

In the roots, sodium contents increased under applied salt stress to threshold limits than that of stems (Figure 6B), up to about 1 mmol g^{-1} DW in all studied genotypes under 400 mM NaCl (the roots of *P. halimoides* were not investigated due to the aforementioned reasons). The most uniform decreasing pattern in K^+ content was observed at root level (Figure 6D); the highest reduction percentage as compared to control plants was recorded in *P. oleracea* L. subsp. *oleracea* (80.94%) contrary to *P. grandiflora* with only a 24.5% reduction when compared to untreated control plants. This translates to the biggest drop in K^+/Na^+ ratios being in *P. oleracea* L. subsp. *oleracea* again (Figure 6F). Chloride contents increased in parallel to applied salt in all genotypes, showed the highest accumulation in *P. oleracea* L. subsp. *oleracea* (Table 2).

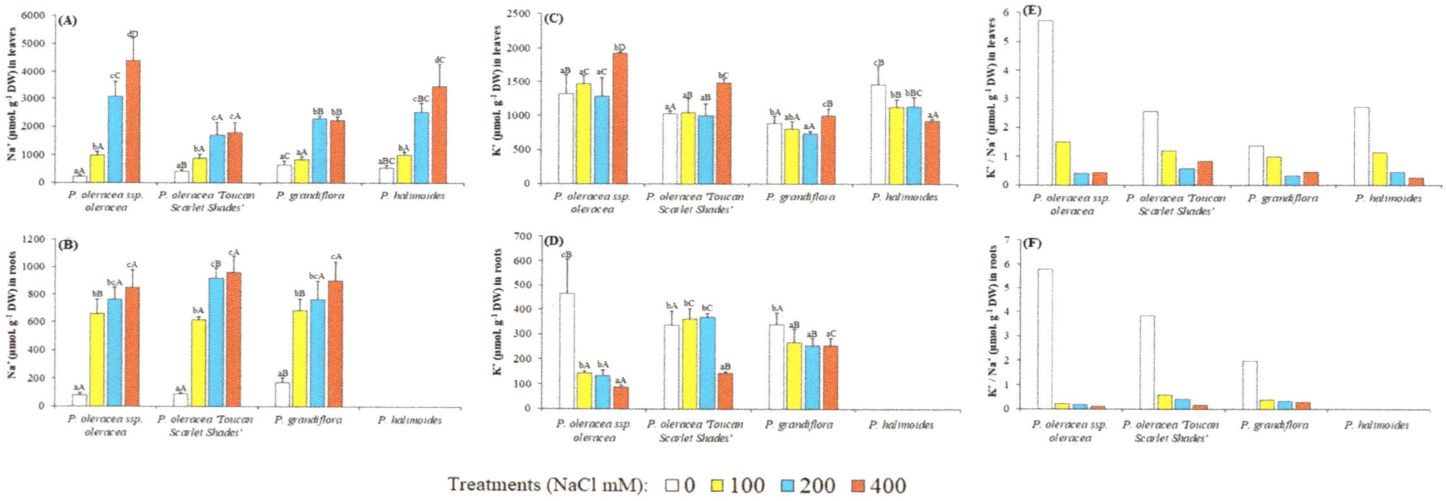

Figure 6. Ion content of leaves and roots of the 4 studied *Portulaca* genotypes undergoing 5 weeks of salt treatments (up to 400 mM NaCl). Sodium and potassium concentrations, as well as potassium over sodium ratios in leaves (**A**,**C**,**E**, respectively) and roots (**B**,**D**,**F**, respectively). The values shown are means ± SE (n = 5). Different lowercase letters above the bars indicate significant differences between the control plants, and different capital letters indicate significant differences between different genotypes.

2.7. Principal Component Analysis

A principal component analysis was performed with all analyzed parameters (except those of total soluble sugars and ions in roots due to the incomplete data set in *P. halimoides*). Five components were found with an Eigenvalue higher than 1, with their cumulative weight accounting for almost 90% of the total variability. The first component, explaining 34.87% of the variability was positively correlated with the levels of total phenolics (TPC) and flavonoids (TF), and K^+ in leaves, and to lesser extent to those of chlorophyll a (Chla) and (Chlb) and negatively with morphological parameters such as length of the leaves (LL), number of shoots (NS), number of flowers (Flo), and stem length (SL) (Figure 7). The second component explaining an additional 29.87% of variability was positively correlated with levels of osmolytes: Proline in roots (ProR) and leaves (ProL), and total soluble sugars in leaves (TSSL), and carotenoids (Caro) and negatively with the water content of roots (WCR) and leaves (WCL), fresh weight of leaves (FWL) and the ratio K^+/Na^+ in leaves (Figure 7). Regarding the projections of the genotypes on the scatterplot, *P. oleracea* susp. *oleracea* (OLE) separated from the other three analyzed, mostly due to high content of TP, TF and K^+ in the salt-stressed plants. Controls of the other three taxa are grouped together. Interestingly, plants from the most stressful treatment of 400 mM NaCl were separated along the first axis, except those of *P. halimoides* (HAL), which was clearly related to the second component. On the other hand, only *P. grandiflora* (GRA) was projected on the left side of the graph together, related with higher leaf length (LL), number of shoots (NS) and flowers (Flo), indicating a better response to salinity (Figure 7).

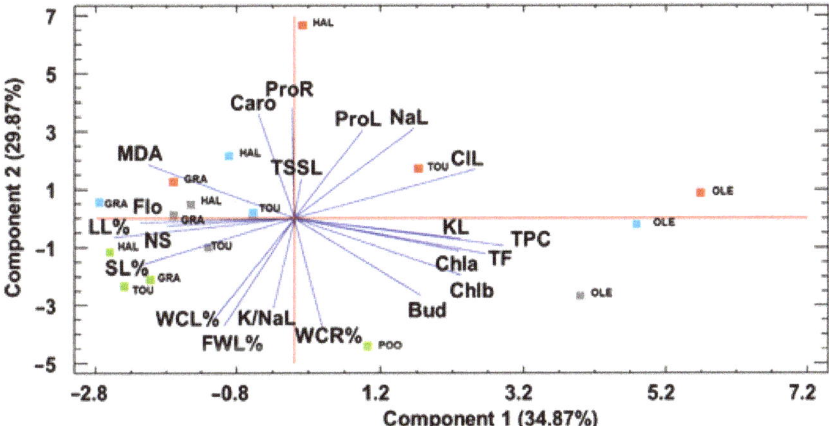

Figure 7. Principal component analysis based on the morphological and biochemical traits measured in plants from control (green), 100 (grey), 200 (blue) and 400 mM (red) NaCl treatments of *Portulaca oleracea* susbp. *oleracea* (OLE), *P. oleracea* cultivar "Toucan Scarlet Shades" (TOU), *P. grandiflora* (GRA) and *P. halimoides* (HAL). Abbreviations of the analyzed parameters: SL%, stem length percentage (control being 100%); NS, number of stems; LL%, leaf length percentage (control being 100%); flower bud, number of flower buds; Flo, number of flowers; FWL%, fresh weight percentage of leaves (control being 100%); WCL%, water content percentage in leaves; WCR%, water content percentage in roots; ProL, proline in leaves; ProR, proline in roots; TSS, total soluble sugars in leaves; MDA, malonaldehyde; TPC, total phenolic compounds; TF, total flavonoids; CL, chloride in leaves; NaL, sodium in leaves; KL, potassium in leaves; K/NaL, potassium over sodium ratio in leaves.

3. Discussion

Purslane (*P. olereacea*) has been proposed as an ideal candidate to be used as a halophytic crop with high nutritional value in the drainage water reuse system [39], and it was reported to accumulate higher biomass under mild salt concentrations and complete its life cycle even at a high salinity level (approximately 350 mM NaCl), maintaining its shoots free of visible salinity induced injury or nutrient deficiency symptoms [39]. This species was reported as moderate salt tolerant in many other studies [40–42]. Besides the accumulation of Na^+ and Cl^-, there are reports regarding its ability to accumulate higher levels of K^+ in shoots in conditions of salt stress, reflected in the K^+/Na^+ selectivity coefficients [39]. Due to its salt tolerance and edible properties purslane was regarded as a good candidate for soil desalinization [43,44], and given its potential in retaining toxic ions, also in heavy metals decontamination [45–47]. In a recent metabolomics study performed on two cultivars of purslane, 132 different metabolites were marked for their significant variation under salinity, notably an increase in proline and other amino acids, neutral and soluble sugars, sugar alcohols, amines, etc. [48]. Growth of both cultivars was strongly affected by the salinity, starting with 100 mM NaCl but metabolic response of the cultivars differed. Also, strong differences in growth parameters, flowering, mineral content or mechanism of salt tolerance were reported when comparing different genotypes of *P. oleracea* [49,50]. In a study analyzing responses to salt stress in a wild and a cultivated variety of *P. oleracea*, simultaneous changes in multiple traits indicated that the two genotypes employed different strategies in allocating resources to cope with saline stress [51]. This great phenotypic plasticity makes the species, and at a larger extent the genus *Portulaca* as a whole ideal for comparative studies focused on the better understanding of mechanism of stress tolerance [52].

Contrary to purslane, which was extensively analyzed in recent years, due to its interest as crop and medicinal plant, only little is known on the other two species included in the study. *P. grandiflora* has been reported as salt tolerant and as an accumulator of proline [53]. Recently, in addition to the C4 photosynthesis common in the genus *Portulaca*, the pathway CAM was found in cotyledons of this species [54]. According to our knowledge, there are no studies regarding the salt tolerance of *P. halimoides*, except our previous reports on germination and early seedling growth in *P. oleracea*, *P. grandiflora* and *P. halimoides* [55] which revealed a greater resilience to drought of the latter, and a seemingly better response to salinity in *P. grandiflora*.

The data presented in this work confirm the salt tolerance of purslane and of congener species. The four *Portulaca* genotypes survived 5 weeks of salt treatments even when 400 mM NaCl was applied. However, growth was affected which was expected since the most common effect of salinity is the inhibition of plant growth [13]. Growth inhibition under stress conditions is strongly related to the redirection of plant resources such as energy and metabolic precursors from their primary metabolisms and growth for the activation of defense mechanisms [14]. Although all *Portulaca* genotypes followed the same pattern of growth response, the smallest reduction of all growth parameters was observed in *P. grandiflora*, which at the end of the salt treatments showed a slight increase of the leaf length and only a small reduction of the shoots fresh weight and water content. This can be explained by a decrease in growth in terms of dry matter while the plants have retained their water content notably in the leaves, around 90% even under the highest applied stress (Figure 3B). This good water retention in the leaves could be explained by the constitutive succulence of the leaves of the *Portulaca* species. Succulence in combination with other adaptive traits, such as reduced leaf surface, leaf thickness, increased cell wall plasticity, or a small number of stomas per unit area, is considered one of the defining features of halophytes as an adaptive structural trait. In addition, the succulence has the effect of diluting the salts that can accumulate in the plant's organs, which allows the plant to cope with a large amount of salts [56]. Previous reports suggest that the osmotic potential of the sap from the leaves of plants grown under saline conditions can change to maintain a constant water uptake from the soil, based also on the osmotic adjustment at the root level [57–59]. A better response in terms of growth parameters to salt stress of this *P. grandiflora*, especially to the highest concentration of 400 mM NaCl, was also reflected in the scatter plot of the principal component analysis. On the other hand, it was observed that relatively small supplements in salt concentrations were enough to reduce vegetative growth and plant development in some genotypes (*P. oleracea* L. subsp. *oleracea* and *P. halimoides*). Greater salt tolerance of *P. grandiflora* was highlighted also by its flowering rate under salt treatments.

Applied salinity is known to have deleterious effects on the photosynthetic pigments in stressed plants, due to activity of ROS which accumulate due to oxidative stress resulting from ionic toxicity and osmotic imbalance [60]. A reduction in the photosynthetic pigments is considered a frequent response of the plants encountering stressful environments [61] being negatively correlated with their relative degree of tolerance to stress conditions [62]. Remarkable decreases in chlorophyll content as a result of drought and salt stress have been observed in various plant species including both trees, and herbaceous species, such as *Paulownia imperialis* [63], *Phaseolus vulgaris* [64], *Carthamus tinctorius* [65], *Salicornia prostrata* [66] or *Inula crithmoides* [67] but with some exceptions among halophytes including *Tecticornia indica* [68], *Sesuvium portulacastrum* [68], *Salvadora persica* [69], *Sarcocornia fruticosa* [70]. In the latter it has been observed that under saline conditions the leaf pigment content varied; in some species the total chlorophyll content was significantly enhanced exhibiting a gradual salt-dependent increase (*T. indica* and *S. portulacastrum*) whereas in others (*S. persica* or *S. fruticosa*) no significant changes were evidenced in chlorophyll a, b, total chlorophyll, and carotenoid contents with increasing salinity [69].

Previous reports indicate that purslane accessions subjected to salt stress generally showed a reduction of their chlorophyll concentration but the relative reductions with respect to the non-stressed controls were maintained below 40%, even at 32 dS m^{-1}, the highest salinity tested [71]. Our findings are in accordance with previous results indicating that a decrease in chlorophylls in some studied genotypes under stress, notably *P. oleracea* L. subsp. *oleracea* and to a lesser extent in *P. grandiflora*. On the other hand, an increase in these pigments was registered in *P. oleracea* "Toucan Scarlet Shades" and *P. halimoides* which is not unusual in plants undergoing low to medium-level stress, as a requirement to enhance the capacity for stress-defense. However, this precedes the expected degradation in case these unfavorable conditions are prolonged [72]. Carotenoids contents increased under the highest concentration of NaCl applied in all studied genotypes except *P. oleracea* subsp. *oleracea*. Some carotenoids such as xanthophylls and β-carotenes are well-known for their radical quenching activities, rendering them invaluable for photoprotection against accumulating ROS [73].

Growth parameters are important stress indicators used to evaluate the impact of different stresses on plants; but when they are combined with "stress biomarkers" which are associated with specific plant responses, they can provide useful information about tolerance mechanisms.

Prior investigations carried out in different plant species confirm that the accumulation and increased production of proline is correlated with stress tolerance. In transgenic tobacco plants, increasing proline content led to an enhanced resistance to drought and salinity [74]. These transgenics over-expressed the P5CS gene—encoding the enzyme that controls the rate-limiting step of proline biosynthesis from glutamate. However, proline accumulation is not always positively correlated with stress tolerance, but could also serve as a symptom of injury in leaves caused by salt [75]. Such were the results obtained especially in food crops rather than halophytes. A previous report reveals that sorghum genotypes differed in their sensitivity to salt [76]. Other studies indicate a negative correlation of proline accumulation with tolerance; for example, in two *Phaseolus* sp. the highest concentrations were measured in the most sensitive genotypes [77]. Similar results have been reported in maize, where proline accumulation in the salt tolerant genotypes (*Zea mays* L. cv. Ceratina) was significantly lower than that in the salt sensitive one (*Zea mays* L. cv. Sacharata) as well as in rice (*Oryza sativa* L.) with lower levels in cultivar IR28 (salt susceptible) than in Pokkari (salt tolerant) [78,79]. The results reported in this work indicate that the concentration of proline in the genotypes analyzed could not be related to their salt tolerance. Although, proline levels significantly increased in all selected genotypes in salt stressed plants, the smallest increment was registered in the most least damaged genotype, *P. grandiflora*.

The accumulation of total soluble sugars (TSS) in plants has been reported to play a role in response to abiotic stresses in many plant species [80]. The role of soluble sugars might be masked by their additional roles in plants as being products of the photosynthesis or components of primary metabolisms, signaling or regulatory networks which make sometimes difficult to clearly define their specific contribution to stress tolerance [81]. In the current research, levels of TSS increased only slightly in response to salinity, or their concentrations even decreased. As such, it is difficult to assess their role in salt tolerance of the analyzed taxa.

As reported for many other plant species, MDA is a product of membrane lipid peroxidation which is considered as a reliable biochemical oxidative stress marker; high MDA level should be correlated with high degree of oxidative stress [82]. However, our results show that MDA levels exhibited only minor variations that could not be correlated with the concentration of salt applied or with the degree of salt tolerance of the genotypes. This reflects a low to no oxidative stress among the investigated four genotypes. It can be concluded that lipid peroxidation due to oxidative stress could not occur within such a short period of applied stress (5 weeks).

Phenols and flavonoids are secondary metabolites which act as antioxidants, and their accumulation in plants can reduce the oxidative damage caused by abiotic stresses [83]. Total phenolic compounds (TFC) and total flavonoids (TF) increased significantly under salt stress mostly in *P. oleracea* subsp. *oleracea*, which also contained high concentrations in control plants, and to a lesser extent in its cultivated variety "Toucan Scarlet Shades", but not in the other two species analyzed.

High level of salt stress alters homeostasis in water potential and ion distribution. Drastic changes caused by the accumulation of Na^+ and Cl^- lead to molecular damage, growth arrest and even plant death due to the induced cytoplasmic toxicity [84]. To achieve tolerance against stress plants are controlling homeostasis by regulating ion transport within the plant, cellular uptake and intercellular distribution of Na+ and other toxic ions. Glycophytes activate mechanisms to reduce the uptake of Na+ and Cl-at root level or block their transport to the aerial parts of the plant, whereas dicotyledonous halophytes accumulate and sequester toxic ions in their vacuoles [84]. Our results show that Na^+ concentrations increased with to the concentration of NaCl applied and were significantly higher in the leaves of the selected *Portulaca* genotypes than in their roots, indicating that they act as sodium accumulators. In general Na^+ accumulation is associated with a decrease of K^+ levels since these two cations are competing for the same binding sites. A significant reduction of K^+ was observed in the roots of all plants from the 400 mM NaCl treatment, but this was associated with an increase of its levels in leaves, indicating that the active transport from roots to leaves is as essential mechanism of salt tolerance in the analyzed *Portulaca* genotypes. The low accumulation of Na^+ in leaves associated with only a small reduction of the ratio between K^+ and Na^+ in leaves, and a small decrease of K^+ in roots in plants of *P. grandiflora* submitted to 400 mM NaCl concentration, are the key elements explaining its better adjustment to high salinity. *P. grandiflora* was also the species with the lowest accumulation of Cl^- in roots and leaves under higher salt concentrations.

4. Materials and Methods

4.1. Plant Material and Stress Treatments

Three species of *Portulaca* were included in the study, purslane (*P. oleracea* L), moss rose (*P. grandiflora* Hook.) and silkcotton purslane (*P. halimoides* L.). Of the first species, two genotypes were compared, the wild *P. olereacea* subsp. *oleracea* and the cultivar *P. oleracea* "Toucan Scarlet Shades". Seeds used in this experiment were received from different Botanical Gardens (Botanicher Garten der Universität Zürich, Botanischen Garten der Universität Potsdam, Hortus Botanicus Bergianus—Bergianska trädgården, Grădina Agro-Botanică USAMV Cluj-Napoca) based on the Agreement on the supply of living plant material for non-commercial purposes while the ornamental variety was purchased from a commercial supplier from the USA.

Seeds of the four *Portulaca* genotypes (*P. oleracea* L. *subsp. oleracea*, *P. grandiflora* Hook., *P. halimoides* L. and *P. oleracea* "Toucan Scarlet Shades") were germinated in growth chambers under a photoperiod of 16 h, and a temperature of 25 °C. After 10 days of growth the seedlings were transferred individually to 0.5 L pots (9 × 9 × 10 cm) with a nutritive mix of peat:vermiculite:perlite (50:25:25) substrate. The plants were grown in the greenhouse of the Institute for Plant Molecular and Cellular Biology, Valencia, Spain (39°28′43.0″ N 0°20′12.1″ W) under the following conditions: Long-day photoperiod (16 h light/8 h dark), with light intensity of 130 µE m^{-2} s^{-1}, temperature (23 °C during the day and 17 °C at night), Humidity ranged between 50–80% during the time course of the treatments. During the growing period, seedlings were watered twice a week by adding 1.5 L by half-strength Hoagland's nutrient solution [85] to each tray (containing 12 pots).

Salt treatments were started after six weeks of growth [86] and applied to nine selected seedlings with homogeneous growth stage per treatment. During salt treatments, plants were treated twice a week with newly-prepared nutritive solutions supplemented with respective salt concentrations (100 mM, 200 mM and 400 mM), applying 150 mL solution for each pot. To avoid osmotic shock of salinity, the watering solution was added to the bottom of the trays providing thus a gradual absorption of the solution. In the meantime, control plants were irrigated in the same regime barring any added salt. Plants were harvested after 5 weeks of treatments. Plant roots, stems, and leaves were sampled separately at harvest.

Electrical conductivity (EC) of the soil (pot substrate) was measured at the beginning (Control) and at the end of the treatments applied to the four studied *Portulaca* genotypes. Soil samples were taken from three different pots of every treatment, air dried, diluted and then passed through a 2-mm opening sieve. The extract ratio used for determining EC was 1:5 soil:water mixture. The suspension was prepared in Milli Q water and stirred for 60 min at 600 u/min at 23 °C. Electrical conductivity was measured with a Crimson Conductometer 522. The values reported were expressed in dS m^{-1}.

4.2. Plant Growth Parameters

To determine the effects of applied salinity on plant growth, five plants per treatment per studied accession were monitored for the following physiological parameters: (a) Stem length, (b) leaf length, (c) number of stems and shoots, (d) number of flower buds and flowers, (e) dry weight and (f) water content.

The length of the main stem was measured every five days during the applied treatment period. The average of the five studied plants per stress treatment for every investigated accession was calculated, before being converted into percentage relative to the observed average measured for each species/cultivar's control at that time point.

Similarly, the length of five leaves per plant was measured for the same aforementioned plants per treatment for each studied accession. The leaves were chosen randomly from three different parts (layers) of the seedling: Lower, middle and upper part. The mean length of the leaves (in cm) was then calculated, and expressed as percentages of the average leaf length of the corresponding non-stressed control, taken as 100%. The total number of shoots per plant developed under stressed and unstressed conditions was recorded at the end of the treatments. In order to evaluate the reproductive traits, the flower bud occurrence was monitored during the treatments. The total number of flower buds and flowers for each treatment were averaged.

In addition to the above-mentioned growth parameters measured during the treatments, stress-induced inhibition of plant growth was also assessed by determining the mean fresh weight (FW), dry weight (DW) and water content (WC) of the leaves after 5 weeks of applied salt stress treatment. The fresh weight (FW) of the plants was measured separately for leaves and roots. After harvest, fresh plant material from both harvested tissues from all genotypes was weighed. The mean was then calculated and expressed in percentages (relative to each genotype's respective control, serving as 100%). After measuring the fresh weight, a part of the fresh material was dried in the oven for four days at 65 °C, until constant weight in order to obtain DW. The fresh weight and dry weight data were then used to calculate the water content percentage of the three harvested tissue per studied plant according to the following formula [87]:

$$WC\ (\%) = [(FW - DW)/FW] \times 100$$

4.3. Biochemical Plant Responses to Salt Stress

After plant harvest, biochemical analyses were carried out to assess the salinity-induced changes in the levels of different biochemical stress markers including: Photosynthetic pigments (Chlorophyll a and b and carotenoids), osmolytes (proline, total soluble sugars), malondialdehyde (MDA), total phenolic (TPC), flavonoids (TFC), and ionic content.

4.3.1. Photosynthetic Pigments

Chlorophyll a (Chl a), chlorophyll b (Chl b), and total carotenoid (Caro) levels were determined via spectrophotometry, using 100 mg fresh leaf material from each of the four studied *Portulaca* genotypes. The collected fresh leaves were ground in a MM 301 mixer mill in the presence of liquid N_2 and then extracted with 1 mL ice-cold 80% acetone. Samples were mixed in the dark overnight on a shaker at 4 °C, before being centrifuged at 4 °C for 15 min (3,000 rpm). The supernatant (100 µL) was transferred into new tubes and diluted with another 900 µL of ice-cold acetone. Samples were mixed thoroughly and absorbance was recorded at 663, 646 and 470 nm. Five replicates were used for each treatment. The amount of pigments in each sample was calculated according to the following equations (80% acetone was used as blank) [88]:

Chl 'a' ($\mu g\ mL^{-1}$) = 12.21 (A663) − 2.81 (A646);
Chl 'b' ($\mu g\ mL^{-1}$) = 20.13 (A646) − 5.03 (A663);
Chl 'a'+'b' ($\mu g\ mL^{-1}$) = 20.13 (A646) − 5.03 (A663).
Caro ($\mu g\ mL^{-1}$) = [1000 × A470 − (3.27×Chl 'a') − (104 × Chl 'b')]/227
Final values of the photosynthetic pigments were expressed as mg. g^{-1} DW.

4.3.2. Osmolytes

Free proline (Pro) content was measured in fresh tissue from the leaves and roots according to the ninhydrin-acetic acid method of Bates et al. [89]. The frozen plant material was homogenized in 3% (*w/v*) aqueous sulfosalicylic acid, mixed with 1 mL acid-ninhydrin and 1 mL of glacial acetic acid. Samples were incubated in boiling water for 1 h at 95 °C. To stop the reaction, samples were cooled down in an ice bath and the reaction mixture was extracted with 2 mL toluene, mixed vigorously and left at room temperature for phase separation. The absorbance of the supernatant was measured at 520 nm using toluene for blank. Proline concentration was determined from a standard curve and expressed in $\mu mol\ g^{-1}$ DW. Total soluble sugar (TSS) content of the *Portulaca* samples was quantified according to the method described by Dubois et al. [90]. Dried material was ground and mixed with 3 mL of 80% methanol on a rocker shaker for 24–48 h. The mixture was filtered and centrifuged at 13 000 rpm for 20 min. Concentrated sulfuric acid and 5% phenol was added to the samples and the absorbance was measured at 490 nm. Glucose solutions of known concentration were used to obtain a standard curve, and TSS contents were expressed as "mg equivalent of glucose" per gram of DW.

4.3.3. Malondialdehyde

Lipid peroxidation levels in the leaves of the four *Portulaca* genotypes were determined measuring MDA content by using thiobarbituric acid-reactive substances (TBARS) assay, as described by Hodges et al. [91]. The ground dry material (0.05 g) was extracted with 400 µL of methanol 80%, by mixing in an orbital shaker for 12 h. After mixing, the sample was filtered through glass wool and then 800 µL of 20% TCA and 800 µL of 0.65% TBA were added to the samples which were mixed vigorously and heated at 95 °C in a block heater for 25 min, cooled and centrifuged at 3000× g for 10 min. The supernatant's absorbance was recorded at 440 nm, 532 nm and 600 nm using a spectrophotometer. The blank was prepared in the same way but without plant extract. MDA concentration was calculated from the absorbance at 440, 532 nm and measurements were corrected for non-specific absorbance by subtracting the absorbance at 600 nm. Malondialdehyde equivalents were expressed as $\mu mol\ MDA\ g^{-1}$ FW.

4.3.4. Total Phenolic and Flavonoid Content

Total phenolic compounds (TPC) were quantified in leaves of the four *Portulaca* genotypes by reaction with Folin–Ciocalteu reagent, according to Blainski et al. [92], using 50 μL of plant extract (extracted from dry matter in 80% methanol). and measuring the absorbance at 765 nm using a spectrophotometer. The amount of TPC was expressed as milligram of Gallic acid (used as standard) equivalents per g of dry weight of the sample (mg eq. GA g^{-1} DW). Total flavonoids (TF) were determined by using the aluminum chloride colorimetric assay according to Zhishen et al. [93] protocol. The absorbance of the samples was measured in comparison to the prepared blank reagent at 510 nm using a spectrophotometer. The total flavonoid content of the leaves was expressed in equivalents of catechin (mg eq. C g^{-1} DW), used as standard.

4.3.5. Ion Content

To perform the ion content measurements, 0.05 g dried and ground plant material (leaves and roots) was suspended in 2 mL distilled water and homogenized. The supernatant was transferred to plastic tubes and boiled at 100 °C for 7 min. The samples were cooled down and centrifuged at 20,000× g for 10 min at 4 °C. The supernatant was transferred into a 1 mL syringe, and then passed through 0.22 μm filters. Na^+, K^+ levels were quantified by using a Flame Photometer Jenway PFP7 while Cl^- were measured using a potentiometer titrator—848 Titrino plus.

4.3.6. Statistical Analyses

All data was analyzed using IBM SPSS Statistics 19 (IBM Corporation, Portsmouth, UK) and Statgraphics Centurion 18 (Statgraphics Technologies, The Plains, VA, USA). One-way analysis of variance (ANOVA) was assessed at 95% confidence level to determine whether there were any statistically significant differences between the means of the treatments within one genotype and for all the genotypes undergoing the same treatment. When the ANOVA null hypothesis was rejected, post-hoc comparisons were performed using Tukey's honestly significant difference test at $P < 0.05$ to determine significant differences between the means. The values shown throughout this manuscript's text, tables and figures, are means ± SE. Furthermore, all parameters recorded in plants from control and salt treatments were correlated by principal component analysis (PCA).

5. Conclusions

The testing of the interspecific variation in salinity in *Portulaca* genotypes allowed a better understanding of the essential mechanisms of salt tolerance in this genus. Based on sustained vegetative growth under applied salt stress, as a criterion for improved adaptability, *P. grandiflora* is assumed to be the least susceptible of the genotypes studied. However, contrary to previous reports in *Portulaca*, which considered proline accumulation as a key response to confer salinity tolerance, the present study revealed only a small increase in proline in the more tolerant *P. grandiflora*. Our study demonstrates that tolerance in the studied taxa depends rather on the maintenance of K^+ homeostasis. When exposed to high salinity, all genotypes accumulated Na^+ and Cl^- in the roots and leaves at substantial levels, acting as Na^+ and Cl^- includers. The role of K^+ in the maintenance of cellular functions is well documented and, as such, its sustained presence in these detrimental conditions ensures the continuation of enzymatic functions and the maintenance of the integrity of photosynthetic pigments, which guarantee greater tolerance. However, slight differences in ion transport have been observed between the four genotypes studied. The significant increase of Na^+ in the leaves of *P. oleracea* subsp. *oleracea*, *P. oleracea* "Toucan Scarlet Shades" and *P. grandiflora* at the highest salt concentration applied and the decrease of the K^+/Na^+ ratio in leaves with increasing salinity suggest that Na^+ was transported in greater proportion than K^+ to this organ in these genotypes.

Author Contributions: Conceptualization, R.E.S. and O.V.; methodology, M.A.H. and M.D.R.; software, C.N.; validation, M.B., R.E.S. and O.V.; formal analysis, O.B. and M.A.H.; investigation, O.B., M.A.H. and M.D.R.; resources, R.E.S. and O.V.; data curation, M.A.H. and C.N.; writing—original draft preparation, O.B. and M.A.H.; writing—review and editing, M.B.; visualization, C.N.; supervision, O.V.; project administration, R.E.S. and O.V.; funding acquisition, O.B., R.E.S. and O.V. All authors have read and agreed to the published version of the manuscript.

Funding: This research and publication was supported by the funds from the National Research Development Projects to finance excellence (PFE)-37/2018–2020 granted by the Romanian Ministry of Research and Innovation.

Acknowledgments: The authors also would like to thank the administrative and technical support provided under the frame of the current undergoing project, grant number POC-A1-A1.1.1-B-2015.

Conflicts of Interest: The authors declare no conflict of interest.

References

1. Grime, J.P. Evidence for the Existence of Three Primary Strategies in Plants and Its Relevance to Ecological and Evolutionary Theory. *Am. Nat.* **1977**, *111*, 1169–1194. [CrossRef]
2. Van Breusegem, F.; Vranová, E.; Dat, J.F.; Inzé, D. The role of active oxygen species in plant signal transduction. *Plant Sci.* **2001**, *161*, 405–414. [CrossRef]
3. Bartels, D.; Sunkar, R. Drought and Salt Tolerance in Plants. *Crit. Rev. Plant Sci.* **2005**, *24*, 23–58. [CrossRef]
4. Katerji, N.; Van Hoorn, J.; Hamdy, A.; Mastrorilli, M. Salinity effect on crop development and yield, analysis of salt tolerance according to several classification methods. *Agric. Water Manag.* **2003**, *62*, 37–66. [CrossRef]
5. Fahad, S.; Bajwa, A.A.; Nazir, U.; Anjum, S.A.; Farooq, A.; Zohaib, A.; Sadia, S.; Nasim, W.; Adkins, S.; Saud, S.; et al. Crop Production under Drought and Heat Stress: Plant Responses and Management Options. *Front. Plant Sci.* **2017**, *8*, 1147. [CrossRef] [PubMed]
6. Wild, A. *Soils, Land and Food: Managing the Land During the Twenty-First Century*; Cambridge University Press: Cambridge, UK, 2003; pp. 192–196.
7. Owens, S. Salt of the Earth. *EMBO Rep.* **2001**, *2*, 877–879. [CrossRef]
8. Joint FAO/WHO Expert Committee on Food Additives. *Meeting, World Health Organization. Evaluation of Certain Food Additives and Contaminants: Fifty-Seventh Report of the Joint FAO/WHO Expert Committee on Food Additives*; World Health Organization: Geneva, Switzerland, 2002.
9. Cuevas, J.; Daliakopoulos, I.N.; Del Moral, F.; Hueso, J.J.; Tsanis, I.K. A Review of Soil-Improving Cropping Systems for Soil Salinization. *Agronomy* **2019**, *9*, 295. [CrossRef]
10. Munns, R. Comparative physiology of salt and water stress. *Plant Cell Environ.* **2002**, *25*, 239–250. [CrossRef]
11. Butcher, K.; Wick, A.F.; DeSutter, T.M.; Chatterjee, A.; Harmon, J. Soil Salinity: A Threat to Global Food Security. *Agron. J.* **2016**, *108*, 2189–2200. [CrossRef]
12. Krasensky, J.; Jonak, C. Drought, salt, and temperature stress-induced metabolic rearrangements and regulatory networks. *J. Exp. Bot.* **2012**, *63*, 1593–1608. [CrossRef]
13. Safdar, H.; Amin, A.; Shafiq, Y.; Ali, A.; Yasin, R.; Shoukat, A.; Hussan, M.U.; Sarwar, M.I. A review: Impact of salinity on plant growth. *Nat. Sci.* **2019**, *17*, 34–40. [CrossRef]
14. Zhu, J.-K. Plant salt tolerance. *Trends Plant Sci.* **2001**, *6*, 66–71. [CrossRef]
15. Parida, A.K.; Das, A.B.; Mittra, B.N. Effects of salt on growth, ion accumulation, photosynthesis and leaf anatomy of the mangrove, *Bruguiera parviflora*. *Trees* **2004**, *18*, 167–174. [CrossRef]
16. He, M.; He, C.-Q.; Ding, N.-Z. Abiotic Stresses: General Defenses of Land Plants and Chances for Engineering Multistress Tolerance. *Front. Plant Sci.* **2018**, *9*, 1771. [CrossRef] [PubMed]
17. Rhodes, D.; Hanson, A.D. Quaternary Ammonium and Tertiary Sulfonium Compounds in Higher Plants. *Annu. Rev. Plant Biol.* **1993**, *44*, 357–384. [CrossRef]
18. Rahnama, H.; Ebrahimzadeh, H. The effect of NaCl on antioxidant enzyme activities in potato seedlings. *Biol. Plant.* **2005**, *49*, 93–97. [CrossRef]
19. Allen, J.A.; Chambers, J.L.; Stine, M. Prospects for increasing the salt tolerance of forest trees: A review. *Tree Physiol.* **1994**, *14*, 843–853. [CrossRef] [PubMed]
20. Munns, R.; Tester, M. Mechanisms of Salinity Tolerance. *Annu. Rev. Plant Biol.* **2008**, *59*, 651–681. [CrossRef]
21. Flowers, T.J.; Colmer, T.D. Salinity tolerance in halophytes. *New Phytol.* **2008**, *179*, 945–963. [CrossRef]
22. Flowers, T.J.; Hajibagheri, M.A.; Clipson, N.J.W. Halophytes. *Q. Rev. Biol.* **1986**, *61*, 313–337. [CrossRef]

23. Van De Wouw, M.; Kik, C.; Van Hintum, T.; Van Treuren, R.; Visser, B. Genetic erosion in crops: Concept, research results and challenges. *Plant Genet. Resour.* **2010**, *8*, 1–15. [CrossRef]

24. Fita, A.; Rodríguez-Burruezo, A.; Boscaiu, M.; Prohens, J.; Vicente, O. Breeding and Domesticating Crops Adapted to Drought and Salinity: A New Paradigm for Increasing Food Production. *Front. Plant Sci.* **2015**, *6*, 978. [CrossRef]

25. Dossa, K.; Mmadi, M.A.; Zhou, R.; Zhang, T.; Su, R.; Zhang, Y.; Wang, L.; You, J.; Dossa, K. Depicting the Core Transcriptome Modulating Multiple Abiotic Stresses Responses in Sesame (*Sesamum indicum* L.). *Int. J. Mol. Sci.* **2019**, *20*, 3930. [CrossRef]

26. Al Hassan, M. Comparative Analyses of Plant Responses to Drought and Salt Stress in Related Taxa: A Useful Approach to Study Stress Tolerance Mechanisms. Ph.D. Thesis, Universitat Politecnica de Valencia, Valencia, Spain, 2016. [CrossRef]

27. Kumari, A.; Das, P.; Parida, A.K.; Agarwal, P.K. Proteomics, metabolomics, and ionomics perspectives of salinity tolerance in halophytes. *Front. Plant Sci.* **2015**, *6*, 537. [CrossRef]

28. Misra, A.N.; Latowski, D.; Strzałka, K. The xanthophyll cycle activity in kidney bean and cabbage leaves under salinity stress. *Russ. J. Plant Physiol.* **2006**, *53*, 102–109. [CrossRef]

29. Qados, A.M.A. Effect of salt stress on plant growth and metabolism of bean plant *Vicia faba* (L.). *J. Saudi Soc. Agric. Sci.* **2011**, *10*, 7–15. [CrossRef]

30. Hasegawa, P.M.; Bressan, R.A.; Zhu, J.K.; Bohnert, H.J. Plant cellular and molecular responses to high salinity. *Annu. Rev. Plant Physiol. Plant Mol. Biol.* **2000**, *51*, 463–499. [CrossRef]

31. Rhodes, D.; Nadolska-Orczyk, A.; Rich, P. Salinity, osmolytes and compatible solutes. In *Salinity: Environment-Plants-Molecules*; Springer: Dordrecht, Germany, 2002; pp. 181–204. [CrossRef]

32. Joshi, R.; Mangu, V.R.; Bedre, R.; Sanchez, L.; Pilcher, W.; Zandkarimi, H.; Baisakh, N. Salt adaptation mechanisms of halophytes: Improvement of salt tolerance in crop plants. In *Elucidation of Abiotic Stress Signaling in Plants*; Springer: New York, NY, USA, 2015; pp. 243–279.

33. Ozgur, R.; Uzilday, B.; Sekmen, A.H.; Turkan, I. Reactive oxygen species regulation and antioxidant defence in halophytes. *Funct. Plant Biol.* **2013**, *40*, 832–847. [CrossRef]

34. Xu, J.; Tian, Y.-S.; Peng, R.-H.; Xiong, A.-S.; Zhu, B.; Jin, X.-F.; Gao, F.; Fu, X.-Y.; Hou, X.; Yao, Q.-H. AtCPK6, a functionally redundant and positive regulator involved in salt/drought stress tolerance in Arabidopsis. *Planta* **2010**, *231*, 1251–1260. [CrossRef]

35. Nakabayashi, R.; Saito, K. Integrated metabolomics for abiotic stress responses in plants. *Curr. Opin. Plant Biol.* **2015**, *24*, 10–16. [CrossRef] [PubMed]

36. Kim, I.; Fisher, D.G. Structural aspects of the leaves of seven species of *Portulaca* growing in Hawaii. *Can. J. Bot.* **1990**, *68*, 1803–1811. [CrossRef]

37. Yazici, I.; Türkan, I.; Sekmen, A.H.; Demiral, T. Salinity tolerance of purslane (*Portulaca oleracea* L.) is achieved by enhanced antioxidative system, lower level of lipid peroxidation and proline accumulation. *Environ. Exp. Bot.* **2007**, *61*, 49–57. [CrossRef]

38. Sdouga, D.; Ben Amor, F.; Ghribi, S.; Kabtni, S.; Tebini, M.; Branca, F.; Trifi-Farah, N.; Marghali, S. An insight from tolerance to salinity stress in halophyte *Portulaca oleracea* L.: Physio-morphological, biochemical and molecular responses. *Ecotoxicol. Environ. Saf.* **2019**, *172*, 45–52. [CrossRef] [PubMed]

39. Grieve, C.; Suarez, D. Purslane (*Portulaca oleracea* L.): A halophytic crop for drainage water reuse systems. *Plant Soil* **1997**, *192*, 277–283. [CrossRef]

40. Kafi, M.; Rahimi, Z. Effect of salinity and silicon on root characteristics, growth, water status, proline content and ion accumulation of purslane (*Portulaca oleracea* L.). *Soil Sci. Plant Nutr.* **2011**, *57*, 341–347. [CrossRef]

41. Uddin, K.; Juraimi, A.S.; Ali, E.; Ismail, M.R. Evaluation of Antioxidant Properties and Mineral Composition of Purslane (*Portulaca oleracea* L.) at Different Growth Stages. *Int. J. Mol. Sci.* **2012**, *13*, 10257–10267. [CrossRef]

42. Alam, A.; Juraimi, A.S.; Rafii, M.Y.; Hamid, A.A.; Aslani, F. Screening of Purslane (*Portulaca oleracea* L.) Accessions for High Salt Tolerance. *Sci. World J.* **2014**, *2014*, 1–12. [CrossRef]

43. Hamidov, A. Environmentally useful technique—*Portulaca oleracea* golden purslane as a salt removal species. *WSEAS Trans. Environ. Dev.* **2007**, *2*, 117–122.

44. Karakaş, S.; Çullu, M.A.; Dikilitaş, M. Comparison of two halophyte species (*Salsola soda* and *Portulaca oleracea*) for salt removal potential under different soil salinity conditions. *Turk. J. Agric. For.* **2017**, *41*, 183–190. [CrossRef]

45. Osma, E.; Ozyigit, I.I.; Demir, G.; Yasar, U. Assessment of some heavy metals in wild type and cultivated purslane (*Portulaca oleracea* L.) and soils in Istanbul, Turkey. *Fresen. Environ. Bull.* **2014**, *23*, 2181–2189.

46. Hammami, H.; Parsa, M.; Mohassel, M.H.R.; Rahimi, S.; Mijani, S. Weeds ability to phytoremediate cadmium-contaminated soil. *Int. J. Phytoremediat.* **2015**, *18*, 48–53. [CrossRef]

47. Negi, S. Heavy metal accumulation in *Portulaca oleracea* Linn. *J. Pharmacogn. Phytochem.* **2018**, *7*, 2978–2982.

48. Zaman, S.; Bilal, M.; Du, H.; Che, S. Morphophysiological and Comparative Metabolic Profiling of Purslane Genotypes (*Portulaca oleracea* L.) under Salt Stress. *BioMed Res. Int.* **2020**, *2020*, 1–17. [CrossRef]

49. Alam, A.; Juraimi, A.S.; Rafii, M.Y.; Hamid, A.A.; Aslani, F.; Hakim, A. Salinity-induced changes in the morphology and major mineral nutrient composition of purslane (*Portulaca oleracea* L.) accessions. *Biol. Res.* **2016**, *49*, 24. [CrossRef]

50. Ozturk, M.; Altay, V.; Güvensen, A. Portulaca oleracea: A vegetable from saline habitats. In *Handbook of Halophytes: From Molecules to Ecosystems towards Biosaline Agriculture*; Grigore, M.N., Ed.; Springer-Multimedia: Cham, Switzerland, 2020; pp. 1–14.

51. Mulry, K.R.; Hanson, B.A.; Dudle, D.A. Alternative Strategies in Response to Saline Stress in Two Varieties of *Portulaca oleracea* (Purslane). *PLoS ONE* **2015**, *10*, e0138723. [CrossRef]

52. Borsai, O.; Al Hassan, M.; Boscaiu, M.; Sestras, R.E.; Vicente, O. The genus *Portulaca* as a suitable model to study the mechanisms of plant tolerance to drought and salinity. *EuroBiotech J.* **2018**, *2*, 104–113. [CrossRef]

53. Kichenaradjou, M.; Jawaharlal, M.; Arulmozhiyan, R.D.; Vijayaraghavan, H. Evaluation of ornamental groundcovers on physiological and quality parameters in salt affected soil condition. *J. Pharmacogn. Phytochem.* **2018**, *7*, 553–557.

54. Guralnick, L.J.; Gilbert, K.E.; Denio, D.; Antico, N. The Development of Crassulacean Acid Metabolism (CAM) Photosynthesis in Cotyledons of the C4 Species, Portulaca grandiflora (*Portulacaceae*). *Plants* **2020**, *9*, 55. [CrossRef] [PubMed]

55. Borsai, O.; Al Hassan, M.; Boscaiu, M.; Sestras, R.E.; Vicente, O. Effects of salt and drought stress on seed germination and seedling growth in Portulaca. *Rom. Biotechnol. Lett.* **2018**, *23*, 13340–13349.

56. Karuppanapandian, T.; Moon, J.C.; Kim, C.; Manoharan, K.; Kim, W. Reactive oxygen species in plants: Their generation, signal transduction, and scavenging mechanisms. *Aust. J. Crop Sci.* **2011**, *5*, 709.

57. Grigore, M.-N.; Toma, C. Succulence. In *Anatomical Adaptations of Halophytes*; Springer: Cham, Switzerland, 2017. [CrossRef]

58. Grigore, M.N.; Toma, C. Kranz Anatomy. In *Anatomical Adaptations of Halophytes*; Springer: Cham, Switzerland, 2017. [CrossRef]

59. Poljakoff-Mayber, A. Morphological and anatomical changes in plants as a response to salinity stress. In *Plants in Saline Environments*; Poljakoff-Mayber, A., Gale, J., Eds.; Springer: New York, NY, USA, 1975; pp. 97–117.

60. Bernstein, L. osmotic adjustment of plants to saline media. II. dynamic phase. *Am. J. Bot.* **1963**, *50*, 360–370. [CrossRef]

61. Dubey, S.; Bhargava, A.; Fuentes, F.; Shukla, S.; Srivastava, S. Effect of salinity stress on yield and quality parameters in flax (*Linum usitatissimum* L.). *Not. Bot. Horti Agrobot. Cluj-Napoca* **2020**, *48*, 954–966. [CrossRef]

62. Hand, M.J.; Taffouo, V.D.; Nouck, A.E.; Nyemene, K.P.; Tonfack, B.; Meguekam, T.L.; Youmbi, E. Effects of Salt Stress on Plant Growth, Nutrient Partitioning, Chlorophyll Content, Leaf Relative Water Content, Accumulation of Osmolytes and Antioxidant Compounds in Pepper (*Capsicum annuum* L.) Cultivars. *Not. Bot. Horti Agrobot. Cluj-Napoca* **2017**, *45*, 481–490. [CrossRef]

63. Ayala-Astorga, G.I.; Lilia, A.-M. Salinity effects on protein content, lipid peroxidation, pigments, and proline in *Paulownia imperialis* (Siebold & Zuccarini) and *Paulownia fortunei* (Seemann & Hemsley) grown In Vitro. *Electron. J. Biotechnol.* **2010**, *13*, 13–14. [CrossRef]

64. Bayuelo-Jiménez, J.S.; Jasso-Plata, N.; Ochoa, I. Growth and Physiological Responses of *Phaseolus* Species to Salinity Stress. *Int. J. Agron.* **2012**, *2012*, 1–13. [CrossRef]

65. Siddiqi, E.H.; Ashraf, M. Can leaf water relation parameters be used as selection criteria for salt tolerance in safflower (*Carthamus tinctorius* L.). *Pak. J. Bot.* **2008**, *40*, 221–228.

66. Akcin, A.; Yalcin, E. Effect of salinity stress on chlorophyll, carotenoid content, and proline in *Salicornia prostrata* Pall. and *Suaeda prostrata* Pall. subsp. *prostrata* (*Amaranthaceae*). *Braz. J. Bot.* **2016**, *39*, 101–106. [CrossRef]

67. Al Hassan, M.; Chaura, J.; López-Gresa, M.P.; Borsai, O.; Daniso, E.; Donat-Torres, M.P.; Mayoral, O.; Vicente, O.; Boscaiu, M. Native-Invasive Plants vs. Halophytes in Mediterranean Salt Marshes: Stress Tolerance Mechanisms in Two Related Species. *Front. Plant Sci.* **2016**, *7*, 473. [CrossRef]

68. Rabhi, M.; Castagna, A.; Remorini, D.; Scattino, C.; Smaoui, A.; Ranieri, A.; Abdelly, C. Photosynthetic responses to salinity in two obligate halophytes: *Sesuvium portulacastrum* and *Tecticornia indica*. *S. Afr. J. Bot.* **2012**, *79*, 39–47. [CrossRef]

69. Rangani, J.; Parida, A.K.; Panda, A.; Kumari, A. Coordinated Changes in Antioxidative Enzymes Protect the Photosynthetic Machinery from Salinity Induced Oxidative Damage and Confer Salt Tolerance in an Extreme Halophyte Salvadora persica L. *Front. Plant Sci.* **2016**, *7*, 50. [CrossRef]

70. Redondo-Gómez, S.; Wharmby, C.; Castillo, J.M.; Mateos-Naranjo, E.; Luque, C.J.; De Cires, A.; Enrique Figueroa, M. Growth and photosynthetic responses to salinity in an extreme halophyte, *Sarcocornia fruticosa*. *Physiol. Plant.* **2006**, *128*, 116–124. [CrossRef]

71. Agathokleous, E.; Feng, Z.; Peñuelas, J. Chlorophyll hormesis: Are chlorophylls major components of stress biology in higher plants? *Sci. Total. Environ.* **2020**, *726*, 138637. [CrossRef] [PubMed]

72. Miller, N.J.; Sampson, J.; Candeias, L.P.; Bramley, P.M.; Rice-Evans, C.A. Antioxidant activities of carotenes and xanthophylls. *FEBS Lett.* **1996**, *384*, 240–242. [CrossRef]

73. García-Caparrós, P.; Hasanuzzaman, M.; Lao, M.T. Oxidative stress and antioxidant defense in plants under salinity. In *Reactive Oxygen, Nitrogen and Sulfur Species in Plants: Production, Metabolism, Signaling and Defense Mechanisms*; Hasanuzzaman, M., Fotopoulos, V., Nahar, K., Fujita, M., Eds.; Wiley: Hoboken, NJ, USA, 2019; pp. 291–309. [CrossRef]

74. Sun, H.; Sun, X.; Wang, H.; Ma, X. Advances in salt tolerance molecular mechanism in tobacco plants. *Hereditas* **2020**, *157*, 1–6. [CrossRef]

75. Lutts, S.; Majerus, V.; Kinet, J.-M. NaCl effects on proline metabolism in rice (*Oryza sativa*) seedlings. *Physiol. Plant.* **1999**, *105*, 450–458. [CrossRef]

76. De Lacerda, C.F.; Cambraia, J.; Oliva, M.A.; Ruiz, H.A. Osmotic adjustment in roots and leaves of two sorghum genotypes under NaCl stress. *Braz. J. Plant Physiol.* **2003**, *15*, 113–118. [CrossRef]

77. Al Hassan, M.; Morosan, M.; López-Gresa, M.P.; Prohens, J.; Vicente, O.; Boscaiu, M. Salinity-Induced Variation in Biochemical Markers Provides Insight into the Mechanisms of Salt Tolerance in Common (*Phaseolus vulgaris*) and Runner (*P. coccineus*) Beans. *Int. J. Mol. Sci.* **2016**, *17*, 1582. [CrossRef]

78. Cha-Um, S.; Kirdmanee, C. Effect of salt stress on proline accumulation, photosynthetic ability and growth characters in two maize cultivars. *Pak. J. Bot.* **2009**, *41*, 87–98.

79. Demiral, T.; Türkan, I. Comparative lipid peroxidation, antioxidant defense systems and proline content in roots of two rice cultivars differing in salt tolerance. *Environ. Exp. Bot.* **2005**, *53*, 247–257. [CrossRef]

80. Gil, R.; Boscaiu, M.; Lull, C.; Bautista, I.; Lidon, A.L.; Vicente, O. Are soluble carbohydrates ecologically relevant for salt tolerance in halophytes? *Funct. Plant Biol.* **2013**, *40*, 805–818. [CrossRef]

81. Lokhande, V.H.; Suprasanna, P. Prospects of Halophytes in Understanding and Managing Abiotic Stress Tolerance. In *Environmental Adaptations and Stress Tolerance of Plants in the Era of Climate Change*; Ahmad, P., Prasad, M.N.V., Eds.; Springer: New York, NY, USA, 2012; pp. 29–56.

82. Kumar, D.; Al Hassan, M.; Naranjo, M.A.; Agrawal, V.; Boscaiu, M.; Vicente, O. Effects of salinity and drought on growth, ionic relations, compatible solutes and activation of antioxidant systems in oleander (*Nerium oleander* L.). *PLoS ONE* **2017**, *12*, e0185017. [CrossRef]

83. Ahmad, P.; Jaleel, C.A.; Salem, M.A.; Nabi, G.; Sharma, S. Roles of enzymatic and nonenzymatic antioxidants in plants during abiotic stress. *Crit. Rev. Biotechnol.* **2010**, *30*, 161–175. [CrossRef]

84. Flowers, T.J.; Munns, R.; Colmer, T.D. Sodium chloride toxicity and the cellular basis of salt tolerance in halophytes. *Ann. Bot.* **2015**, *115*, 419–431. [CrossRef]

85. Hoagland, D.R.; Arnon, D.I. The water-culture method for growing plants without soil. In *California Agricultural Experiment Station Publications Series*; College of Agriculture, University of California: Davis, CA, USA, 1950.

86. Rahdari, P.; Tavakoli, S.; Hosseini, S.M. Studying of salinity stress effect on germination, proline, sugar, protein, lipid and chlorophyll content in purslane (*Portulaca oleracea* L.) leaves. *J. Stress. Physiol. Biochem.* **2011**, *8*, 182–193.

87. Li, W.; Zhang, H.; Zeng, Y.; Xiang, L.; Lei, Z.; Huang, Q.; Li, T.; Shen, F.; Cheng, Q. A Salt Tolerance Evaluation Method for Sunflower (*Helianthus annuus* L.) at the Seed Germination Stage. *Sci. Rep.* **2020**, *10*, 1–9. [CrossRef]

88. Lichtenthaler, H.K.; Wellburn, A.R. Determinations of total carotenoids and chlorophylls a and b of leaf extracts in different solvents. *Biochem. Soc. Trans.* **1983**, *11*, 591–592. [CrossRef]

89. Bates, L.S.; Waldren, R.P.; Teare, I.D. Rapid determination of free proline for water-stress studies. *Plant Soil* **1973**, *39*, 205–207. [CrossRef]

90. Dubois, M.; Gilles, K.A.; Hamilton, J.K.; Rebers, P.A.; Smith, F. Colorimetric Method for Determination of Sugars and Related Substances. *Anal. Chem.* **1956**, *28*, 350–356. [CrossRef]

91. Hodges, D.M.; Delong, J.M.; Forney, C.F.; Prange, R.K. Improving the thiobarbituric acid-reactive-substances assay for estimating lipid peroxidation in plant tissues containing anthocyanin and other interfering compounds. *Planta* **1999**, *207*, 604–611. [CrossRef]

92. Blainski, A.; Lopes, G.C.; De Mello, J.C.P. Application and Analysis of the Folin Ciocalteu Method for the Determination of the Total Phenolic Content from *Limonium Brasiliense* L. *Molecules* **2013**, *18*, 6852–6865. [CrossRef]

93. Zhishen, J.; Mengcheng, T.; Jianming, W. The determination of flavonoid contents in mulberry and their scavenging effects on superoxide radicals. *Food Chem.* **1999**, *64*, 555–559. [CrossRef]

Publisher's Note: MDPI stays neutral with regard to jurisdictional claims in published maps and institutional affiliations.

Article

The Influence of Soil Acidity on the Physiological Responses of Two Bread Wheat Cultivars

Brigitta Tóth [1,*], Csaba Juhász [2], Maryke Labuschagne [3] and Makoena Joyce Moloi [3]

[1] Institute of Food Science, University of Debrecen, 138 Böszörményi St., 4032 Debrecen, Hungary
[2] Arid Land Research Centre, University of Debrecen, 138 Böszörményi St., 4032 Debrecen, Hungary; juhasz@agr.unideb.hu
[3] Department of Plant Sciences, University of the Free State-Main Campus, P.O. Box 339, Bloemfontein 9300, South Africa; LabuscM@ufs.ac.za (M.L.); MoloiMJ@ufs.ac.za (M.J.M.)
* Correspondence: btoth@agr.unideb.hu; Tel.: +3630/2738842

Received: 23 September 2020; Accepted: 30 October 2020; Published: 31 October 2020

Abstract: The recent study was conducted to examine the influence of acidic soil on the activities of ascorbate (APX) and guaiacol peroxidase (POD), proline, protein as well as malon-dialdehyde (MDA) content, in two commercial spring wheat cultivars (PAN3497 and SST806) at different growth stages (tillering and grain filling). A cultivar effect was significant only for MDA content, while the treatment effect was highly significant for proline, protein, and MDA. The sampling time effect was significant for most characteristics. MDA, antioxidative capacity, as well as protein content increased with maturity. At grain filling, MDA and proline contents were significantly higher at pH 5 than pH 6 and 7 for both cultivars, with the highest content in SST806. Similarly, SST806 had significantly higher APX and POD when growing at pH 5. There were no significant differences in protein content at grain filling between either genotype or treatments affected by low pH. This study showed that growth stage and soil pH influence the rate of lipid peroxidation as well as the antioxidative capacity of wheat, with a larger effect at grain filling, at pH 5. Although SST806 had higher proline, POD, and APX content than PAN3497 at this growth stage, this coincided with a very high MDA content. This shows that the high antioxidative capacity observed here, was not associated with a reduction of lipid peroxidation under low soil pH. Further research should, therefore, be done to establish the role of the induced antioxidant system in association with growth and yield in wheat.

Keywords: antioxidative enzyme activity; low pH; proline; protein; wheat

1. Introduction

Climate change has severe consequences on the natural environment and agricultural production, leading to poverty and food insecurity [1]. The indirect impact of such alterations include a shift in cropland distribution, while a direct effect involves, among others, sulfidic wetland drying (because of climate change-induced drought). As a result, climate change has a catalytic effect on the changes in soil quality [2], specifically soil acidification, which is a worldwide problem, but especially so in Africa and Europe. Africa contributes a higher proportion (16.7%) of global acidic soils compared to Europe (9.9%) [3,4]. In parts of South Africa, for example, the KwaZulu-Natal province, 85% of soils have a pH of less than five [5], which has a significant impact on crop production, leading to yield losses of up to 70% [6]. The extent of yield reduction depends on the level of hydrogen ions (which affects plant growth and development) [7–9], climatic conditions and genetic background of the cultivar. In addition, the combination of H^+-toxicity, nutrient deficiency and reduced water uptake influences growth negatively [10]. Many studies associate soil acidification with nitrogen fertilizer over-application [11–13], acid rain [13–15], changes in soil physical properties [16], plant residues [17], and changes in soil chemistry [18] as the main common reasons worldwide.

Acid soil affects phosphorus uptake, root length, and mean root diameter [19]. Furthermore, low soil pH could influence the levels of reactive oxygen species (ROS) [20] such as singlet oxygen ($^1O_2$), superoxide anion (O_{2-}), hydrogen peroxide (H_2O_2), and hydroxyl radical (OH). These radicals are able to oxidize vital cellular ultrastructure and give rise to oxidative damage and devastation of cellular organelles [21]. For instance, Zhang et al. [22] reported that lipid peroxidation and H_2O_2 concentration increased in rice seedling when H^+ concentration was high. Similarly, growing *Plantago* in pH 4 soil contributed to enhanced lipid peroxidation [23]. Yang et al. [24] indicated that low soil pH enhanced the membrane permeability in Eucalyptus leaves. Contrary to this, pH 4 did not affect H_2O_2, malondialdehyde (MDA), electrolyte leakage, and protein oxidation in the roots and leaves of *Plantago algarbiensis* and *P. almogravensis* [25].

To compensate for the unfavorable influence of free radicals, plants possess an antioxidative system (non-enzymatic and enzymatic) aimed at reducing the effect of oxidative stress [26,27]. The activity of antioxidant enzymes was associated with low soil pH in cucumber, citrus, and rice [22,28,29]. Soil acidity selectively increased ascorbate peroxidase (APX) while it reduced superoxide dismutase (SOD) and catalase (CAT) activity in rice roots [22]. Such selective induction of antioxidative enzymes (monodehydroascorbate reductase, guaiacol peroxidase, APX, and glutathione reductase) was also observed in cucumber [28]. In some instances, however, increase in the antioxidative enzyme activities was not sufficient to protect plants against oxidative damage [23]. Results of a study in the leaves and roots of *P. algarbiensis* and *P. almogravensis* contradicted such findings, where the antioxidative enzymes were not affected by low soil pH [25,30].

To date, most studies have focused on the influence of soil acidity on plant growth and yield. There is limited knowledge of the influences of low soil pH on the physiological aspects of wheat, especially on leaves. Therefore, our experiment was set to investigate the influence of low soil pH on lipid peroxidation, protein and proline contents as well as on antioxidant enzymes (ascorbate oxidase and guaiacol peroxidase) activities, in the leaves of two South African hard spring wheat cultivars, at two growth stages, tillering and grain filling.

2. Results

The rate of lipid peroxidation was measured by the amount of MDA used in plants grown in low pH soil. Compared to the pH 7 treatment, the highest differences in MDA content were observed at pH 5 for PAN3497 (34%) and SST806 (39%) at the grain filling stage, while at tillering, significantly higher levels of MDA was recorded in PAN3497 only. The levels of lipid peroxidation were higher at grain filling than the tillering stage for both genotypes. For both growth stages and genotypes, the pH 6 treatment was not significantly different from pH 7 (Figure 1).

At grain filling, cultivars responded differently to pH treatments, where PAN3497 had a significant increase in APX activity at pH 6 while SST806 had highly significantly increased activity at pH 5. In contrast to this, at the tillering stage, APX activity did not change significantly for any of the pH treatments and genotypes (Figure 2).

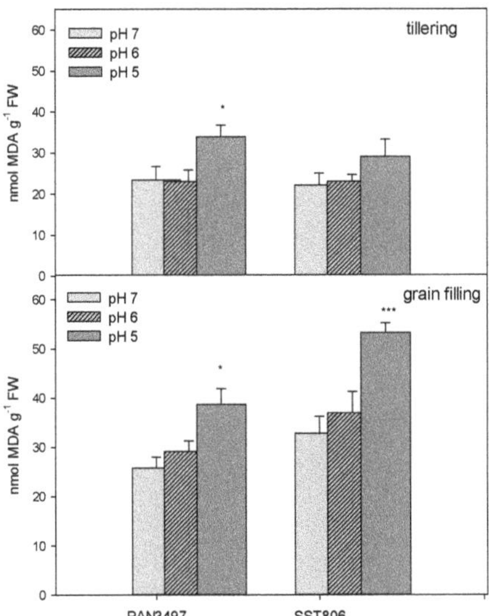

Figure 1. The MDA content of PAN3497 and SST806 grown at three soil pH levels (pH 7, 6, and 5), with two sampling times (tillering and grain filling). Values are the averages of three biological and technical repetitions ± SE. Significant difference compared to pH 7: * $p < 0.05$, *** $p < 0.001$.

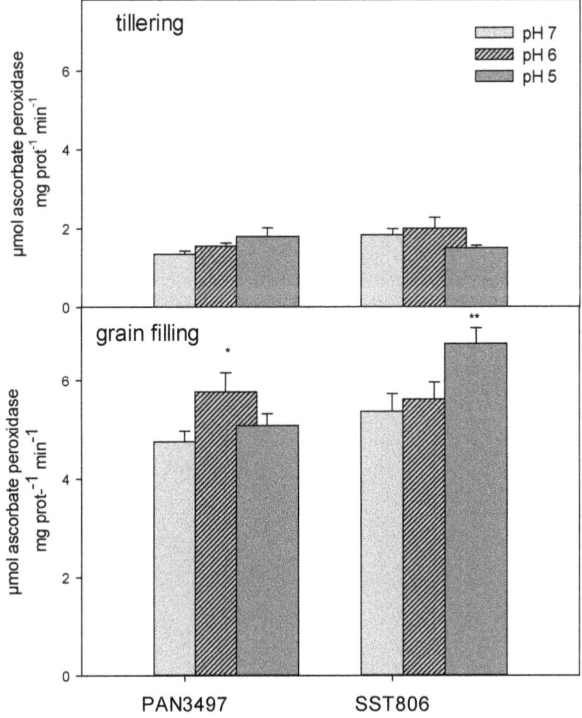

Figure 2. The APX activity of PAN3497 and SST806 grown at three soil pH levels (pH 7, 6, and 5), with two sampling times (tillering and grain filling). Values are the averages of three biological and technical repetitions ± SE. Significant difference compared to pH 7: * $p < 0.05$, ** $p < 0.01$.

Higher tetraguaiacol content (representing high peroxidase activity) was measured during grain filling than tillering (Figure 3). Peroxidase activity for PAN3794 showed no significant changes for either growth stage. Contrary to this, a highly significant reduction in peroxidase activity was observed for SST806 at pH 5 during tillering. However, at grain filling, activity was significantly higher at pH 5.

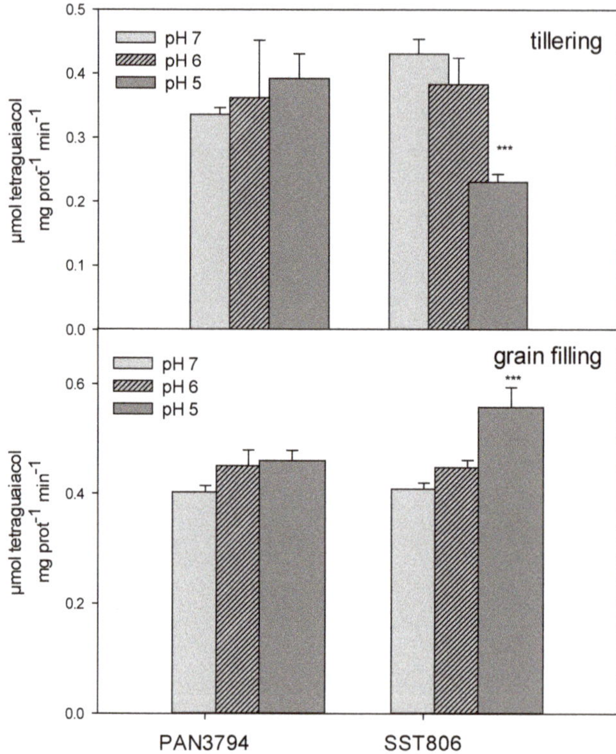

Figure 3. The peroxidase (POD) activity of PAN3497 and SST806 grown at three soil pH levels (pH 7, 6, and 5), with two sampling times (tillering and grain filling). Values are the averages of three biological and technical repetitions ± SE. Significant difference compared to pH 7: *** $p < 0.001$.

Proline content increased with a decrease in the pH for both genotypes at grain filling. This increase, however, was significant only at pH 5. The pattern, however, changed at the tillering stage where pH 6 influenced proline accumulation negatively. For PAN3497, there were no significant differences between pH 7 and 5. SST806 displayed differences between the treatments, however, the differences were not significant (Figure 4).

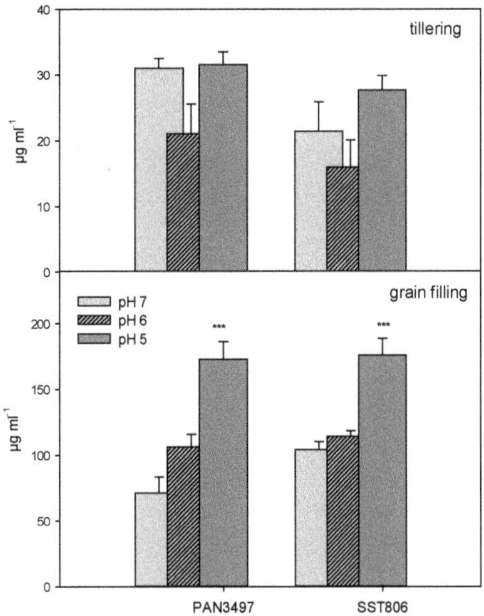

Figure 4. Proline content in leaves of PAN3497 and SST806 grown at three soil pH levels (pH 7, 6, and 5), with two sampling times (tillering and grain filling). Values are the averages of three biological and technical repetitions ± SE. Significant difference compared to pH 7: *** $p < 0.001$.

Although protein content increased with a decrease in pH for SST806, the treatments were not significantly different at the tillering stage. For PAN3497 at tillering, protein content was significantly decreased at pH 5. In contrast, at grain filling, there were no significant differences between treatments or genotypes (Figure 5).

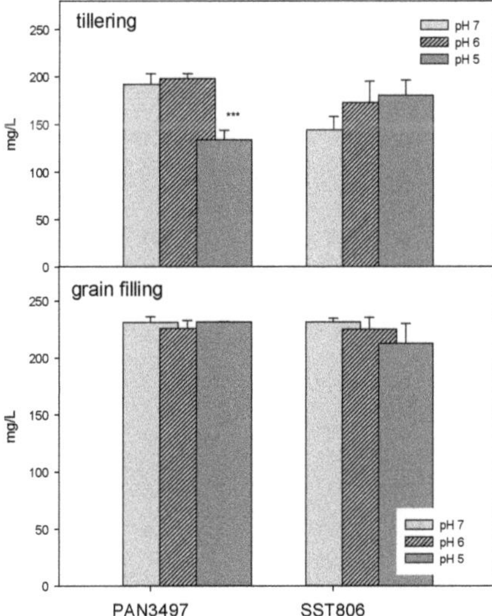

Figure 5. Protein content in leaves of PAN3497 and SST806 grown at three soil pH levels (pH 7, 6, and 5), with two sampling times (tillering and grain filling). Each value is the mean ± standard error of three biological and technical replicates. Significant difference compared to pH 7: ** $p < 0.001$.

The cultivar effect was only significant ($p \leq 0.01$) for MDA content. Based on ANOVA the influence of treatments were significant ($p \leq 0.001$) on MDA, proline, and protein content and insignificant on APX and POD. Highly significant sampling time effects were measured for all characteristics. POD was the least influenced by the sampling time. The cumulative effects of cultivar and pH treatments significantly influenced the proline and protein content. In addition, a highly significant interaction was observed between cultivar and sampling times for MDA content. Treatment and sampling time interactions were significant for APX, POD, proline, and protein content (Table 1).

Table 1. The mean squares from the combined analysis of variance for ascorbate peroxidase activity, malondialdehyde content, peroxidase activity, proline, and protein content for two wheat cultivars, with three treatments over two sampling times.

	Cultivar (C)	Treatment (T)	Sampling Times (S)	CxT	CxS	TxS	CxTxS
APX	1.91	1.58	281.91 ***	0.75	0.03	4.17 **	0.68
MDA	441.39 **	1055.19 ***	1144.73 ***	20.67	571.61 **	181.34	28.36
POD	0.00	0.01	0.30 ***	0.01	0.01	0.06 **	0.01
Proline	612.26	17,677.81 ***	185,844.69 ***	1038.73 *	2003.23	15,533.44 *	833.11
Protein	485.68	1905.05 *	62,151.63 ***	2011.45 **	679.57	4362.51 ***	3423.66 ***

$* \ p < 0.05$; $** \ p < 0.01$; $*** \ p < 0.001$; APX, ascorbate peroxidase; MDA, malondialdehyde; POD, peroxidase.

Table 2 shows the average values of the two genotypes. The pH treatments had no effect on POD. APX slightly increased with decreasing soil pH. Decreasing soil pH inversely affected MDA content, which was highest for pH 5. Proline content significantly increased in all treatments, with the largest effect at pH 5. Protein content was the least for pH 5 treatment.

Table 2. Average values of two genotypes and two sampling times for measured parameters for three treatments.

	pH 7 Mean ± S.D.	pH 6 Mean ± S.D.	pH 5 Mean ± S.D.	LSD (0.05)
APX	3.32 ± 2.03	3.71 ± 2.30	3.78 ± 2.56	0.38
MDA	26.51 ± 4.32	28.05 ± 6.60	38.76 ± 10.42	3.71
POD	0.40 ± 0.03	0.41 ± 0.05	0.41 ± 0.14	0.03
Proline	57.10 ± 38.22	64.46 ± 53.17	102.17 ± 83.80	8.54
Protein	199.60 ± 41.41	206.00 ± 23.03	189.10 ± 42.93	9.65

Values in columns are means ± standard deviation; APX, ascorbate peroxidase; MDA, malondialdehyde; POD, peroxidase.

The genotype had no effect on POD. APX activity was insignificantly higher for SST806 compared to PAN3497. PAN3497 had a significantly lower MDA content than SST806, whilst the proline content was notably higher for SST806 compared to PAN3497 (Table 3). Protein content was significantly higher for PAN3497 compared to SST806.

Table 3. Average values for measured parameters in two genotypes with three treatments over two sampling times.

	PAN3497	SST806	LSD (0.05)
APX	3.36 ± 2.04	3.84 ± 0.11	0.31
MDA	29.03 ± 6.26	33.19 ± 11.14	3.03
POD	0.40 ± 0.05	0.41 ± 0.11	0.03
Proline	72.58 ± 58.97	76.57 ± 65.08	6.97
Protein	202.00 ± 37.36	194.97 ± 33.64	7.88

Values in columns are means ± standard deviation; APX, ascorbate peroxidase; MDA, malondialdehyde; POD, peroxidase.

3. Discussion

Exposure to abiotic stressors such as acidic soil conditions commonly results in the overproduction of reactive oxygen species (ROS) [31]. The increased activity of antioxidant enzymes is one of the first reactions as a response of plants to oxidative stress [10].

The few studies that have reported on the influence of oxidative stress catalyzed by acidic soil conditions were contradictory. Long et al. [29] stated that low pH influenced ROS more in roots than in leaves. Bhuyan et al. [32] stated that under acidic growing conditions, as well as during other abiotic stresses such as cold stress [33] wheat MDA content was higher. In this experiment, MDA content, APX and POD activity, proline, and protein content of wheat leaves significantly changed when plants were grown in pH 5 soil. Moreover, a very strong cultivar effect was found. Significant changes were observed only at pH 5 in the case of SST806, while MDA concentration and POD activity significantly changed in PAN3497 at soil pH 6.

Turhan et al. [34] found that the growth stage of a plant influences its response to abiotic stress, which was confirmed by the higher levels of proline, protein, and MDA content at grain filling than at the tillering stage in the current study. Firscher [35] also stated that the most sensitive development stages of wheat are stem elongation, after flowering, and grain filling. Dreccer et al. [36] showed that high temperatures and water imbalance influenced the production of barley, wheat, chickpea, and canola at tillering, stem elongation, and grain filling.

The effect of cultivar on MDA content was negligible, although SST806 had slightly higher values during grain filling. The sampling time effect was significant for the different traits. Nikolaeva et al. [37] stated that the activity of APX, proline, and MDA content of wheat leaves were influenced by cultivar, time of stress, and the phase of leaf development. This study showed similar results. The effects of the treatments were highly significant for MDA, proline, and protein content in this experiment. MDA and proline content was notably higher in the pH 5 treatment, whilst protein content was reduced compared to pH 7 conditions in this study. MDA content was higher by 95% when wheat was grown at pH 5.5 compared to pH 7.0 conditions [32].

Proline is a so-called non-enzymatic antioxidant. Its accumulation is caused by increased synthesis or moderated deterioration [38], which is one of the plants' reactions to abiotic stressors [39]. In this study, proline content was significantly higher at pH 5.0 compared to pH 7.0 in both cultivars. A higher concentration of proline is an indication that the plants are exposed to stressors [40]. The expression levels of the antioxidant (SOD, CAT, and POX) and proline genes responsible for scavenging or neutralizing ROS were identified in plants. The tolerance of plants to low pH soil is based on the expression of genes related to proline and antioxidant production [41].

Interestingly, proline may enhance the activity of some antioxidant enzymes, for instance, peroxidase [42]. APX activity was also stated to be enhanced under several stress conditions; e.g., Shi et al. [28] found that the activity of APX was higher when cucumber was grown at pH 4.5 compared to pH 6.5. However, the current study did now show significant changes in APX activity. Furthermore, Zhang et al. [22] also published enhanced APX activity in rice roots at pH 2.5. They suggested that the activation of APX is a principal part in scavenging ROS and aids the adaptation of plants to acidic pH.

According to Sanmartin et al. [43], oxidative stress can catalyze the expression of ascorbate oxidase (AOX), suggesting its role in regulating oxidative stress [44]. In the current study, the APX was significantly higher at 6 pH in PAN3497 and also significantly higher in SST806 at pH 5.0 during the grain filling stage.

In conclusion, growth stages influence the rate of lipid peroxidation, as well as the antioxidative capacity of bread wheat cultivars with a larger effect at grain filling, making this a very important time for studying the effects of acidity stress at the biochemical level. Although SST806 had higher proline, POD, and APX levels, this coincided with a very high MDA content. This shows that the high antioxidative capacity observed here cannot be associated with the reduction of membrane damage under low soil pH. This study should therefore be advanced to establish the role of the induced antioxidant system in association with the yield performance of these cultivars. The current research findings were obtained under controlled conditions in a glasshouse. Nitrogen leaching or other circumstances can affect results in field conditions.

4. Materials and Methods

4.1. Green House Trial

Two South-African bread wheat genotypes, PAN3497 and SST806, were sown in 2 L pots filled with 2 kg soil in the greenhouse (temperature was 18 °C during the night and 22–24 °C during the day) in a randomized complete block design with three factors (pH treatments, genotypes, and sampling times) and three repetitions (20 pots per repetition where each pot contained three plants). The soil was collected at Bainsvlei (GPS S 29.05° S 26.11667°, Bloemfontein, South Africa) from 1.5 m deep subsoil. The data of the soil analysis are shown in Table 4. Soil nitrogen content was determined with the classic Kjeldahl-method [45]. Soil phosphorus availability was measured with Bray 1 [46]. The amount of extractable cations and microelements were determined with the use of ammonium acetate, where 2.5 g soil and 25 mL 1 M ammonium acetate was applied for the extraction. The concentration of cations was measured with an atomic adsorption spectrophotometer, based on Miller et al. [47]. The total organic carbon content of the soil was measured by the Walkley–Black method [48]. The soil texture (sand, clay, and silt %) was determined with the Bouyoucos method [49]. To determine the soil density, the weight of 10 cm^3 of dry soil was divided by its volume.

Table 4. Main parameters of the used soil.

Parameters	Value
pH_{H2O}	6.8
sand (%)	96
clay (%)	4
silt (%)	0
CEC (cmol $_C$ kg^{-1})	3.27
density (g cm^{-1})	1.18
C (%)	0.04
N (mg kg^{-1})	<0.004
P (mg kg^{-1})	26.5
K (mg kg^{-1})	163.6
Ca (mg kg^{-1})	13,630 *
S (mg kg^{-1})	1.62
Mg (mg kg^{-1})	38.914 *
Na (mg kg^{-1})	1.80 *
Fe (mg kg^{-1})	4.82
Zn (mg kg^{-1})	0.71
Cu (mg kg^{-1})	0.60
Mn (mg kg^{-1})	21.39

* result as a % of CEC. CEC, cation exchange capacity.

The experiment was conducted from June to the end of October 2018.

The following macro and micronutrient fertilization was applied for all treatments: 261 mg L^{-1} KNO_3, 210 mg L^{-1} K_2SO_4, 87 mg L^{-1} $NH_4H_2PO_4$, 758 mg L^{-1} $Ca(NO_3)_2$, 348 mg L^{-1} $MgSO_4$, 3.45 mg L^{-1} $C_{10}H_{13}FeN_2O_8$, 0.30 mg L^{-1} $MnSO_4$, 0.13 mg L^{-1} $ZnSO_4$, 0.62 mg L^{-1} H_3BO_3, 0.05 mg L^{-1} $CuSO_4$, 0.02 mg L^{-1} Na_2MoO_4.

4.2. Treatments

For low soil pH treatments, sulfur (S) was applied at different concentrations. The control did not receive any S. The S treatments were as follows: in order to the lower pH to 6, 0.155 g of S was added per kg of soil; to lower the pH to 5, 0.387 g of S was added per kg of soil.

After the treatments, the soil was kept moist and incubated for four weeks with weekly pH measurements to ensure that the desired pH was reached. Sowing commenced after the soil incubation period.

4.3. Sampling

The youngest leaves were collected from the main stem at the tillering and grain filling stages. Thereafter, they were placed in liquid nitrogen and stored in an ultra-freezer at 70 °C until the determination of the measured parameters.

4.4. Enzymes Assays

The experiment followed the process from Pukacka and Ratajczak [50] to make the enzyme extracts. Samples (1 g) of each treatment were weighed, homogenized with liquid nitrogen and 5 mL of 50 mM potassium phosphate buffer (pH 7.0) was added. The buffer contained the following chemicals: 1 mM EDTA, 2% PVPP, 0.1% Triton X-100 and 1 mM ascorbate. To obtain the supernatant used for examination, the resulting extract was centrifuged at 15,000× g for 20 min at 4 °C.

The APX assay was carried out based on Mishra et al. [51] with some changes to the composition of the reaction mixture. Each 1 mL of mixture contained 470 μL 50 mM phosphate buffer (pH 7.0), 250 μL 0.1 mM H_2O_2, 200 μL 0.5 mM sodium ascorbate, 50 μL 0.1 mM EDTA and 30 μL enzyme. The APX activity was measured from the decrease in absorbance due to ascorbate oxidation. The absorbance was measured at 290 nm for 5 min at 20 °C against a blank, which contained 50 mM phosphate buffer instead of the enzyme. Each sample was measured three times. To calculate the enzyme activity, an extinction coefficient of 2.8 $mM^{-1}cm^{-1}$ was applied.

To measure the POD activity, the study used a method published by Zieslin and Ben-Zaken [52] with some modifications to the amount of chemicals used in the reaction mixture. The modified mixture had 50 μL 0.2 M H_2O_2, 100 μL 50 mM guaiacol, 340 μL distilled H_2O, 500 μL 80 mM phosphate buffer (pH 5.5), and 10 μL enzyme. The POD activity was determined based on the concentration of generated tetraguaiacol. The absorbance of the reaction mixture was read at 470 nm for 3 min at 30 °C. All the above mentioned chemicals were used for the blank, but 50 mM phosphate buffer was used instead of the enzyme. To calculate the concentration of tetraguaiacol created, the extinction coefficient was 26.6 $mM^{-1}cm^{-1}$.

The study used the method of Bradford [53] to measure protein content for the enzyme extract.

The rate of lipid peroxidation was deduced by measuring the quantity of malondialdehyde (MDA) generated in the chemical assay, using the method described by Heath and Packer [54]. Wheat leaves (100 mg) were homogenized with liquid nitrogen, and 1 mL 0.25% thiobarbituric acid (TBA) and 10% trichloroacetic acid (TCA) was added. The samples were centrifuged at 10,800× g for 25 min at 4 °C and the supernatants were used. The reaction was generated by the use of 0.2 mL supernatant and 0.8 mL 20% TCA, then 0.5% TBA was added to a clean Eppendorf tube. The mixture was vortexed, heated at 95 °C for 30 min and immediately cooled on ice. This was followed by centrifugation at 10,800× g for 10 min at 4 °C. The absorbance was read at 532 nm and 600 nm. The amount of malondialdehyde was computed with the use of an extinction coefficient of 155 $mM^{-1}cm^{-1}$.

The proline content in the leaves was measured based on Carillo and Gibon [55]. Fresh wheat leaves (0.1 g) were homogenized with liquid nitrogen and 2 ml 70% (*v/v*) ethanol was added [56]. The 1 mL reaction mix (1% ninhydrin in 60% (*v/v*) acetic acid) was added to 500 μL ethanolic extract and placed into 1.5 mL Eppendorf tubes, which were shaken and kept in a 95 °C water bath for 20 min. Afterward, samples were cooled on ice and centrifuged at 12,000× g for 1 min. The absorbance of the samples was measured at 520 nm, and the amount of proline was calculated from the proline standard curve. The color of our reaction with ninhydrin was yellowish to deep yellow/pinkish, which indicated the presence of proline. We established a proline standard curve using different concentrations of proline in this experiment to verify the content of proline in our extracts. Similarly, the reaction of proline standards were also yellowish and pinkish (depending on the concentration of proline).

4.5. Statistical Analysis

Analysis of variance (ANOVA) was carried out on the data for both genotypes, the four treatments, and two years, as a three-factor analysis [57]. ANOVA was also used for the two cultivars separately, as well as for the two sampling times combined, in order to observe the influence of treatments on the APX, POD, MDA, proline, and protein contents within each cultivar. Differences were inspected at a $p < 0.05$ level of significance. The Tukey test and least significant difference (LSD) were used for means separation.

Author Contributions: B.T.: designed the experiment, investigation, completed the statistical analysis, writing–original draft preparation. C.J.: completed the statistical analysis, writing–original draft preparation, resources. M.L.: conceptualization, completed the statistical analysis, supervision, writing–review and editing, resources. M.J.M.: investigation, writing–original draft preparation. All authors have read and agreed to the published version of the manuscript.

Funding: This research did not receive any specific grant from funding agencies in the public, commercial, or not-for-profit sectors.

Acknowledgments: This research was funded by the National Research Foundation of South Africa, through the SARChI chair initiative.

Conflicts of Interest: The authors declare that they have no conflict of interest.

References

1. Arraiza, M.P.; Santamarta, J.C.; Ioras, F.; García-Rodríguez, J.L.; Abrudan, I.V.; Korjas, H.; Borála, G. (Eds.) *Climate Change and Restoration of Degraded Land*; Colegio de Ingenieros de Montes: Madrid, Spain, 2014.
2. Schoenholtz, S.; Miegroet, H.V.; Burger, J. A review of chemical and physiological properties as indicators of forest soil quality. Challenges and opportunities. *Ecol. Manag.* **2000**, *138*, 335–356.
3. Von Uexkull, H.R.; Mutert, E. Global extent, development and economic impact of acid soils. *Plant Soil* **1995**, *171*, 1–15. [CrossRef]
4. Venter, A.; Herselman, J.E.; VanderMerwe, G.M.E.; Steyn, C.; Beukes, D.J. Developing soil acidity maps for South Africa. In Proceedings of the 5th International Symposium on Soil Plant Interactions at Low pH, Bergville, South Africa, 12–16 March 2001; p. 21.
5. Roberts, V.G.; Smeda, Z. The distribution of soil fertility constraints in KwaZulu-Natal, South Africa. In Proceedings of the 5th International Symposium on Plant Soil Interactions at low pH, Bergville, South Africa, 12–16 March 2001; p. 12.
6. Dewi-Hayati, P.K.; Sutoyo, A.; Syarif, A.; Prasetyo, T. Performance of maize single-cross hybrids evaluated on acidic soils. *Int. J. Adv. Sci. Eng. Inf. Technol.* **2014**, *4*, 30–33.
7. Koyama, H.; Toda, T.; Hara, T. Brief exposure to low-pH stress causes irreversible damage to the growing root in Arabidopsis thaliana: Pectin-Ca interaction may play an important role in proton rhizotoxicity. *J. Exp. Bot.* **2001**, *52*, 361–368. [PubMed]
8. Ghiimire, R.; Machado, S.; Bista, P. Soil pH, soil organic matter, and crop yields in winter wheat-summer fallow systems. *Agron. J.* **2017**, *109*, 707–717. [CrossRef]
9. Anderson, G.; Bell, R. Wheat grain-yield response to lime application: Relationships with soil pH and aluminium in Western Australia. *Crop Pasture Sci.* **2019**, *70*, 295–305. [CrossRef]
10. Long, A.; Zhang, J.; Yang, L.T.; Ye, X.; Lai, N.W.; Tan, L.L.; Lin, D.; Chen, L.S. Effects of low pH on photosynthesis, related physiological parameters and nutrient profile of *Citrus*. *Front. Plant Sci.* **2017**, *8*, 185. [CrossRef]
11. Schroeder, J.L.; Zhang, H.; Girma, K.; Raun, W.R.; Penn, C.J. Soil acidification from long-term use of nitrogen fertilizers on winter wheat. *Soil Sci. Soc. Am. J.* **2011**, *75*, 957–964. [CrossRef]
12. Tian, D.; Niu, S. A global analysis of soil acidification caused by nitrogen addition. *Environ. Res. Lett.* **2015**, *10*, 024019. [CrossRef]
13. Oertel, C.; Matschullat, J.; Zurba, K.; Zimmermann, F.; Erasmi, S. Greenhouse gas emissions from soils—A review. *Geochemistry* **2016**, *76*, 327–352. [CrossRef]
14. Bini, C.; Breslin, F. Soil acidification by acid rain in forest ecosystems: A case study in northern Italy. *Sci. Total Environ.* **1998**, *222*, 1–15. [CrossRef]

15. Russell, A.E.; Laird, D.A.; Mallarino, A.P. Nitrogen fertilizer and cropping system impacts on soil quality in mid-western Mollisols. *Soil Sci. Soc. Am. J.* **2006**, *70*, 249–255.

16. Matsumoto, S.; Ogata, S.; Shimada, H.; Sasaoka, T.; Hamanaka, A.; Kusuma, G.J. Effects of pH-induced changes in soil physical characteristics on the development of soil water erosion. *Geoscience* **2018**, *8*, 134. [CrossRef]

17. Meda, A.R.; Cassidato, M.E.; Pavan, M.A.; Miyazawa, M. Alleviating soil acidity through plant organic compounds. *Braz. Arch. Biol. Technol.* **2001**, *44*, 185–189. [CrossRef]

18. Blake, L.; Goulding, K.W.T.; Mott, C.J.B.; Johnston, A.E. Changes in soil chemistry accompanying acidification over more than 100 years under woodland and grass at Rothamsted Experimental Station, UK. *Eur. J. Soil Sci.* **1999**, *50*, 401–412. [CrossRef]

19. Robles-Aguilar, A.A.; Pang, J.; Postma, J.A.; Schrey, S.D.; Lambers, H.; Jablonowski, N.D. The effect of pH on morphological and physiological root traits of *Lupinus angustifolius* treated with struvite as a recycled phosphorus source. *Plant Soil* **2019**, *434*, 65–78. [CrossRef]

20. Song, H.; Xu, X.; Wang, H.; Tao, Y. Protein carbonylation in barley seedling roots caused by aluminum and proton toxicity is suppressed by salicylic acid. *Russ. J. Plant Physiol.* **2011**, *58*, 653–659. [CrossRef]

21. Sharma, P.; Jha, A.B.; Dubey, R.S.; Pessarakli, M. Reactive oxygen species, oxidative damage, and antioxidative defence mechanisms in plants under stressful conditions. *J. Bot.* **2012**, 217037. [CrossRef]

22. Zhang, Y.K.; Zhu, D.F.; Zhang, Y.P.; Chen, H.Z.; Xiang, J.; Lin, X.Q. Low pH-induced changes of antioxidant enzyme and ATPase activities in the roots of rice (*Oryza sativa* L.) seedlings. *PLoS ONE* **2015**, *10*, e0116971. [CrossRef]

23. Martins, N.; Gonçalves, S.; Romano, A. Metabolism and aluminum accumulation in *Plantago almogravensis* and *P. algarbiensis* in response to low pH and aluminum stress. *Biol. Plant.* **2013**, *57*, 325–331. [CrossRef]

24. Yang, M.; Huang, S.X.; Fang, S.Z.; Huang, X.L. Response of seedling growth of four *Eucalyptus* clones to acid and aluminum stress. *Plant Nutr. Fertil. Sci.* **2011**, *17*, 195–201.

25. Martins, N.; Osório, M.L.; Gonçalves, S.; Osório, J.; Romano, A. Differences in Al tolerance between *Plantago algarbiensis* and *P. almogravensis* reflect their ability to respond to oxidative stress. *Biometals* **2013**, *26*, 427–437. [CrossRef] [PubMed]

26. Liu, E.U.; Liu, C.P. Effects of simulated acid rain on the antioxidative system in *Cinnamomum philippinense* seedlings. *Water Air Soil Poll.* **2011**, *215*, 127–135. [CrossRef]

27. Chen, J.; Wang, W.H.; Liu, T.W.; Wu, F.H.; Zheng, H.L. Photosynthetic and antioxidant responses of *Liquidambar formosana* and *Schima superba* seedlings to sulfuric-rich and nitric-rich simulated acid rain. *Plant Physiol. Biochem.* **2013**, *64*, 41–51. [CrossRef] [PubMed]

28. Shi, Q.H.; Zhu, Z.J.; Juan, L.I.; Qian, Q.Q. Combined effects of excess Mn and low pH on oxidative stress and antioxidant enzymes in cucumber roots. *Agric. Sci. China* **2006**, *5*, 767–772. [CrossRef]

29. Long, A.; Huang, W.L.; Qi, Y.P.; Yang, L.T.; Lai, N.W.M.; Guo, J.X.; Chen, L.S. 2019. Low pH effects on reactive oxygen species and methylglyoxal metabolisms in citrus roots and leaves. *BMC Plant Biol.* **2019**, *19*, 477. [CrossRef]

30. Martins, N.; Gonçalves, S.; Palma, T.; Romano, A. The influence of low pH on *in vitro* growth and biochemical parameters of *Plantago almogravensis* and *P. algarbiensis*. *Plant Cell Tissue Organ Cult.* **2011**, *107*, 113–121. [CrossRef]

31. Hasanuzzaman, M.; Nahar, K.; Hossain, M.S.; Mahmud, J.A.; Rahman, A.; Inafuku, M.; Oku, H.; Fujita, M. Coordinated actions of glyoxalase and antioxidant defense systems in conferring abiotic stress tolerance in plants. *Int. J. Mol. Sci.* **2017**, *18*, 200. [CrossRef]

32. Bhuyan, M.H.M.B.; Hasanuzzaman, M.; Mahmud, J.A.; Hossain, S.; Bhuiyan, T.F.; Fujita, M. Unraveling morphophysiological and biochemical responses of *Triticum aestivum* L. to extreme pH: Coordinated actions of antioxidant defence and glyoxalase systems. *Plants* **2018**, *8*, 24. [CrossRef]

33. Hou, Y.; Guo, Z.; Yi, Y.; Li, H.-N.; Li, H.-G.; Chen, L.; Ma, H.; Zhang, L.; Lin, J.W.; Zhong, M. Effects of cold acclimation and exogenous phytohormones abscisic acid treatment on physiological indicators of winterness wheat. *J. Plant Sci.* **2010**, *5*, 125–136.

34. Turhan, A.; Ozmen, N.; Kuscu, H.; Serbeci, S.M.; Seniz, V. Influence of rootstocks on yield and fruits characteristics and quality of watermelon. *Hortic. Environ. Biotechnol.* **2012**, *53*, 336–341. [CrossRef]

35. Firscher, R.A. Number of kernels in wheat crops and the influence of solar radiation and temperature. *J. Agric. Sci.* **1985**, *105*, 447–461. [CrossRef]

36. Dreccer, M.F.; Fainges, J.; Whish, J.; Ogbonnaya, F.C. Comparison of sensitive stages of wheat, barley, canola, chickpea and field pea to temperature and water stress across Australia. *Agric. For. Meteorol.* **2018**, *248*, 275–294. [CrossRef]

37. Nikolaeva, M.K.; Maevskaya, S.N.; Shugaev, A.G.; Bukhov, N.G. Effect of drought on chlorophyll content and antioxidant enzyme activities in leaves of three wheat cultivars varying in productivity. *Russ. J. Plant. Physiol.* **2010**, *57*, 87–95. [CrossRef]

38. Verbruggen, N.; Hermans, C. Proline accumulation in plants: A review. *Amino Acids* **2008**, *35*, 753–759. [CrossRef]

39. Hayat, S.; Hayat, Q.; Alyemeni, M.N.; Wani, A.S.; Pichtel, J.; Ahmad, A. Role of proline under changing environments. *Plant Signal. Behav.* **2012**, *7*, 1–11. [CrossRef] [PubMed]

40. Ogagaoghene, A.J. pH level, ascorbic acid, proline and soluble sugar as bio-indicators for pollution. *Chem. Search J.* **2017**, *8*, 41–49.

41. Naing, A.H.; Lee, D.B.; Ai, T.N.; Lim, K.B.; Kim, C.K. Enhancement of low pH stress tolerance in anthocyanin-enriched transgenetic Petunia overexpressing *RsMYB1* gene. *Front. Plant Sci.* **2018**, *9*, 1124. [CrossRef] [PubMed]

42. Hoque, M.A.; Banu, M.N.A.; Okuma, E.; Amako, K.; Nakamura, Y.; Shimoishi, Y.; Murata, Y. Exogenous proline and glycinebetaine increase NaCl-induced ascorbate-glutathione cycle enzyme activities, and proline improves salt tolerance more than glycinebetaine in tobacco Bright Yellow-2 suspension–cultures cells. *J. Plant Physiol.* **2007**, *164*, 1457–1468. [CrossRef]

43. Sanmartin, M.; Drogoudi, P.A.; Lyons, T.; Pateraki, I.; Barnes, J.; Kanellis, A.K. Over-expression of ascorbate oxidase in the apoplast to transgenic tobacco results in altered ascorbate and glutathione redox states and increased sensitivity to ozone. *Planta* **2003**, *216*, 918–928. [CrossRef]

44. De Tullio, M.C. Antioxidants and redox regulation: Changing notions in a changing world. *Plant Physiol. Biochem.* **2010**, *48*, 289–291. [CrossRef] [PubMed]

45. Bremner, J.M. Determination of nitrogen in soil by the Kjeldahl method. *J. Agric. Sci.* **1960**, *55*, 11–33. [CrossRef]

46. Bray, R.H.; Kurtz, L.T. Determination of total, organic, and available forms of phosphorus in soils. *Soil Sci.* **1945**, *59*, 39–45. [CrossRef]

47. Miller, R.O.; Gavlak, R.; Horneck, D. *Plant, Soil and Water Reference Methods for the Western Region*; WREP: Western Rural Development Center: Corvallis, OR, USA, 2013; Volume 125.

48. Walkley, A.; Black, I.A. An examination of the Degtjareff method for determining ssoil organic matter, and a proposed modification of the chromi acid titration method. *Soil Sci.* **1934**, *37*, 29–38. [CrossRef]

49. Bouyoucos, G.J. Directions for Making Mechanical Analysis of Soils by the Hydrometer Method. *Soil Sci.* **1936**, *4*, 225–228. [CrossRef]

50. Pukacka, S.; Ratajczak, E. Production and scavenging of reactive oxygen species in *Fagus sylvatica* seeds during storage at varied temperature and humidity. *J. Plant Physiol.* **2005**, *162*, 873–885. [CrossRef]

51. Mishra, N.P.; Mishra, R.K.; Singhal, G.S. Changes in the activities of anti-oxidant enzymes during exposure of intact wheat leaves to strong visible light at different temperatures in the presence of protein synthesis inhibitors. *Plant Physiol.* **1993**, *102*, 903–910. [CrossRef]

52. Zeislin, N.; Ben-Zaken, R. Peroxidases, phenylalanine ammonia-lyase and lignification in peduncles of rose flowers. *Plant Physiol. Biochem.* **1991**, *29*, 147–151.

53. Bradford, M.M. A rapid and sensitive method for the quantitation of microgram quantities of protein utilizing the principle of protein-dye binding. *Anal. Biochem.* **1976**, *72*, 248–254. [CrossRef]

54. Heath, R.L.; Packer, L. Photoperoxidation in isolated chloroplasts. I. Kinetics and stoichiometry of fatty acid peroxidation. *Arch. Biochem. Biophys.* **1968**, *125*, 189–198. [CrossRef]

55. Carillo, P.; Gibon, Y. Protocol: Extraction and Determination of Proline. 2011. Available online: https://www.researchgate.net/publication/211353600_PROTOCOL_Extraction_and_determination_of_proline (accessed on 1 September 2020).

56. Hummel, I.; Pantin, F.; Sulpice, R.; Piques, M.; Rolland, G.; Dauzat, M.; Christophe, A.; Pervent, M.; Bouteillé, M.; Stitt, M.; et al. *Arabidopsis* plants acclimate to water deficit at low cost through changes of carbon usage: An integrated perspective using growth, metabolite, enzyme, and gene expression analysis. *Plant Physiol.* **2010**, *154*, 357–372. [CrossRef]

57. Agrobase. *Agrobase Generation II, Agronomix*; Agrobase: Winnipeg, MB, Canada, 2014.

Publisher's Note: MDPI stays neutral with regard to jurisdictional claims in published maps and institutional affiliations.

 *plants*

Article

Responses of Upland Cotton (*Gossypium hirsutum* L.) Lines to Irrigated and Rainfed Conditions of Texas High Plains

Addissu. G. Ayele [1], Jane K. Dever [2,*], Carol M. Kelly [2], Monica Sheehan [2], Valerie Morgan [2] and Paxton Payton [3]

[1] Department of Agricultural Sciences, College of Agriculture, Family Sciences, and Technology, Fort Valley State University, Fort Valley, GA 31030, USA; Addissu.ayele@fvsu.edu
[2] Texas A&M AgriLife Research and Extension Center, Lubbock, TX 79403, USA; cmkelly@ag.tamu.edu (C.M.K.); m-bellow@tamu.edu (M.S.); vmorgan@ag.tamu.edu (V.M.)
[3] USDA-ARS Cropping Systems Research Laboratory, Lubbock, TX 79415, USA; paxton.payton@ars.usda.gov
*   Correspondence: jdever@ag.tamu.edu; Tel.: +1-(806)-746-6101

Received: 28 September 2020; Accepted: 16 November 2020; Published: 18 November 2020

Abstract: Understanding drought stress responses and the identification of phenotypic traits associated with drought are key factors in breeding for sustainable cotton production in limited irrigation water of semi-arid environments. The objective of this study was to evaluate the responses of upland cotton lines to rainfed and irrigated conditions. We compared selected agronomic traits over time, final yield and fiber quality of cotton lines grown in irrigated and rainfed trials. Under rainfed conditions, the average number of squares per plant sharply declined during weeks 10 to 14 while the average number of bolls per plant significantly reduced during weeks 13 to 15 after planting. Therefore, weeks 10 to 14 and weeks 13 to 15 are critical plant growth stages to differentiate among upland cotton lines for square and boll set, respectively, under drought stress. Variation in square and boll set during this stage may translate into variable lint percent, lint yield and fiber properties under water-limited conditions. Lint yield and fiber quality were markedly affected under rainfed conditions in all cotton lines tested. Despite significantly reduced lint yield in rainfed trials, some cotton lines including 11-21-703S, 06-46-153P, CS 50, L23, FM 989 and DP 491 performed relatively well under stress compared to other cotton lines. The results also reveal that cotton lines show variable responses for fiber properties under irrigated and rainfed trials. Breeding line 12-8-103S produced long, uniform and strong fibers under both irrigated and rainfed conditions. The significant variation observed among cotton genotypes for agronomic characteristics, yield and fiber quality under rainfed conditions indicate potential to breed cotton for improved drought tolerance.

Keywords: rainfed; irrigated; *Gossypium hirsutum*

---

1. Introduction

Climatic variability and elevated levels of greenhouse gases could cause the induction of flooding, heat waves and drought stress [1]. Among these environmental factors, water scarcity, which leads to drought stress, is the major limitation for crop production. Water availability is a key element for sustainable cotton production and its limitation adversely affects the biochemical and physiological process of a plant, leading to a reduction of yield and fiber quality [1]. The productivity of agricultural land is seriously affected by the change in patterns of temperature, the amount and distributions of rainfall and climate change. These changes are likely to remain critical barriers to keep up with food and fiber production in the future [2,3]. As water is a limited resource, and drought frequency and intensity show increasing trends [4], appropriate use of irrigation water is expected to balance food demands with the anticipated increase of the world population growth. To succeed with the subsequent estimates of water shortages, measures aimed at reorganizing and optimizing the efficiency of water consumption in the agricultural sector are critical.

Cotton (*Gossypium* spp.) is one of the world's most important crops, accounting for around 35% of all-natural and man-made fibers produced in the world [5]. Production of cotton in many regions of the U.S. cotton belt is limited by insufficient irrigation water and erratic rainfall patterns. In the Texas High Plains, the Ogallala aquifer historically provides irrigation water for cotton production. However, the Ogallala aquifer water table has declined by more than 50%, mainly due to the intensification of irrigated crop production [6–8]. Depletion of groundwater and high energy costs associated with pumping water to the surface affect cotton production, which makes the selection for drought tolerance a primary objective of cotton breeding in the high plains of Texas.

Studies indicate that when a cotton population is subjected to abiotic stresses, particularly drought and high-temperature stresses, more than 50% yield reduction occurs as compared to irrigated plants with a similar genetic background [9]. Most crops, including cotton, are sensitive to drought stress, particularly during flowering through seed developmental stages [10]. Cook and El-Zik, 1992, [11] suggested drought stress during anthesis can result in a reduction of lint yield due to shedding squares and young bolls. Drought reduces yield and fiber quality, costing producers millions of dollars each year. Wang, et al., 2016, [12] observed when soil moisture decreased from approximately 60 to 45% of field capacity, yield reduction was doubled (from approximately 30 to 60%), and fiber quality, particularly fiber length and strength, was reduced. Periodic drought also increased within-plant variability of fiber maturity and fiber length of upland cotton cultivars, being more pronounced when boll setting was in the higher fruiting branches [13].

Water availability has a direct effect on plant growth. Drought stress decreases plant turgor potential, inhibiting normal plant functions [14], and reduces both cotton yield and fiber quality [15]. Though plants with fewer bolls can compensate to some degree by producing larger bolls, the number of bolls per unit area is the most significant yield component [16]. The impact of drought stress on cotton is complex. Therefore, research is needed to better understand the responses of upland cotton to drought stress on reproductive growth, yield and fiber quality, and how they can be improved. We hypothesized that differences in response to drought stress among upland cotton genotypes exist and can lead to the identification of novel sources of germplasm that could be used for introgression of enhanced stress tolerance alleles by conventional breeding. The main objective of this study was to investigate the response of upland cotton genotypes to rainfed and irrigated conditions with respect to fiber quality, yield and reproductive growth over time, and to characterize agronomic traits useful in plant selection under drought conditions.

2. Results

2.1. The Response of Cotton Genotypes for Yield and Agronomic Traits in Irrigated and Rainfed Trials

A significant interaction effect between genotype × week was observed for the number of the squares (NSQR), number of bolls per plant (NB) and number of flowers (NF) in the irrigated trial. The NB and NF were also affected by genotype × week interaction in rainfed trial, indicating that the response of some genotypes was not the same across weeks for these agronomic traits (Table 1). Conversely, no significant effect of genotype × week interaction was observed for the number of main-stem nodes (NN) and plant height (PH) in both irrigated and rainfed trials. Genotypes showed consistent variability across weeks in rainfed conditions for NSQR (Table 2). The results revealed high variability among genotypes for NB, NSQR and PH. However, no significant differences were observed among genotypes for NF and NN in both irrigated and rainfed conditions.

Table 1. Analysis of variance for the number of bolls (NB), number of squares (NSQR), number of flowers (NF), plant height (PH), and number of nodes (NN) for upland cotton lines grown under irrigated and rainfed conditions. DF—degrees of freedom.

Sources of Variance	DF	Irrigated					Rainfed				
		NSQR	NF	NN	NB	PH	NSQR	NF	NN	NB	PH
Genotypes	8	<0.0001 *	0.2164	<0.0001 *	<0.0001 *	<0.0001 *	<0.0001 *	0.3662	0.1144	<0.0001 *	<0.0001 *
Week	8	<0.0001 *	<0.0001 *	<0.0001 *	<0.0001 *	<0.0001 *	<0.0001 *	<0.0001 *	<0.0001 *	<0.0001 *	<0.0001 *
Genotypes * Week	64	0.0004 *	0.000 *	0.9938	0.0267 *	0.9909	0.6795	0.0232 *	0.6795	0.026 *	1.000

* Significant at the <0.05 probability level. Notes. Week represents the period across cotton plant growth stages. Data for agronomic traits were collected for nine consecutive weeks starting from 50 days after planting. Square root transformations were used for all counted traits.

Table 2. Least square means for the number of bolls per plant (NB), boll size, plant height (PH), lint turnout (LT %), and lint yield of cotton lines grown under irrigated and rainfed conditions.

Genotypes	Irrigated					Rainfed				
	Sqrt (NB)	PH (cm)	Boll Size (g)	LT (%)	Yield (kg ha^{-1})	Sqrt (NB)	PH (cm)	Boll Size (g)	LT (%)	Yield (kg ha^{-1})
06-46-153P	3.2 a [†]	41 cd	5.5 a	22.9 e	972 cd	2.5 a	36 bcd	4.3 ab	23.6 cd	436 ab
11-21-703S	3.1 a	40 d	5.4 ab	27.2 cd	1156 b	1.9 abc	34 e	4.6 a	26.7 ab	465 a
12-8-103S	3.1 a	40 d	4.8 ab	23.1 e	1057 abc	1.8 abc	35 de	3.6 bc	22.2 d	387 bc
CS 50	2.7 ab	42 cd	3.9 c	29.9 a	1121 abc	2.2 ab	37 bcd	3.0 c	29.2 a	446 ab
DP 491	3.1 a	41 cd	4.9 ab	26.1 d	1101 bc	1.9 abc	36 b	4.1 ab	26.6 b	460 a
FM 989	3.0 a	43 c	5.3 ab	28.2 bc	1267 a	1.7 bc	37 bcd	3.8 b	26.3 b	434 b
L23	2.7 ab	46 b	4.7 b	28.8 ab	1082 bc	1.5 c	38 bc	3.9 abc	27.6 b	410 ab
TX 1151	2.2 b	49 a	4.8 ab	21.4 f	652 e	1.5 c	42 a	3.5 bc	19.7 e	217 d
TX 62	2.1 c	41 cd	5.4 ab	23.7 e	826 d	1.4 c	38 b	4.6 a	24.1 c	341 c

Notes. Sqrt (NB): number of bolls-square root transformation was applied. PH: plant height cm; LT: lint turnout %; boll size: seed cotton weight in grams boll^{-1}. [†] Means with the same letters are not significantly different at $p < 0.05$.

The least-square means indicate high variability among cotton lines for final yield, lint percent (LT %), boll size, plant height, and number of bolls in both irrigated and rainfed trials. Under rainfed conditions cotton lines 06-46-153P, 11-21-703S, 12-8-103S, CS 50, L 23, DP 491, and FM 989 produced a similar number of bolls per plant (Table 2). Relatively, a low average number of bolls were recorded for TX 62, TX 1151, and L23 in rainfed trials. TX 62 consistently showed poor boll setting under both irrigated and rainfed conditions. Cotton line L23 produced a good number of bolls under irrigated conditions, with poor boll set under rainfed trials. The smallest boll size was obtained from CS 50 in irrigated and rainfed trials.

Under rainfed trials, some cotton lines such as, 11-21-703S, 06-46-153P, CS 50, FM 989, and DP 491, which produced a relatively high number of bolls, tended to produce better lint yield (Table 3). The TX 1151 cotton line was the tallest plant among the cotton lines evaluated. However, TX 1151 produced the lowest NB, boll size, lint percent and lint yield both in irrigated and rainfed trials.

Figure 1 depicts the effect of irrigation and rainfed treatment on the number of squares and boll retention capacity of cotton lines across weeks and growth stages. No significant differences were observed between rainfed and irrigated plots during weeks seven to 10 for the average number of squares produced per plant. In rainfed trials, the average number of squares produced per plant was significantly reduced between weeks 10 (70 DAP) to 12 (84 DAP), while in the irrigated trials, the number of squares remains constant between weeks 10 (70 DAP) to 11 (77 DAP). During week 13 (91 DAP) to 15 (105 DAP), square production sharply declined and the difference between rainfed and irrigated plots was negligible. Boll setting began around week nine after planting (Figure 1) and continuously increased until it reached a plateau during weeks 12 to 13 in rainfed, and weeks 13 to 14 after planting in the irrigated trials. The average boll setting and retention capacity of upland cotton in both irrigated and rainfed did not show differences until weeks 12 (84 DAP).

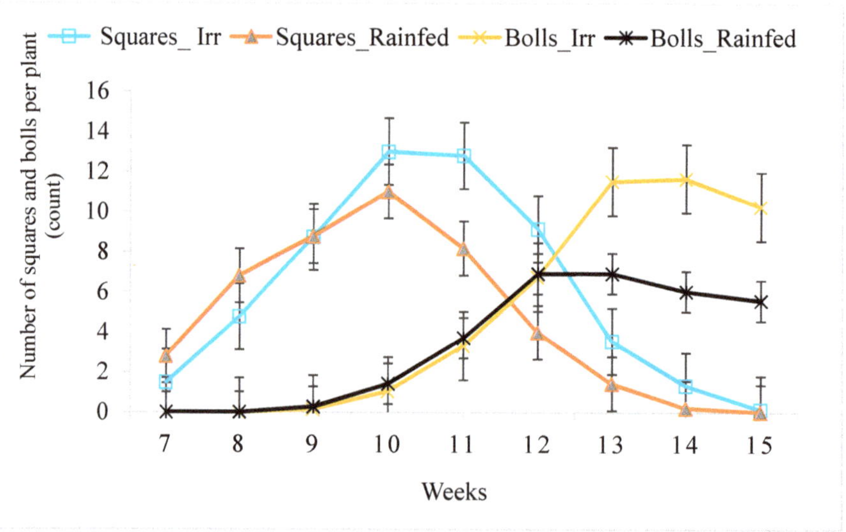

Figure 1. The average number of squares and bolls per plant distribution by week for upland cotton lines grown under irrigated and rainfed conditions. Data for agronomic traits were collected for nine consecutive weeks (weeks 7 to 15). Data averaged for three years, nine genotypes, and over four replications. Standard error (SE) bars were used to show the variations between irrigated and rainfed trials. Overlapping SE bars show no significant differences between irrigated and rainfed treatments.

Table 3. Least square means for micronaire (no unit), upper half mean length (mm), length uniformity (%), strength (kN m kg -1), and elongation (%) of upland cotton lines grown under irrigated and rainfed conditions.

Genotypes	Irrigated					Rainfed				
	Micronaire	Length (mm)	Uniformity (%)	Strength (kN. m kg^{-1})	Elongation (%)	Micronaire	Length (mm)	Uniformity (%)	Strength (kN. m kg^{-1})	Elongation (%)
06-46-153P	3.8 ab [†]	29.3 b	81.2 bcd	316.0 a	7.2 abcd	4.0 a	27.0 bc	79.8 ab	292.0 abc	6.8 abcd
11-21-703S	4.1 ab	29.3 b	81.7 ab	317.4 a	6.6 d	3.7 abc	27.1 bc	79.7 ab	291.4 abc	6.61 d
12-8-103S	3.8 ab	31.2 a	81.9 ab	328.0 a	7.3 abc	3.7 bc	29.2 a	79.6 b	319.1 a	7.3 ab
CS 50	3.9 ab	28.4 bc	80.7 d	303.2 b	7.0 bcd	4.1 a	26.1 cd	80.1 a	278.0 cd	6.7 bcd
DP 491	4.0 ab	27.8 c	81.3 bcd	309.6 ab	7.3 abc	3.9 ab	26.4 cd	78.8 bc	286.1 bcd	7.2 abc
FM 989	4.2 a	28.2 bc	81.2 bcd	315.4 ab	6.7 cd	3.9 ab	26.0 cd	79.5 ab	293.3 abc	6.9 abc
L23	3.9 ab	28.8 bc	80.9 cd	326.2 a	7.5 ab	4.0 ab	27.2 bc	79.8 ab	306.7 ab	7.3 abc
TX 1151	3.7 b	29.4 b	82.3 a	311.8 ab	6.6 cd	3.5 c	28.0 ab	78.8 bc	292.5 abc	6.6 c
TX 62	3.8 ab	27.6 c	79.5 c	276.5 c	7.7 a	3.6 bc	25.7 d	78.2 c	258.5 d	7.4 a

[†] Means with the same letters are not significantly different at $p < 0.05$.

Figure 2 illustrates the average number of bolls distribution per plant for different cotton lines from weeks 12 to 15 which corresponds to 84 to 105 days after planting (DAP). Cotton lines responded differently to irrigated and rainfed conditions for boll setting and retention capacity during weeks 13 (91 DAP), 14 (98 DAP) and 15 (105 DAP) plant growth stages. In rainfed trials, during weeks 13 to 15 after planting, boll production and retention capacity of CS 50 were significantly higher compared to other cotton lines. Cotton lines 06-46-153P, L23, 12-8-103S, and 11-21-703S produced a relatively higher number of bolls that were stable across weeks 13 to 15 after planting. Drought stress during weeks 13 to 15 significantly affected boll setting and retention capacity that led to variable responses among upland cotton lines. Genotypes also showed variable responses for the number of bolls per plant under rainfed conditions during weeks 13 to 15 after planting.

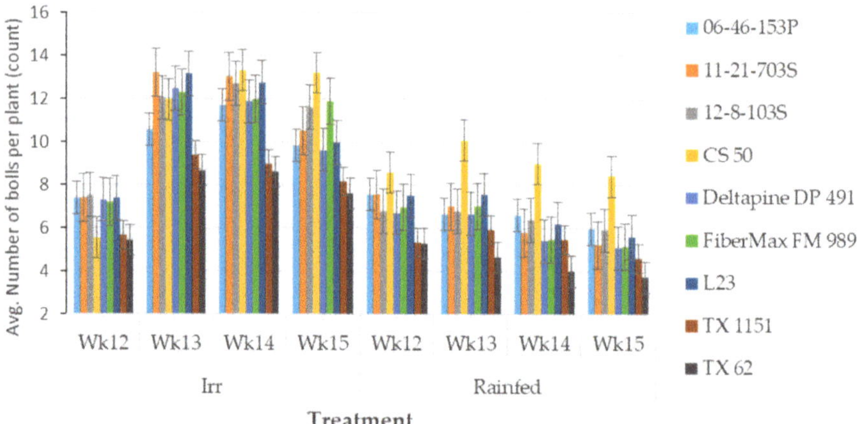

Figure 2. Variation in average number bolls per plant across weeks for upland cotton grown under irrigated and rainfed conditions. Wk., weeks; Irr., irrigated. Average number of bolls compared by subsampling of weeks 12 (Wk12) to weeks 15 (Wk15). Standard error (SE) bars were used to show the variations among genotypes in irrigated and rainfed trials. Overlapping SE bars show no differences between genotypes in irrigated and rainfed conditions.

2.2. Responses of Cotton Genotypes for Fiber Quality under Rainfed and Irrigated Conditions

All high-volume instrument (HVI) measured fiber properties tested were significantly affected due to differences among cotton lines. All genotypes produced stable fiber properties relative to other genotypes in both irrigated and rainfed trials. Best performing cotton lines for fiber quality in irrigated trials were also best in rainfed trials. Cotton lines that produced poor fiber quality under rainfed produced poor fiber quality under irrigated conditions. The least-square means analysis indicates genotypes showed high variability for HVI fiber properties in both rainfed and irrigated conditions (Table 3). Compared to other cotton lines, breeding line 12-8-103S produced significantly longer and stronger fibers under both irrigated and rainfed conditions. The 12-8-103S cotton line was developed in the Texas A&M AgriLife breeding program and selected for salt tolerance. Cotton lines TX 1151, 11-21-703S and 12-8-103S produced relatively uniform fibers, while L23 and 12-8-103S produced strong fibers. TX 62, L23, 12-8-103S and DP 491 produced relatively higher elongation as compared to other cotton lines evaluated under both irrigated and rainfed trials (Table 3).

2.3. Correlation Analyses

Correlation results for this set of cotton lines revealed a significant and positive relationship between yield and HVI fiber properties, including micronaire (r = 0.50), fiber length (r = 0.30), length uniformity (r = 0.51), strength (r = 0.53) and elongation (r = 0.40) under rainfed conditions

(Table 4). In the irrigated trial, fiber length (r = 0.29) and strength (r = 0.38) positively correlated with lint yield. The genotypes selected for this study showed a positive correlation between yield and fiber quality across growing seasons. Relatively poor yielding NCGC accession TX 62 (Table 2), selected for differential response in previous (unpublished) root development screening studies, also produced lower quality fiber (Table 3). Two breeding lines selected for salt tolerance showed relatively good yield in irrigated (12-8-103S) and rainfed (11-21-703S) trials (Table 2) and produced higher fiber quality (Table 3). Inference on correlation between yield and fiber quality was restricted to this set of lines, though results showed promise for developing lines that maintain fiber value in limited water environments. Compared to irrigated trials, a strong and positive relationship was observed between yield, agronomic traits and fiber properties of cotton produced under rainfed conditions. In the rainfed trial, a significant and positive relationship was observed between yield and final plot average agronomic characteristic such as BS (r = 0.41), NSQR (r = 0.45), NN (r = 0.57), PH (r = 0.72), and NB (r = 0.74). Plants with more NSQR, NN and NB were more productive, which indicates that these agronomic traits at certain growth stages could be used as reliable selection criteria to develop drought-tolerant cotton lines.

3. Discussion

Plants respond continuously to changes in various abiotic factors [17], of which the response of plants to limited water availability is considerably high. Studies show drought stress can prevent crops from reaching their genetic potential for yield, quality and agronomically valuable traits [3]. Our results also showed the potential of cotton lines to produce squares, flowers, bolls and main-stem nodes was negatively affected under rainfed conditions. Drought stress limits cotton physiological traits including photosynthesis rate and stomatal conductance leading to reduced productivity by adversely affecting valuable agronomic properties and yield of cotton [18,19]. Our results revealed high variability among selected cotton lines for agronomic characteristics, including the number of squares, bolls and flowers at different plant growth stages, and the number of main-stem nodes, leading to differences in yield and fiber quality under rainfed and irrigated conditions. The variability among agronomic characteristics, including growth and reproductive traits in upland cotton has been observed when water is a limiting factor [20–24]. However, limited information is available that indicates at what particular plant growth stages water stress could contribute to maximum variability among cotton genotypes for the agronomic traits most likely to affect the productivity of cotton under water-limited environments. Our results indicate that drought stress significantly affected the performances of upland cotton lines to produce squares, flowers and bolls during weeks 10 (70 DAP) through weeks 15 (105 DAP), which resulted in reduced boll size, lint turnout and lint yield.

When soil moisture is depleted, young bolls tend to shed [25]. In this study, drought stress tended to accelerate square shedding of cotton lines particularly during weeks 10 to 12 after planting. Similarly, the number of bolls set was significantly reduced during weeks 13 to 15 after planting in the rainfed trials, compared to the number of bolls set in the same period under irrigated conditions. The negative impact imposed by drought stress on the valuable agronomic traits at these critical plant growth stages may result in reduced lint yield and fiber quality.

Studies indicate that the number of bolls formed under drought is less than the number of bolls produced during the nonstress growing seasons [23,26]. We observed that drought stress not only reduced the number of bolls overall but also revealed potential variability among cotton lines at some plant growth stages. Under rainfed trials, the maximum variability of cotton boll production among cotton lines was recorded during weeks 13 to 15 after planting. For example, the CS 50 cotton line produced a relatively higher number of bolls in rainfed trials. Under drought stress conditions, cotton lines 06-46-153P, L23, 12-8-103S, and 11-21-703S produced a relatively higher number of bolls that were stable across weeks 13 to 15 after planting. Some cotton lines, such as TX 1151 and TX 62 produced a significantly low number of bolls under rainfed conditions. Among cotton lines evaluated, 11-21-703S, 06-46-153P, CS 50, L23, FM 989, and DP 491 produced higher lint yield under rainfed conditions.

Table 4. Pearson's correlation coefficient (r) showing the relationship between yield, agronomic traits and selected fiber properties of upland cotton lines grown under irrigated and rainfed conditions. Note: the upper half of the correlation table shows the relationship between different traits in rainfed trial, while the lower half of the correlation table shows the relationship between different traits in irrigated conditions.

	Traits	Yield	BS	NSQR	NN	PH	NB	NF	Micronaire	Length	Uniformity	Strength	Elongation
							Rainfed						
	Yield	1	0.41 **	0.45 **	0.57 **	0.72 ***	0.74 ***	−0.16	0.50 **	0.30 *	0.51 **	0.53 **	0.40 **
	BS	−0.04	1	0.12	0.22	0.39*	0.34 *	−0.17	0.31 *	0.09	0.25	0.12	0.31 *
	NSQR	0.28 *	−0.28 *	1	0.56 **	0.40 **	0.44 **	0.27 *	0.50 **	0.17	0.33 *	0.40 **	0.01
	NN	0.33 *	−0.41 **	0.61 **	1	0.68 **	0.70 ***	−0.22	0.21	0.11	0.35 *	0.31 *	0.34 *
Irrigated	PH	−0.03	−0.34 *	0.11	0.59 **	1	0.75 **	−0.30 *	0.37 *	0.31 *	0.48 **	0.42 **	0.47 **
	NB	0.16	−0.26 *	0.13	0.48 **	0.75 ***	1	−0.22	0.16	0.23	0.42 **	0.36 **	0.46 **
	NF	0.15	−0.35 *	0.38 *	0.43 **	0.18	0.33 *	1	0.25 *	−0.05	−0.15	−0.01	−0.36 *
	Micronaire	0.00	0.40 **	−0.33 *	−0.26 *	−0.27 *	−0.10	−0.34 *	1	0.16	0.33 *	0.35 *	−0.04
	Length	0.29 *	−0.04	0.29 *	0.20	−0.04	−0.05	0.17	−0.38 *	1	0.67 ***	0.79 ***	0.09
	Uniformity	0.13	0.34 *	−0.02	0.04	−0.05	0.01	−0.06	0.18	0.48 **	1	0.73	0.21
	Strength	0.38 *	−0.23	0.34 *	0.45 **	0.26 *	0.23	0.29 *	−0.38 *	0.66 **	0.34 *	1	0.12
	Elongation	−0.21	0.23	−0.41	−0.11	0.24	0.31 *	−0.17	0.29 *	−0.31	−0.07	−0.32	1

BS: boll size; NSQR: number of squares; NN: number of nodes; PH cm: plant height; NB: number of bolls; NF: number of flowers,* significance at $p < 0.05$; ** significant at $p < 0.01$; *** significant at $p < 0.001$.

Our findings, like many others, indicate that, generally, differences in yield loss observed among cotton genotypes may be attributed to the severity and duration of drought at critical plant growth stages The variability observed among genotypes for the number of squares and bolls set per plant at 10–15 weeks after planting, boll size and lint turnout tended to follow the variability observed in lint yield. Therefore, evaluating breeding nurseries based on the relative number of squares and bolls set at weeks 10 to 14 and weeks 13 to 15, respectively, may be predictive of differences in yield potential among genotypes under water-limited environments of the Texas high plains.

All HVI fiber properties showed significant variability in response to irrigated and rainfed trials. Genotypes showed consistent variability for fiber quality in both irrigated and rainfed conditions, which means that cotton lines with better performance under irrigation also performed well in rainfed trials. Similarly, genotypes with low fiber quality in the irrigated trials also produced low fiber quality under rainfed trials.

Studies indicate that drought stress has negative effects on fiber properties, including fiber length, fiber fineness, fiber strength and fiber elongation [25,27,28]. Cotton lines show variable responses for fiber properties under irrigated and rainfed trials. For example, breeding line 12-8-103S produced long, uniform and strong fibers in both rainfed and irrigated trials and is a good candidate for further research on improving fiber quality.

As in the results obtained by [29], we observed a significant and positive relationship between lint yield and other agronomic properties. The author of [23] also observed that yield components and agronomic traits were positively associated with yield in a drought-stressed condition. In this study, boll size, number of bolls and number of nodes showed a significant and positive association with lint yield. For cotton lines included in this study, the results did not necessarily agree with the previous studies that indicated fiber quality traits are negatively associated with fiber yield [30,31]. However, a significant and positive relationship between yield and fiber length (r = 0.61), length uniformity (r = 0.64), and strength (r = 0.59) was observed among the lines selected for this study, which indicates the possibility of simultaneous improvement of cotton for yield and fiber quality. In addition to evaluating cotton for square and boll development at critical plant growth stages to help select for yield potential under drought stress conditions, it is important to select for high fiber quality so that cotton fiber value can be retained in limited water production. The variation observed among genotypes for different fiber properties in rainfed conditions reveals the possibility of selection for genotypes that can produce adequate fiber properties for water-limited cotton production in the high plains of Texas.

4. Materials and Methods

4.1. Plant Materials

Field trials were conducted during the 2014, 2015 and 2016 growing seasons at the Texas A&M AgriLife Research and Extension Center at Lubbock (LREC) on Olton Clay loam soil (fine, mixed, superactive, thermic Aridic Paleustolls). LREC is located at 33°41′ N, 101°49′ W, the elevation is 997 m above sea level and the average annual rainfall is 472 mm. Lubbock is characterized by a semiarid climate, resulting in dry conditions with low precipitation (Table 5), which provides a good environment to study crop drought response. From an initial screening of several genotypes, nine cotton lines were selected to evaluate phenotypic response under irrigated and rainfed conditions: three LREC breeding lines (06-46-153P, 11-21-703S, 12-8-103S), cultivars Deltapine DP 491 (PI 618609), FiberMax FM 989 (PI 639508), CS 50 and SIOKRA L23 [32], and two accessions from the National Cotton Germplasm Collection (NCGC), TEX 1151 (PI 529967) and TX 62 (PI 154096). DP 491 and FM 989 are cultivars that have been successfully grown in commercial production in Texas. The breeding line 06-46-153P was developed in the LREC cotton breeding program and registered as CA 4007 [33]. For this study, breeding lines and cultivars were selected based on differential response to multilocation performance testing over years in irrigated and rainfed trials in West Texas. NCGC accessions were selected for a variable response to greenhouse-observed root development. The selected cotton lines represented variation in maturity, plant height, yield, and fiber quality.

Table 5. Total rainfall, growing degree days and the amount of irrigation water used in 2014, 2015 and 2016 growing seasons.

Months	2014		2015		2016	
	Total Rainfall	$GDD_{15.6}$	Total Rainfall	$GDD_{15.6}$	Total Rainfall	$GDD_{15.6}$
	mm	°C	mm	°C	mm	°C
May	15	177	26	177	32	177
June	66	300	26	282	55	308
July	67	341	15	376	101	452
August	14	357	77	350	6	319
September	176	151	37	234	13	180
October	10	103	27	116	109	174
DD (May–October)		1429		1535		1610
Total Rainfall	348		208		316	
Total Irr	157		338		320	
Irr + Rainfall	505		546		636	

Notes. Irr: Irrigation; GDD15.6: growing degree days at 15.6 °C threshold for cotton. GDD15.6 were calculated based on the National Weather Service data of Lubbock, TX. using means of each maximum and minimum daily temperature for each month during the growing seasons of the cotton.

4.2. Experimental Design and Agronomic Practices

Within the irrigated and rainfed field trials, upland cotton genotypes were arranged in a randomized complete block design (RCBD) with four field replications. Seeds were planted in 4-row 12.19 m long plots on 1.02 m wide centers, each with 10.16 cm spacing between plants. Plants were managed under two conditions, rainfed and irrigated. Plants under irrigated conditions were managed within LREC irrigation capacity to apply water to sustain the growth and development of irrigated cotton on the Texas high plains. Irrigation water was usually applied on a monthly interval if there was no rain during the growing season. Irrigation water was delivered until the furrows were full and the duration of each irrigation time was recorded. The irrigation volumes and depth were estimated using the known flow rate of the irrigation pump. Note that rainfall in West Texas is not evenly distributed and varies from year to year. Under the rainfed conditions, plants did not receive any supplemental irrigation throughout the growing season. However, in both growth conditions, pre-irrigation was applied for all growing seasons to initiate seed germination. In 2014, total water applied, including precipitation to the irrigated trial, was 505 mm, while rainfed trials received 348 mm from precipitation. In 2015, the total volume of water applied to irrigated trials was 546 mm, while rainfed received 208 mm from precipitation. In 2016, the total volume of water applied to the irrigated trial was 636 mm while the rainfed trial received 316 mm water obtained as rainfall (Table 5). In 2014, cotton was planted on 19 May and harvested on 20 November. In 2015, cotton was planted on 27 May and harvested on 3 November for the rainfed and 10 November for the irrigated trials. In 2016, cotton was planted on 13 May and harvested on 15 November. Fertilizer was applied preplant incorporated at the rate of 80-0-0 kg N-P2O5-K2O ha^{-1} for both irrigated and rainfed experimental plots throughout the growing seasons. All other in-season agronomic inputs, such as applications of herbicide and insecticide, were applied following agronomic practices typical for cotton production for Lubbock County. Accumulated growing degree days (GDD15.6) were calculated as the average of the daily maximum and minimum air temperatures less than the base temperature of 15.6 °C [34].

4.3. Data Collection

In the 2014 and 2015 growing seasons, plant height (PH), number of squares (NSQR), number of flowers (NF), number of main-stem nodes (NN) and number of bolls (NB) per plant were recorded for nine consecutive weeks to understand the response of genotypes at different growth stages under irrigated and rainfed conditions. In 2014 and 2015, the first week of data recorded in the growth stages began around 50 DAP (weeks 7) and was completed at 107 DAP (weeks 15). In both growing seasons, a set of 10 plants in each plot were tagged to measure all agronomic traits for nine consecutive weeks. Plant height was recorded from the base of the plant to the meristematic leaf at the apical bud using a measuring tape. The number of bolls per plant was an average of bolls from 10 plants. In 2016, the field data collection strategy was modified based on the 2014 and 2015 evaluations. In 2016, data for different agronomic traits were collected for three consecutive weeks (14 July through 28 July) corresponding to squaring, flowering and boll setting growth stage. The aim of reducing the data collection period was to determine a potential developmental window useful to practically evaluate the agronomic performance of numerous cotton candidate lines under irrigated and rainfed conditions in a breeding nursery.

Just before harvest, final plant height and the total number of main-stem nodes were measured from 10 plants in each plot. A random sample of 25 bolls was picked before harvest from each plot. Boll size was calculated by dividing seed cotton weight in grams by 25, and lint percent calculated by dividing lint weight by seed cotton weight. All tests were mechanically harvested using a two-row cotton stripper modified for small-plot harvesting with no burr extractor. Harvest weights were recorded for each plot, and a 600-g subsample of burr cotton was collected and ginned on a 10-saw laboratory gin with a stick machine, feeder-extractor and lint cleaner. Weighed lint percentage (gin turnout) from the subsample was used to calculate lint yield estimates from the plot harvest weights. A 10-g lint sample was collected from each ginned subsample and sent to the Texas Tech University Fiber and Biopolymer Research Institute for High Volume Instrument (HVI) fiber quality analysis.

4.4. Statistical Analysis

Data from irrigated and rainfed trials were analyzed separately using the general linear model (GLM) and mixed procedures of SAS, version 9.4. The GLM procedure was run with fixed effects to determine the relative magnitude of the main effect of genotype and year × genotype interactions. Because year × genotype interaction was not significant in GLM analysis, the SAS PROC MIXED procedure was applied to multiyear agronomic properties and fiber quality data by using replications and years as random effects.

For agronomic traits data analysis, weeks in different growth stages were included in the model both as the fixed effect and random effect to account for the agronomic data collected as a repeated measurement over time. Including week as a fixed effect in the model indicated how the average of the outcome changed over each week, while the random effects emphasized how much variability of the outcome that the plants had at each week. To validate normality and homoscedasticity of all measured variables, Shapiro–Wilk's and Brown–Forsythe's and Levene's tests were used. When the data met the criteria for normality and homoscedasticity assumptions, all agronomic traits data, yield, and fiber quality traits were analyzed with SAS 9.4 (SAS Institute Inc). For non-normally-distributed count data, square root transformation was applied. Tukey's HSD test was used to determine differences among genotypes for different traits at the $p \leq 0.05$ level of significance. The least-square means were calculated using JMP Genomics 6 (JMP, 2013), where year and replication were treated as random effects. Correlation analysis was performed using the restricted maximum likelihood method to evaluate the relationship between different traits of interest.

Author Contributions: Conceptualization, J.K.D. and P.P.; methodology, J.K.D.; software, V.M.; validation, C.M.K., P.P., and A.G.A.; formal analysis, A.G.A.; investigation, M.S.; resources, J.K.D. and P.P.; data curation, V.M.; writing—original draft preparation, A.G.A.; writing—review and editing, J.K.D.; visualization, C.M.K.; supervision, J.K.D.; project administration, J.K.D.; funding acquisition, J.K.D. and P.P. All authors have read and agreed to the published version of the manuscript.

Funding: This research was funded by USDA-ARS consortium, The Ogallala Aquifer Program, with partial assistance from USDA CSREES Hatch TEX09297 and USDA CRIS 6208-21000-018-00D.

Acknowledgments: The authors sincerely appreciate Shuling Liu, for the valuable advice in statistical data analysis.

Conflicts of Interest: The authors have no conflict of interest.

References

1. Rizza, F.; Badeck, F.W.; Cattivelli, L.; Lidestri, O.; Di Fonzo, N.; Stanca, A.M. Use of a water stress index to identify barley genotypes adapted to rainfed and irrigated conditions. *Crop Sci.* **2004**, *44*, 2127–2137. [CrossRef]

2. Flexas, J.; Bota, J.; Galmés, J.; Medrano, H.; Ribas-Carbó, M. Keeping a positive carbon balance under adverse conditions: Responses of photosynthesis and respiration to water stress. *Phys. Plant* **2006**, *127*, 343–352. [CrossRef]

3. Parida, A.K.; Dagaonkar, V.S.; Phalak, M.S.; Umalkar, G.; Aurangabadkar, L.P. Alterations in photosynthetic pigments, protein, and osmotic components in cotton genotypes subjected to short-term drought stress followed by recovery. *Plant Biotechnol. Rep.* **2007**, *1*, 37–48. [CrossRef]

4. Mancosu, N.; Snyder, L.R.; Kyriakakis, G.; Spano, D. Water Scarcity and Future Challenges for Food Production. *Water* **2015**, *7*, 975–992. [CrossRef]

5. Agricultural Marketing Resource Center (AgMRC). Cotton. 2016. Available online: https://www.agmrc.org/commodities-products/fiber/cotton/ (accessed on 28 September 2020).

6. Colaizzi, P.D.; Gowda, P.H.; Marek, T.H.; Porter, D.O. Irrigation in the Texas High Plains: A brief history and potential reductions in demand. *Irrig. Drain.* **2009**, *58*, 257–274. [CrossRef]

7. Konikow, L. *Groundwater Depletion in the United States (1900–2008)*; U.S. Geological Survey: Reston, VA, USA, 2013.

8. Schaefer, C.R.; Ritchie, G.L.; Bordovsky, J.P.; Lewis, K.; Kelly, B. Irrigation timing and rate affect cotton boll distribution and fiber quality. *Agron. J.* **2018**, *110*, 922–931. [CrossRef]

9. Khan, A.; Pan, X.; Najeeb, U.; Tan, D.K.Y.; Fahad, S.; Zahoor, R.; Luo, H. Coping with drought: Stress and adaptive mechanisms, and management through cultural and molecular alternatives in cotton as vital constituents for plant stress resilience and fitness. *Biol. Res.* **2018**, *51*, 47. [CrossRef]

10. Basal, H.; Dagdelen, N.; Unay, A.; Yilmaz, E. Effects of deficit drip irrigation ratios on cotton (*Gossypium hirsutum* L.) yield and fiber quality. *J. Agron. Crop. Sci.* **2009**, *195*, 19–29. [CrossRef]

11. Cook, C.G.; El-Zik, K.M. Cotton seedling, and first-bloom plant characteristics: Relationships with drought-influenced boll abscission and lint yield. *Crop Sci.* **1992**, *32*, 1464–1467. [CrossRef]

12. Wang, R.; Ji, S.; Zhang, P.; Meng, Y.; Wang, Y.; Chen, B.; Zhou, Z. Drought Effects on Cotton Yield and Fiber Quality on Different Fruiting Branches. *Crop Sci.* **2016**, *56*, 1265–1276. [CrossRef]

13. Ayele, A.; Hequet, E.; Kelly, B. Evaluating within-plant variability of cotton fiber length and maturity. *Agron. J.* **2018**, *110*, 47–55. [CrossRef]

14. Muller, B.; Pantin, F.; Génard, M.; Turc, O.; Freixes, M.S.; Piques, M.; Gibon, Y. Water deficits uncouple growth from photosynthesis, increase C content, and modify the relationships between C and growth in sink organs. *J. Exp. Bot.* **2011**, *62*, 1715–1729. [CrossRef]

15. Liu, R.X.; Zhou, Z.G.; Guo, W.Q.; Chen, B.L.; Oosterhuis, D.M. Effects of N fertilization on root development and activity of water-stressed cotton *Gossypium hirsutum* L. plants. *Agric. Water Manag.* **2008**, *95*, 1261–1270. [CrossRef]

16. Wells, R.; Meredith, W.R. Comparative growth of obsolete and modern cotton cultivars. III. Relationship of yield to observed growth characteristics. *Crop Sci.* **1984**, *24*, 868–872. [CrossRef]

17. Zheng, J.; Zhao, J.; Tao, Y.; Wang, J.; Liu, Y.; Fu, J.; Jin, Y.; Gao, P.; Zhang, J.; Bai, Y.; et al. Isolation and analysis of water stress-induced genes in maize seedlings by subtractive PCR and cDNA macroarray. *Plant Mol. Biol.* **2004**, *55*, 807–823. [CrossRef]

18. Grimes, D.W.; El-Zik, K.M. Cotton. In *Irrigation of Agricultural Crops. Stewart*; BA Nielsen, D.R., Ed.; ASA-CSSASSSA: Madison, WI, USA, 1990; pp. 741–773.

19. Loka, D.A.; Oosterhuis, D.M.; Ritchie, G.L. Water-deficit stress in cotton. In *Stress Physiology in Cotton*; Oosterhuis, D.M., Ed.; The Cotton Foundation: Cordova, TN, USA, 2011; pp. 37–72.

20. Pettigrew, W.T. Physiological consequences of moisture deficit stress in cotton. *Crop Sci.* **2004**, *44*, 1265–1272. [CrossRef]

21. Ullah, I.; Ur-Rahman, M.; Ashraf, M.; Zafar, Y. Genotypic variation for drought tolerance in cotton (Gossypium hirsutum L.): Leaf gas exchange and productivity. *Flora* **2008**, *203*, 1050–1115. [CrossRef]

22. Iqbal, K.; Azhar, F.M.; Khan, I.A.; Ullah, E. Variability for drought tolerance in cotton (*Gossypium hirsutum*) and its genetic basis. *Int. J. Agric. Biol.* **2011**, *1*, 61–66.

23. Dahab, A.A.; Elhag, B.; Husnain, T.; Muhammad, S. Variability for drought tolerance in cotton (*Gossypium Hirsutum* L.) for growth and productivity traits using the selection index. *Afr. J. Agric. Res.* **2012**, *7*, 4934–4942. [CrossRef]

24. Mvula, J.; James, M.; Kabambe, V.; Banda, M.H.P. Screening cotton (*Gossypium hirsutum* L.) genotypes for drought tolerance under screen house conditions in Malawi. *J. Plant Breed. Crop Sci.* **2018**, *10*, 48–57. [CrossRef]

25. Snowden, M.C.; Ritchie, G.L.; Simao, F.R.; Bordovsky, J.P. Timing of Episodic Drought can be Critical in Cotton. *Agron. J.* **2014**, *106*, 452–458. [CrossRef]

26. Ritchie, G.L.; Whitaker, J.R.; Bednarz, C.W.; Hook, J.E. Subsurface drip and overhead irrigation: A comparison of plant boll distribution in upland cotton. *Agron. J.* **2009**, *101*, 1336–1344. [CrossRef]

27. Karademir, C.; Remzi, E.; Berekatoğlu, K. Yield and fiber properties of cotton (*Gossypium hirsutum* L.) under water stress and non-stress conditions. *Afr. J. Biotechnol.* **2011**, *10*, 12575–12583. [CrossRef]

28. Feng, L.; Mathis, G.; Ritchie, G.; Han, Y.; Li, Y.; Wang, G.; Zhi, X.; Bednarz, C.W. Optimizing Irrigation and Plant Density for Improved Cotton Yield and Fiber Quality. *Agron. J.* **2014**, *106*, 1111–1118. [CrossRef]

29. Zeng, L.; Meredith, W.R., Jr. Associations among lint yield, yield components, and fiber properties in an introgressed population of cotton. *Crop Sci.* **2009**, *49*, 1647–1654. [CrossRef]

30. Culp, T.W.; Harrell, D.C.; Kerr, T. Some genetic implications in the transfer of high fiber strength genes to upland cotton. *Crop Sci.* **1979**, *19*, 481–484. [CrossRef]

31. Ulloa, M. Heritability, and correlations of agronomic and fiber traits in an okra-leaf upland cotton population. *Crop Sci.* **2006**, *46*, 1508–1514. [CrossRef]

32. Reid, P.E. 'CS 50' and 'Siokra L23'. *Plant Var. J.* **2001**, *14*, 81.

33. Dever, J.K.; Kelly, C.; Ayele, A.; Zwonitzer, J.; Payton, P.; Jones, D. Registration of CA 4007 cotton germplasm line for water-limited production. *J. Plant Regist.* **2020**, *14*, 49–56. [CrossRef]

34. Peng, S.; Krieg, D.R.; Hicks, S.K. Cotton lint yield response to accumulated heat units and soil water supply. *Field Crops Res.* **1989**, *19*, 253–262. [CrossRef]

Publisher's Note: MDPI stays neutral with regard to jurisdictional claims in published maps and institutional affiliations.

Communication

Decomposition of Calcium Oxalate Crystals in *Colobanthus quitensis* under CO_2 Limiting Conditions

Olman Gómez-Espinoza [1,2], Daniel González-Ramírez [2], Panagiota Bresta [3], George Karabourniotis [3] and León A. Bravo [1,*]

[1] Laboratorio de Fisiología y Biología Molecular Vegetal, Instituto de Agroindustria, Departamento de Ciencias Agronómicas y Recursos Naturales, Facultad de Ciencias Agropecuarias y Forestales & Center of Plant, Soil Interaction and Natural Resources Biotechnology, Scientific and Technological Bioresource Nucleus, Universidad de La Frontera, 1145 Temuco, Chile; o.gomez01@ufromail.cl or oespinoza@itcr.ac.cr

[2] Centro de Investigación en Biotecnología, Escuela de Biología, Instituto Tecnológico de Costa Rica, Cartago 30101, Costa Rica; daniel.agr13@estudiantec.cr

[3] Laboratory of Plant Physiology and Morphology, Faculty of Crop Science, Agricultural University of Athens, 118 55 Athens, Greece; brestapan@aua.gr (P.B.); karab@aua.gr (G.K.)

* Correspondence: leon.bravo@ufrontera.cl; Tel.: +56-45-259-2821

Received: 1 September 2020; Accepted: 21 September 2020; Published: 2 October 2020

Abstract: Calcium oxalate (CaOx) crystals are widespread among plant species. Their functions are not yet completely understood; however, they can provide tolerance against multiple environmental stress factors. Recent evidence suggested that CaOx crystals function as carbon reservoirs since its decomposition provides CO_2 that may be used as carbon source for photosynthesis. This might be advantageous in plants with reduced mesophyll conductance, such as the Antarctic plant *Colobanthus quitensis*, which have shown CO_2 diffusion limitations. In this study, we evaluate the effect of two CO_2 concentrations in the CaOx crystals decomposition and chlorophyll fluorescence of *C. quitensis*. Plants were exposed to airflows with 400 ppm and 11.5 ppm CO_2 and the number and relative size of crystals, electron transport rate (ETR), and oxalate oxidase (OxO) activity were monitored along time (10 h). Here we showed that leaf crystal area decreases over time in plants with 11.5 ppm CO_2, which was accompanied by increased OxO activity and only a slight decrease in the ETR. These results suggested a relation between CO_2 limiting conditions and the CaOx crystals decomposition in *C. quitensis*. Hence, crystal decomposition could be a complementary endogenous mechanism for CO_2 supply in plants facing the Antarctic stressful habitat.

Keywords: alarm photosynthesis; Antarctic; oxalate oxidase

1. Introduction

Calcium oxalate (CaOX) is a salt of oxalic acid ($C_2H_2O_4$) and calcium (Ca^{2+}) that forms insoluble crystals of diverse morphology [1]. Detected in at least 215 families, CaOx crystals are widespread among plant kingdom [2,3]. Crystals occur in roots, stems, leaves, flowers, fruits and seeds, and within epidermal, ground, and vascular tissues [4]. They are formed in the vacuoles of specialized cells called crystal idioblasts, which possess distinct structure and content from the surrounding cells. In leaves, crystals idioblasts are commonly located within mesophyll and/or bundle sheath extensions [5]. Some plants (including mainly succulents) like *Cactus senilis*, accumulate CaOx by as much as 85% by dry weight [6,7]. The huge variation in distribution among organs, tissues, and cells among plant species suggests that crystals may have independent origins of formation and multiple functions [8,9].

Recent evidence showed that indeed CaOx crystals represent multifunctional tools which are essential especially under stress conditions [8]. They are dynamic storage systems supplying calcium and oxalate ions upon demand. Both parts of these inclusions serve vital functions. The Ca part controls the levels of cytosolic concentration and immobilizes the excess quantities of this element, taking into account that plants do not have an excretory system. The oxalate part produced in the root can take part in nutrient acquisition, metal detoxification, mineral weathering, and selection of beneficial bacterial populations, whereas oxalate in the leaves can function as a dynamic carbon reservoir, providing CO_2 in a process called alarm photosynthesis [10]. Moreover, oxalate of all organ and tissues can take part in defense reactions upon pathogen and/or herbivore attack [8,11].

Regarding alarm photosynthesis, CaOx crystals located within mesophyll, function as dynamic carbon reservoirs: Crystal decomposition releases CO_2, which is further used for photosynthesis in plants exposed to CO_2 limiting conditions, such as total or partial stomata closure during drought stress [10,12]. However, the function of CaOx as a source of CO_2 for photosynthesis seems to be restricted to specific plant species or situations related to stressful environments, especially to water stress conditions [8,13]. The study of this process at the interspecific level in different climatic regions of the planet is still needed [10].

Antarctic is considered one of the territories with hardest conditions for plant species survival [14]. This is due to multiple extreme environmental traits of this ecosystem. In addition to the well-known low temperatures and sporadic high irradiance, plants inhabiting the Antarctic continent face short growing seasons, windiest climate, and high vapor pressure deficit, leading to leaf dryness [15]. Therefore, these plants become a proper model to research on this mechanism.

Deschampsia antarctica Desv. (Poaceae) and *Colobanthus quitensis* (Kunth) Bartl. (Caryophyllaceae) are the only two plants that naturally have colonized parts of the maritime Antarctic [16]. *D. antarctica* is characterized by the absence of carbon calcium inclusions typical of the members of the Poaceae family [17,18], and therefore is not an appropriate candidate for this study. In contrast, *C. quitensis* has an abundant amount of CaOx crystals in its leaves. Moreover, *C. quitensis*, apart from its tolerance to extreme environmental conditions [19], represents a highly suitable plant model for the evaluation of the AP mechanism disposing three basic features that have been associated with the AP process: (1) a considerable amount of crystals with dimensions appropriate for accurate measurements of crystal properties (large idioblasts about 50 μm in diameter [17,18]), (2) the presence of transcripts with a high similarity to germin-like proteins (OxO enzymes) in the transcriptome [20,21], and (3) high CO_2 diffusion limitations [22,23].

It has been shown that *C. quitensis* CO_2 assimilation is highly limited by CO_2 diffusion; this is partially due to leaf anatomical traits, such as mesophyll and chloroplast thickness [22]. This species has unusually low values of mesophyll conductance (g_m) causing a constraint in the CO_2 diffusion. Therefore, it becomes the principal restriction process for CO_2 acquisition in the plant [23]. It is remarkable that this plant species can achieve high rates of photosynthesis with a very low g_m. Authors suggest that some biochemical components might compensate this low CO_2 diffusion, and therefore facilitate the CO_2 availability, for instance, gas transport aquaporins or carbonic anhydrase and a robust enzymatic machinery [22,24]. Considering the recent findings regarding AP, an intriguing question arises: is this biochemical appendance of the photosynthetic machinery implicated in the photosynthetic function of *C. quitensis*?

For *C. quitensis*, AP might play a role as a complementary endogenous mechanism that could facilitate the supply of CO_2, given the reported limitations in the diffusion of CO_2. Therefore, this study aims to evaluate the dynamics of CaOx crystals in *C. quitensis* leaves under a CO_2 limitation. We hypothesize that the exposure to a low external $[CO_2]$ causes decomposition of CaOx crystals in the leaves of *C. quitensis*, this will provide internal CO_2 for a baseline level of photosynthesis, which will drain electrons from photosystem II even below the CO_2 compensation point. The obtained results will give us novel data about CaOx crystals functions in plants and will add new evidence on the AP mechanism in a plant species from an extreme environment.

2. Results and Discussion

2.1. Calcium Oxalate Crystal Decomposition

A CO_2 restriction experiment was performed with *C. quitensis* plants to test whether a CO_2 limiting condition might trigger the decomposition of CaOx crystals. The experimental setup allowed to compare the CaOx crystal dynamics under a limiting CO_2 condition on *C. quitensis* plants, with CO_2 concentration close to 11.5 ppm, which is below the CO_2 compensation point for this species (25–30 ppm) [22] and ambient 400 ppm CO_2 (control). There were significant main effects for both CO_2 concentration and time. There was a statistically significant interaction between CO_2 concentration and time, where the relative area of crystals decreased as time passed under low [CO_2]. Tukey's post hoc test showed that, after ten hours of treatment, the mean of total crystal area per leaf was significantly lower in the 11.5 ppm-CO_2-concentration group compared to the control group; a significant crystal decomposition was evident (Figure 1a, Figure 2 and Figure S1).

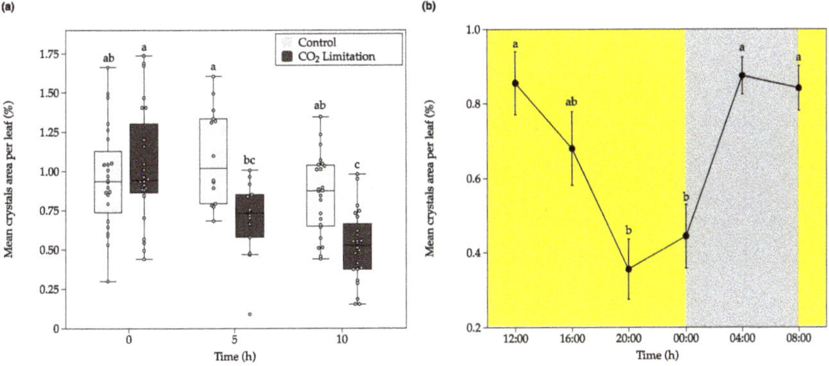

Figure 1. Fluctuations in total mean crystals area per leaf of *C. quitensis* plants under ambient or low CO_2 concentration. (**a**), Boxplots representing the effect of the CO_2 limitation treatment on leaf crystals area per leaf (%). The horizontal line indicates the mean and length of each whisker indicates the interquartile range (IQR); $n = 14$, different letters represent statistically significant differences (Two-way ANOVA; $p < 0.05$). In (**a**), plants were kept over time under constant light in airtight chambers injected either with ambient air (Control, 400 ppm CO_2, white boxes) or filtered air with soda lime (CO_2 limitation, 11.5 ppm CO_2, gray boxes). (**b**), Diurnal fluctuations in CaOx crystal area in *C. quitensis* leaves. Plants were grown under optimal growth conditions. Yellow background denotes light hours, while gray background dark hours. Error bars denote SE of mean; $n = 15$. Different letters represent statistically significant differences between time of the day (one-way ANOVA; $p < 0.05$).

The imposed carbon limitation, generated by the low [CO_2], triggered a significant reduction of CaOx crystals areas in the *C. quitensis* leaves, which is in accordance with our proposed hypothesis. The results are also in agreement with those reported by Tooulakou et al. (2016) [10], where a condition that boosted the stomatal closure and limited the availability of CO_2 (e.g., exogenous application of abscisic acid or drought stress), increased the CaOX crystal decomposition in the leaves of *Amaranthus hybridus*. Furthermore, the number of crystals per leaf area decreased significantly in the low [CO_2] treatment after 10 h (Figure S2). Therefore, the observed reduction of total crystal area per leaf (%) was the sum of both the reduction in size and number of crystals per leaf area. Consequently, given that complete crystal formation (maximum size) and maximum number of crystals per leaf are observed early in leaf development [25]; the observed differences in the crystals per leaf area should be a result of full crystal decomposition and not of differences in the number of idioblasts.

The CaOx crystal decomposition obtained here was also similar to that observed by Tooulakou et al. (2019) [26], where they showed that *A. hybridus* plants that were grown under CO_2

restrictive conditions exhibited a considerable reduction in the leaf CaOx crystal volume over time, compared to the control group. Both Tooulakou et al. reports (2016; 2019) consider that leaf CaOx crystals act as dynamic carbon reservoirs, capable of providing CO_2 for photosynthesis when the entry of atmospheric CO_2 into the mesophyll is limited by an environmental factor. Therefore, the CaOx crystal decomposition that is observed during the day might be associated to carbon requirements for photosynthesis [10,12,26].

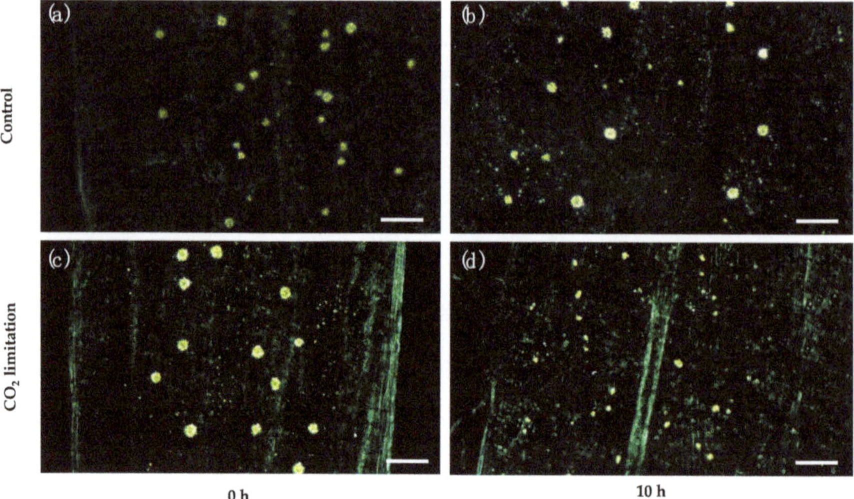

Figure 2. Paradermal view of the chlorine-bleached leaves under polarized light. CaOx crystals are visible as bright spots. Note the obvious differences in size and distribution of crystals between 0 h and 10 h under CO_2 limitation. (**a**) Control leaf, time = 0 h, 400 ppm CO_2. (**b**) Control leaf, time = 10 h, 400 ppm CO_2. (**c**) Treatment leaf, time = 0 h, 11.5 ppm CO_2. (**d**) Treatment leaf, time = 10 h, 11.5 ppm CO_2. Bars = 200 μm.

Moreover, the observation of the leaves crystals for 24 h, under optimal growth conditions, showed that these structures display diurnal fluctuations, similar to those reported in *A. hybridus* and *Oxalis corniculate* [12,27]. According to Figure 1b, during the first hours of light the mean crystal area per leaf kept stable. However, by 20:00 to 00:00 h the mean crystal area decreased significantly, while during the dark hours (00:00 to 08:00), these structures undergo a full recovery. During light time, and therefore the period where the plant is photosynthetically active, the crystals undergo a decomposition process. This process could be associated with supplementing—through the supply of subsidiary CO_2 released from crystals—the CO_2 requirement of this plant, since this species particularly suffers from a strong limitation in the acquisition of environmental CO_2 [22].

Throughout dark hours, as there are no electrons (from light-dependent reactions) to fix carbon, the plants could restore the crystals. The recovery of the oxalate can occur through several metabolic pathways; however, its origin is exclusively biological [8]. Diurnal fluctuations evidence that crystals decomposition is not always associated to an environmental stress per se in *Colobanthus quitensis*, and more a complementary process that could be suppling subsidiary CO_2 to its daily cycle. The obtained results suggest that the CaOX crystals in *C. quitensis* are a dynamic system, which respond to environmental stimuli, such as limitation of CO_2 (Figure 1a), and fluctuates in a daily course (Figure 1b).

2.2. Chlorophyll Fluorescence and Oxalate Oxidase Activity Measurements

Tooulakou et al. (2016) also showed that crystal decomposition was accompanied by boosted oxalate oxidase enzymatic activity (OXO; transforms oxalate into CO_2) [10]. The enzymatic analysis for oxalate oxidase activity on *C. quitensis* leaves showed that there were significant main effects for both CO_2 concentration and Time, and also there was a statistically significant interaction between both effects on the oxalate oxidase activity. Tukey's post hoc test showed that there were no significant differences between treatments at 0 h or 5 h of the test. However, statistically significant differences between the two studied groups were observed after 10 h (Figure 3a). The observed OXO activity was similar to that reported for *Podophyllum peltatum* after water stress [10]. The obtained results allow us to observe an association between the increase in oxalate oxidase activity and the decomposition of the CaOx crystals.

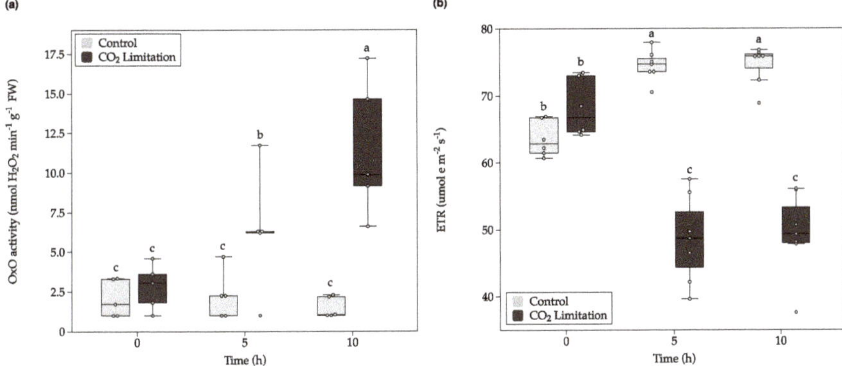

Figure 3. Fluctuations in the oxalate oxidase activity (OxO) (**a**), and electron transport rate (ETR) (**b**) of *C. quitensis* plants under ambient or low CO_2 concentration. Plants were kept over time under constant light in airtight chambers injected either with ambient air (Control, 400 ppm CO_2, white boxes) or filtered air with soda lime (CO_2 limitation, 11.5 ppm CO_2, gray boxes). The horizontal line indicates the mean and length of each whisker indicates the interquartile range (IQR); $n = 5$. Different letters represent statistically significant differences (two-way ANOVA; $p < 0.05$).

Chlorophyll fluorescence (ChlF) measurements showed that the electron transport rate (ETR) of the CO_2-limited plants decreased significantly compared to the control group (Figure 3b). There were significant main effects for CO_2 concentration, but not for Time. In addition, there was a statistically significant interaction between both effects on the ETR. However, despite the statistical differences, the plants exposed to low [CO_2] for 10 h still maintain high ETR values (~50 μmol e m^{-2} s^{-1}). This ETR values are similar to those reported before for this species under different temperatures and light intensity [22,28]. Following the AP hypothesis, a plant that does not have the AP mechanism and is exposed to a low [CO_2], should experience an inhibition of photosynthesis also evident by a reduced ETR. It is known that in C3 plants exposed to low [CO_2] the rate of carboxylation of Rubisco is reduced and consequently, the net photosynthetic rates are also affected due to substrate limitations; a situation that boosts photorespiration rates [29]. However, the obtained results showed just a 25% reduction of ETR in *C. quitensis* plants exposed to low [CO_2] (Figure S3).

The linearity between ETR and net CO_2 assimilation is commonly absent in C3 species, especially due the existence of alternative electron sinks [30]. Furthermore, under excess light, reducing equivalents from photosynthetic electron transport (NADPH) are exported from the chloroplasts to the cytosol, via malate/oxaloacetate shuttle, and the mitochondrial non-phosphorylating pathways may facilitate the dissipation of these excess reductants in the cell [31]. Therefore, the occurrence of sufficient ETR values alone is not a satisfactory indication that AP is responsible for the use of electrons because there are other alternative electron sinks, such as photorespiration and mitochondrial respiratory chain. In *C. quitensis*, the photosynthetic electron transport is insensitive to variations in oxygen concentration under non-photorespiratory conditions, indicating that electron transport to oxygen (Mehler reaction) is negligible [32]. However, it has also been shown that under low CO_2 availability, the relationship between ETR and Gross photosynthesis (A_G) in *C. quitensis* leaves present high values, indicative of enhanced photorespiration rates [22].

2.3. Calcium Oxalate Crystal Decomposition under Non-Photorespiratory Conditions

In order to eliminate the effect of photorespiration as an alternative electron sink on the observed level of ETR under the 11.5 ppm [CO_2] treatment (50 µmol e m^{-2} s^{-1}) (Figure 3b), as well to reduce the mitochondrial respiratory CO_2 efflux, and the putative contribution of mitochondria electron chain consuming chloroplast redox power, a second individual experiment under non-photorespiratory conditions (100% N_2) was performed. The *C. quitensis* plants showed a significant reduction in the crystals area per leaf between the beginning and the end of the treatment after 10 h under 100%-N_2 atmosphere in the glass container (Figure 4a), as well as a high percentage of crystal decomposition (Figure S1). However, under non-photorespiratory conditions, the ETR values decreased intensely after 2 h of treatment but no further reduction was observed, and until the end of the treatment, the ETR values were kept constant close to ~20 µmol e m^{-2} s^{-1} (Figure 4b). The difference of ETR between low CO_2 and 100% N_2 was about 30 µmol e m^{-2} s^{-1}, about 40% of total ETR; this is the putative contribution of photorespiration, mitochondrial respiration and Mehler reaction as alternative electron sinks to ETR in *C. quitensis* leaves. This is consistent with a high contribution of oxygen as an alternative electron sinks observed in other plant species [33,34]. Therefore, as previously reported by Saez et al. (2017) [22], it seems that photorespiration is enhanced in this species, which may help to counteract the harmful consequences that are generated as a result of a limited carbon assimilation. Photorespiration could also play a beneficial role in the dynamic and fast response of photosynthetic metabolism under CO_2 limitations, as has been observed in *Arabidopsis thaliana* [35].

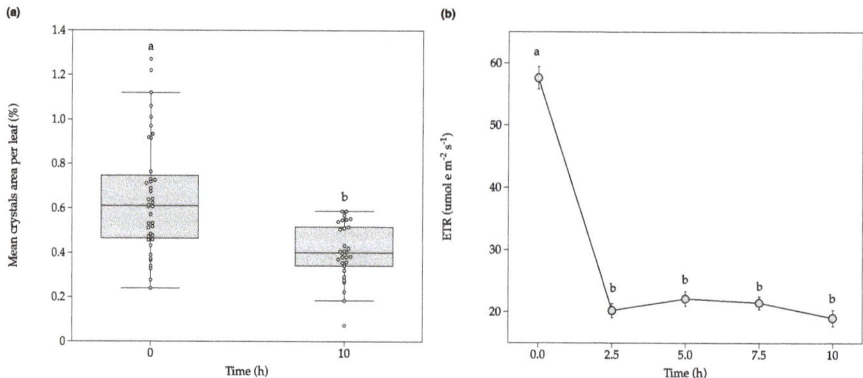

Figure 4. Fluctuations in total mean crystals area per leaf (**a**); and electron transport rate (ETR) (**b**) of *C. quitensis* plants under non-photorespiratory conditions (100% N_2). Plants were kept over time under constant light in airtight chambers injected with N_2 100%. In (**a**), the horizontal line indicates the mean and length of each whisker indicates the interquartile range (IQR); $n = 35$; different letters represent statistically significant differences (Independent samples *t*-test, $p < 0.01$). In (**b**), error bars denote SE of mean; $n = 15$. Different letters represent statistically significant differences between time (One-way ANOVA; $p < 0.05$).

Despite having neither atmospheric CO_2 nor O_2 as final electron acceptors, *C. quitensis* plants were able to maintain a stable ETR value for about 10 h of stress. Theoretically, four electrons are required to reduce 2NADP$^+$ to NADPH to fix one CO_2 [36]. In practice, because not all the electron really flows linearly and alternative electron sinks are active, about 8–12 electrons are required per CO_2 fixed. Therefore, with around ~20 μmol e m^{-2} s^{-1} of ETR, the *C. quitensis* plants could hypothetically fix about 2 μmol CO_2 m^{-2} s^{-1}, or even higher (about 5 μmol CO_2 m^{-2} s^{-1}) if we consider that under low oxygen the main alternative electron sinks, photorespiration, and Mehler reaction are constraint. If these CO_2 molecules are effectively being supplied by the CaOx crystals decomposition needs to be probed.

Alarm photosynthesis could be a process that may allow plants maintain a baseline level of photosynthesis when face stress situations. This mechanism is advantageous as a quenching regulator for the energy excess accumulated from the electron transport chain, when the light-dependent reactions are not in pace with photosynthetic CO_2 assimilation from the atmosphere [10]. To this respect, it is possible that some unknown molecules could be involved as final electron acceptors in *C. quitensis* plants under a CO_2 limiting condition, among them, the CO_2 released from the CaOx crystals decomposition.

The multiple functional role(s) of CaOx crystals in plants is not yet well understood and even some researchers have doubts about the functionality of the bioavailable calcium stocks in plants [13]. However, the obtained results generate a contribution on the understanding of the functions that have been attributed to CaOx crystals, and their relation with the AP mechanism, even though more evidence is still required to ensure that CO_2 release by CaOx decomposition is being used for photosynthesis in *C. quitensis*.

In this study, we found that *Colobanthus quitensis* plants exposed to a CO_2 limitation significantly increased the CaOx crystal decomposition, as well as the oxalate oxidase activity in its leaves. This means that under stress conditions, the crystal decomposition could provide CO_2 molecules to the mesophyll tissue [8]. In parallel, ETR decreased but remained stable when compared to the control group. Moreover, under non-photorespiratory conditions a significant CaOx crystal decomposition was also observed, whereas ETR decreased around 40% but was still adequate for the maintenance of a baseline level of photosynthesis if required.

For the Antarctic plant *C. quitensis*, alarm photosynthesis could play an important role as a complementary endogenous mechanism that could facilitate a CO_2 supply given the limitations in the CO_2 diffusion that have been widely studied. Alarm photosynthesis is considered as a process that may enable the prevention of water losses when plants are under stressful conditions such as drought or strong winds coupled with low relative humidity. Further investigations on other extreme environment plant species is envisioned besides the genetic mechanisms of alarm photosynthesis, especially with focus on the CaOx genetic process of decomposition and regeneration (biosynthetic–degradation pathways), as well as the role of calcium ion during crystal recycling cycles, will allow a more detailed understanding of plant responses to intense drought scenarios. Future research should also focus on further understanding the diurnal fluctuations that have been observed in this plant, its relationship with daily hours and exposure to light. Likewise, attention should be paid to the description and characterization of the oxalate oxidase protein/gene and its regulation, not only because of their participation in the crystal decomposition processes, but also because of their association with stress tolerance processes.

3. Materials and Methods

3.1. Plant Material and Growth Conditions

Colobanthus quitensis plants were collected in King George Island near to Henryk Arctowski Polish Antarctic station (62°09′34″ S; 58°28′19″ W) during March 2018. *C. quitensis* plants were reproduced vegetatively in plastic pots (5 × 5 × 5 cm) using a soil/peat/vermiculite mixture (3:1:2) and maintained in a greenhouse until having a regular-size cushion. Plants were fertilized with 0.2 g L^{-1} Phostrogen® solution once a month. Before the experiment, *C. quitensis* plants were grown for at least 21 days (acclimation period) in a growth room (photoperiod 16/8 h, temperature 16 °C, photon flux density (photosynthetically active radiation) of 200 μmol m^{-2} s^{-2} and ambient air conditions (approximately 400 ppm CO$_2$)).

3.2. Experimental Design and Sample Collection

The individual plant cushions were placed inside a transparent airtight borosilicate container and were supplied with air (7.0 L/min) either at ambient (400 ppm, control) or low [CO$_2$] (11.5 ppm, treatment) (Figure S4). A soda lime scrubber was used to achieve the low [CO$_2$], which was continuously monitored by an infrared gas analyzer (IRGA-LI−6400XT, LI-COR Inc., Lincoln, NE, USA). During the experiment (10 h, from 08:00 to 18:00) plants were kept under 200 μmol m^{-2} s^{-2} PAR and 16 °C temperature. Leaf samples (24) were collected for crystal decomposition measurements. Additionally, 20 mg of leaves from 5 plants were collected at time 0, 5, and 10 h for oxalate oxidase activity determination.

A second independent group of plants was kept at optimal growth conditions (temperature 16 °C, PAR of 200 μmol m^{-2} s^{-2} and ambient air conditions (approximately 400 ppm CO$_2$)) for monitoring the leaves crystals for 24 h. Plants were exposed to 16 h of light (from 08:00 to 23:59) and 8 h of dark (00:00 to 07:59). Leaf samples (15) were collected for crystal decomposition measurements every four hours.

3.3. Chlorophyll Fluorescence

The electron transport rate (ETR) of *C. quitensis* plants exposed to both [CO$_2$] was measured in vivo during the CO$_2$-restrictive experiment. Chlorophyll fluorescence measurements were performed using a Maxi-Imaging-PAM Chlorophyll Fluorimeter (Walz, Effeltrich, Germany). Ten areas of interest (AOI) were selected from each cushion and the ETR was calculated according: ETR = Φ_{PSII}·PAR·αL·(PSII/PSI); where Φ_{PSII} is the quantum efficiency of the photosystem II (PSII), PAR is the photosynthetically active radiation, αL the leaf absorptance (0.73 for control leaves, 0.68 for CO$_2$ Limitation), and PSII/PSI the distribution of absorbed energy between the two photosystems (assumed to be 0.5). Φ_{PSII} was measured with a Saturation Pulse (6000 μmol photons m^{-2} s^{-1}, 800 ms) applied after 3 min illumination at 750 μmol quanta m^{-2} s^{-1} of actinic light (AL) (Figure S5). The leaf absorptance was directly measured in the plants using the Maxi-Imaging PAM as described by Saéz et al., 2018 [37]. Briefly, successive illumination of the samples with red (R) and near infrared (NIR) light and the capture of each remission image were used by the equipment software to calculate pixel by pixel as follows: Abs = 1 − R/NIR. The minimum number of cushions used for the analysis (*n*) was at least 5.

3.4. Measurements of Crystal Degradation

The collected *C. quitensis* leaves were bleached in sodium hypochlorite solution (5% p/p) according to Tooulakou et al. (2016) [10]. Briefly, whole mature leaves were put in an aqueous solution of commercial bleach for 48 h until full depigmentation. Depigmented leaves were rinsed with abundant distillated water and then put between two microscope slides. Samples were observed under an optical microscope adapted with a polarizing filter at 10× magnification Leica DM750-Camera Leica ICC50W (Leica Microsystems, Wetzlar, Hesse, Germany). Several images were taken per leaf, covering the total leaf area. The area of each crystal was calculated by digital image analysis (ImageJ-Fiji v 2.0.0-rc69/1.52i) [38]. For each individual leaf, several images were taken comprising the total leaf area. Each individual image was analyzed as follows: (1) image was converted to 8 bits, (2) 8 bits image was converted to Mask, and (3) the tool "Analyzing Particles" was used to count and measure all crystals area in the picture using the following parameters: Size 400–5000 pixel^2, Circularity 0.35–1.00. The total counts (crystals area) of all images from an individual leaf were sum together to obtain the total area of crystals per leaf. Then, this value is divided by the total leaf area to obtain a ratio area crystals/area leaf. The minimum number of leaves used for the analysis at each time (*n*) was at least 14.

3.5. Oxalate Oxidase Activity Determination

The activity of the oxalate oxidase enzyme was determined at time 0, 5, and 10 h of the CO$_2$ restrictive experiment using the Oxalate Oxidase Activity Assay Kit (BioVision, Inc., Milpitas, CA, USA).

Assays were performed following kit protocol using 20 mg of *C. quitensis* fresh leaves from 5 plants ($n = 5$). In this test, the decomposition of oxalate by an oxalate oxidase release hydrogen peroxide, which generates a fluorescent signal directly proportional to the amount of active oxalate oxidase present in samples.

3.6. Statistical Analysis

Two-way analysis of variance (ANOVA) at a 95% level of significance ($p < 0.05$) were applied using JASP software (Version 0.13.1) [39] to assess the effects of both time and CO_2 concentration. Tukey Post Hoc Test was carried out in those cases where ANOVA was significant. One-way ANOVA was used to analyze one factor multiple comparation and Independent samples *t*-test were applied for simple comparison. The assumption of data normality, and homoscedasticity were tested with the Shapiro–Wilk and Levenes's test (data regarding all the statistical tests can be found in Tables S1–S8)

Supplementary Materials: The following are available online at http://www.mdpi.com/2223-7747/9/10/1307/s1, Figure S1: Decomposition percentage of CaOx crystals in *C. quitensis* leaves under adequate (Control, 400 ppm CO_2), low CO_2 (CO_2 limitation, 11 ppm CO_2) and non-photorespiratory conditions (100% N_2, 2 ppm CO_2) at the end of the treatment. The horizontal line indicates the mean and length of each whisker indicates the interquartile range (IQR); $n = 5$. Different letters denote significant differences between groups (one-way ANOVA; $p < 0.05$). Figure S2: CaOx crystals number counted in whole *C. quitensis* leaves under adequate (Control, 400 ppm CO_2) or low (Treatment, 11 ppm CO_2) CO_2 concentration. The horizontal line indicates the mean and length of each whisker indicates the interquartile range (IQR); $n = 14$. Different letters represent statistically significant differences (two-way ANOVA; $p < 0.05$). Figure S3: Light responses curves (ETR/ETR$_{max}$ %) of *Colobanthus quitensis* under adequate (Control, 400 ppm CO_2) and low CO_2 (CO_2 limitation, 11 ppm CO_2). Measurements were performed at 16 °C. Error bars denote SD of mean; $n = 6$. Figure S4: Experimental setup: System used to run the CO_2 restriction experiment on *C. quitensis* plants. Figure S5: Quantum yield of PSII (φPS II) in *C. quitensis* leaves under adequate (Control, 400 ppm CO_2) or low (Treatment, 11 ppm CO_2) CO_2 concentration. Error bars denote SD of mean; $n = 15$. Table S1: Two-way independent ANOVA results for Figure 1a data. Table S2: One-way independent ANOVA results for Figure 1b data. Table S3: Two-way independent ANOVA results for Figure 3a data. Table S4: Two-way independent ANOVA results for Figure 3b data. Table S5: Independent samples *t*-test results for Figure 4a data. Table S6: One-way independent ANOVA results for Figure 4b data. Table S7: One-way independent ANOVA results for Figure S1 data. Table S8: Two-way independent ANOVA results for Figure S2 data.

Author Contributions: O.G.-E., G.K., P.B., and L.A.B. designed the experiments. O.G.-E. and D.G.-R. grew the plant material and performed the experiments. O.G.-E., G.K., P.B., and L.A.B. analyzed the results. O.G.-E. prepared the manuscript. G.K., P.B., and L.A.B. revised the manuscript. All authors have read and agreed to the published version of the manuscript.

Funding: This research was funded by the Instituto Antártico Chileno (INACH), Grant DG_10_18; the Chilean National Commission for Scientific and Technological Research (CONICYT-PFCHA/Doctorado Nacional/ 2017-21170265 to O.G-E) and the Network for Extreme Environments Research, (NEXER) Grant NXR 17-0002.

Acknowledgments: We would like to thank Jorge Farías Avendaño for facilitating the use of the microscope in his laboratory. We are also thankful to Eng. Lorena Sandoval for her valuable work in plant maintenance and fertilization. Finally, we thank Kattia Núñez-Montero for the graphical abstract creation (Created with BioRender.com).

Conflicts of Interest: The authors declare no conflict of interest.

References

1. Prasad, R.; Shivay, Y.S. Oxalic acid/oxalates in plants: From self-defence to phytoremediation. *Curr. Sci.* **2017**, *112*, 1665–1667. [CrossRef]

2. Raman, V.; Horner, H.T.; Khan, I.A. New and unusual forms of calcium oxalate raphide crystals in the plant kingdom. *J. Plant Res.* **2014**, *127*, 721–730. [CrossRef] [PubMed]

3. Franceschi, V. Calcium oxalate in plants. *Trends Plant Sci.* **2001**, *6*, 331. [CrossRef]

4. Webb, M.A. Cell-mediated crystallization of calcium oxalate in plants. *Plant Cell* **1999**, *11*, 751–761. [CrossRef]

5. Dickison, W.C. *Integrative Plant Anatomy*; Academic Press: Cambridge, MA, USA, 2000; ISBN 978-0-12-215170-5.

6. Franceschi, V.R.; Nakata, P.A. Calcium Oxalate in Plants: Formation and Function. *Annu. Rev. Plant Biol.* **2005**, *56*, 41–71. [CrossRef] [PubMed]

7. Nakata, P.A. Advances in our understanding of calcium oxalate crystal formation and function in plants. *Plant Sci.* **2003**, *164*, 901–909. [CrossRef]

8. Karabourniotis, G.; Horner, H.T.; Bresta, P.; Nikolopoulos, D.; Liakopoulos, G. New insights into the functions of carbon–calcium inclusions in plants. *New Phytol.* **2020**, nph.16763. [CrossRef]

9. Horner, H.; Wagner, B. Calcium oxalate formation in higher plants. In *Calcium Oxalate in Biological Systems*; Khan, S., Ed.; CRC Press: Boca Raton, FL, USA, 1995; pp. 53–72.

10. Tooulakou, G.; Giannopoulos, A.; Nikolopoulos, D.; Bresta, P.; Dotsika, E.; Orkoula, M.G.; Kontoyiannis, C.G.; Fasseas, C.; Liakopoulos, G.; Klapa, M.I.; et al. "Alarm photosynthesis": Calcium oxalate crystals as an internal CO_2 source in plants. *Plant Physiol.* **2016**, *171*, 2577–2585. [CrossRef]

11. Gaberščik, A.; Grašič, M.; Vogel-Mikuš, K.; Germ, M.; Golob, A. Water shortage strongly alters formation of calcium oxalate druse crystals and leaf traits in *Fagopyrum esculentum*. *Plants* **2020**, *9*, 917. [CrossRef]

12. Tooulakou, G.; Giannopoulos, A.; Nikolopoulos, D.; Bresta, P.; Dotsika, E.; Orkoula, M.G.; Kontoyannis, C.G.; Fasseas, C.; Liakopoulos, G.; Klapa, M.I.; et al. Reevaluation of the plant "gemstones": Calcium oxalate crystals sustain photosynthesis under drought conditions. *Plant Signal. Behav.* **2016**, *11*, e1215793. [CrossRef]

13. Paiva, E.A.S. Are calcium oxalate crystals a dynamic calcium store in plants? *New Phytol.* **2019**, *223*, 1707–1711. [CrossRef] [PubMed]

14. Fuentes-Lillo, E.; Cuba-Díaz, M.; Rifo, S. Morpho-physiological response of *Colobanthus quitensis* and *Juncus bufonius* under different simulations of climate change. *Polar Sci.* **2017**, *11*, 11–18. [CrossRef]

15. Convey, P.; Gibson, J.A.E.; Hillenbrand, C.D.; Hodgson, D.A.; Pugh, P.J.A.; Smellie, J.L.; Stevens, M.I. Antarctic terrestrial life—Challenging the history of the frozen continent? *Biol. Rev.* **2008**, *83*, 103–117. [CrossRef]

16. Cavieres, L.A.; Sáez, P.; Sanhueza, C.; Sierra-Almeida, A.; Rabert, C.; Corcuera, L.J.; Alberdi, M.; Bravo, L.A. Ecophysiological traits of Antarctic vascular plants: Their importance in the responses to climate change. *Plant Ecol.* **2016**, *217*, 343–358. [CrossRef]

17. Prychid, C.J.; Rudall, P.J. Calcium oxalate crystals in monocotyledons: A review of their structure and systematics. *Ann. Bot.* **1999**. [CrossRef]

18. Brönnimann, D.; Ismail-Meyer, K.; Rentzel, P.; Pümpin, C.; Lisá, L. Excrements of Herbivores. In *Archaeological Soil and Sediment Micromorphology*; John Wiley & Sons, Ltd.: Chichester, UK, 2017; pp. 55–65.

19. Cuba-Díaz, M.; Cerda, G.; Rivera, C.; Gómez, A. Genome size comparison in *Colobanthus quitensis* populations show differences in species ploidy. *Polar Biol.* **2017**, *40*, 1475–1480. [CrossRef]

20. Arthofer, W.; Bertini, L.; Caruso, C.; Cicconardi, F.; Delph, L.F.; Fields, P.D.; Ikeda, M.; Minegishi, Y.; Proietti, S.; Ritthammer, H.; et al. Genomic Resources Notes accepted 1 February 2015–31 March 2015. *Mol. Ecol. Resour.* **2015**, *15*, 1014–1015. [CrossRef]

21. Cho, S.M.; Lee, H.; Jo, H.; Lee, H.; Kang, Y.; Park, H.; Lee, J. Comparative transcriptome analysis of field- and chamber-grown samples of *Colobanthus quitensis* (Kunth) Bartl, an Antarctic flowering plant. *Sci. Rep.* **2018**, *8*, 11049. [CrossRef]

22. Sáez, P.L.; Bravo, L.A.; Cavieres, L.A.; Vallejos, V.; Sanhueza, C.; Font-Carrascosa, M.; Gil-Pelegrín, E.; Javier Peguero-Pina, J.; Galmés, J. Photosynthetic limitations in two Antarctic vascular plants: Importance of leaf anatomical traits and Rubisco kinetic parameters. *J. Exp. Bot.* **2017**, *68*, 2871–2883. [CrossRef]

23. Sáez, P.L.; Galmés, J.; Ramírez, C.F.; Poblete, L.; Rivera, B.K.; Cavieres, L.A.; Clemente-Moreno, M.J.; Flexas, J.; Bravo, L.A. Mesophyll conductance to CO_2 is the most significant limitation to photosynthesis at different temperatures and water availabilities in Antarctic vascular species. *Environ. Exp. Bot.* **2018**, *156*, 279–287. [CrossRef]

24. Pérez-Torres, E.; Bascuñán, L.; Sierra, A.; Bravo, L.A.; Corcuera, L.J. Robustness of activity of Calvin cycle enzymes after high light and low temperature conditions in Antarctic vascular plants. *Polar Biol.* **2006**, *29*, 909–916. [CrossRef]

25. Giannopoulos, A.; Bresta, P.; Nikolopoulos, D.; Liakopoulos, G.; Fasseas, C.; Karabourniotis, G. Changes in the properties of calcium-carbon inclusions during leaf development and their possible relationship with leaf functional maturation in three inclusion-bearing species. *Protoplasma* **2019**, *256*, 349–358. [CrossRef] [PubMed]

26. Tooulakou, G.; Nikolopoulos, D.; Dotsika, E.; Orkoula, M.G.; Kontoyannis, C.G.; Liakopoulos, G.; Klapa, M.I.; Karabourniotis, G. Changes in size and composition of pigweed (*Amaranthus hybridus* L.) calcium oxalate crystals under CO_2 starvation conditions. *Physiol. Plant* **2019**, *166*, 862–872. [CrossRef] [PubMed]

27. Seal, S.N.; Sen, S.P. The photosynthetic production of oxalic acid in *Oxalis corniculata*. *Plant Cell Physiol.* **1970**, *11*, 119–128. [CrossRef]

28. Casanova-Katny, M.A.; Bravo, L.A.; Molina-Montenegro, M.; Corcuera, L.J.; Cavieres, L.A. Photosynthetic performance of *Colobanthus quitensis* (Kunth) Bartl. (Caryophyllaceae) in a high-elevation site of the Andes of central Chile. *Rev. Chil. Hist. Nat.* **2006**, *79*, 41–53. [CrossRef]

29. Gerhart, L.M.; Ward, J.K. Plant responses to low [CO_2] of the past. *New Phytol.* **2010**, *188*, 674–695. [CrossRef]

30. Kalaji, H.M.; Schansker, G.; Brestic, M.; Bussotti, F.; Calatayud, A.; Ferroni, L.; Goltsev, V.; Guidi, L.; Jajoo, A.; Li, P.; et al. Frequently asked questions about chlorophyll fluorescence, the sequel. *Photosynth. Res.* **2017**, *132*, 13–66. [CrossRef]

31. Noguchi, K.; Yoshida, K. Interaction between photosynthesis and respiration in illuminated leaves. *Mitochondrion* **2008**, *8*, 87–99. [CrossRef]

32. Pérez-Torres, E.; Bravo, L.A.; Corcuera, L.J.; Johnson, G.N. Is electron transport to oxygen an important mechanism in photoprotection? Contrasting responses from Antarctic vascular plants. *Physiol. Plant* **2007**, *130*, 185–194. [CrossRef]

33. Ort, D.R.; Baker, N.R. A photoprotective role for O_2 as an alternative electron sink in photosynthesis? *Curr. Opin. Plant Biol.* **2002**, *5*, 193–198. [CrossRef]

34. Fernández-Marín, B.; Gulías, J.; Figueroa, C.M.; Iñiguez, C.; Clemente-Moreno, M.J.; Nunes-Nesi, A.; Fernie, A.R.; Cavieres, L.A.; Bravo, L.A.; García-Plazaola, J.I.; et al. How do vascular plants perform photosynthesis in extreme environments? An integrative ecophysiological and biochemical story. *Plant J.* **2020**, *101*, 979–1000. [CrossRef] [PubMed]

35. Eisenhut, M.; Bräutigam, A.; Timm, S.; Florian, A.; Tohge, T.; Fernie, A.R.; Bauwe, H.; Weber, A.P.M. Photorespiration Is Crucial for Dynamic Response of Photosynthetic Metabolism and Stomatal Movement to Altered CO_2 Availability. *Mol. Plant* **2017**, *10*, 47–61. [CrossRef] [PubMed]

36. Zeinalov, Y. Mechanisms of Photosynthetic Oxygen Evolution and Fundamental Hypotheses of Photosynthesis. In *Handbook of Photosynthesis*, 2nd ed.; Pessarakli, M., Ed.; Taylor & Francis Group, LLC.: Boca Raton, FL, USA, 2005; pp. 23–25. ISBN 9781420027877.

37. Sáez, P.L.; Cavieres, L.A.; Galmés, J.; Gil-Pelegrín, E.; Peguero-Pina, J.J.; Sancho-Knapik, D.; Vivas, M.; Sanhueza, C.; Ramirez, C.F.; Riviera, B.K.; et al. In situ warming in the Antarctic: Effects on growth and photosynthesis in Antarctic vascular plants. *New Phytol.* **2018**. [CrossRef] [PubMed]

38. Schindelin, J.; Arganda-Carreras, I.; Frise, E.; Kaynig, V.; Longair, M.; Pietzsch, T.; Preibisch, S.; Rueden, C.; Saalfeld, S.; Schmid, B.; et al. Fiji: An open-source platform for biological-image analysis. *Nat. Methods* **2012**, *9*, 676–682. [CrossRef]

39. JASP Team. *JASP*, version 0.13.1; Computer software. 2020.

Article

Chlorophyll a Fluorescence Transient and 2-Dimensional Electrophoresis Analyses Reveal Response Characteristics of Photosynthesis to Heat Stress in *Malus*. 'Prairifire'

Tao Wang [1,2], Siqian Luo [1], Yingli Ma [1], Lingyu Li [1], Yinfeng Xie [1,*] and Wangxiang Zhang [1]

[1]   Co-Innovation Center for Sustainable Forestry in Southern China, College of Biology and the Environment, Nanjing Forestry University, Nanjing 210037, China; johnwt@cnbg.net (T.W.); lq8429996@gmail.com (S.L.); yli_ma@sina.com (Y.M.); li13505153082@sina.com (L.L.); wangxiangzh2002@sina.com.cn (W.Z.)

[2]   Institute of Botany, Jiangsu Province and Chinese Academy of Sciences (Nanjing Botanical Garden Mem. Sun Yat-Sen), Nanjing 210014, China

*   Correspondence: xxyyff@njfu.edu.cn

Received: 18 July 2020; Accepted: 13 August 2020; Published: 15 August 2020

Abstract: Flowering crabapples are a series of precious ornamental woody plants. However, their growth and development are inhibited in the subtropical regions due to the weak photosynthesis under high-temperature environment in the summer. Chlorophyll a fluorescence transient and 2-dimensional electrophoresis (2-DE) analyses were conducted to investigate the response characteristics of photosynthesis under simulated 38 °C heat stress in leaves of *Malus*. 'Prairifire', a spring-red leaf cultivar of flowering crabapple with strong thermal adaptability. In the present study, the net photosynthetic rate (Pn) was significantly decreased during the heat shock process, which showed a similar trend to the stomatal conductance (Gs), indicating a sensitive stomatal behavior to heat stress. Moreover, an efficient reaction center in photosystem II (PSII), and a functionally intact oxygen-evolving complex (OEC) conferred strong photosynthetic adaptability under heat stress. The higher level of transketolase (TK) under 48-h heat shock treatment was considered a protective mechanism of photosynthetic apparatus. However, heat stress inhibited the functions of light harvesting complex II (LHCII), electron transport in PSII, and the levels of key enzymes in the Calvin cycle, which were considered as the reasons causing an increase in the proportion of non-stomatal restrictions.

Keywords: heat stress; *Malus*. 'Prairifire'; photosynthetic characteristics; chlorophyll a fluorescence; 2-dimensional electrophoresis

1. Introduction

Heat stress is one of the main environmental factors affecting plant growth and reproduction [1]. With the intensification of greenhouse gas emissions, extremely high temperatures occur frequently with longer duration, which is becoming a serious challenge for the entire agroforestry system worldwide [2].

As the basis of yield and quality in plants, photosynthesis is regarded as one of the most sensitive processes in response to heat stress [3] and has been reported to be the key to reveal the thermal adaptability of plants [4,5]. The impact of heat stress on photosynthesis can be mainly summarized as follows: in the early stage, heat stress causes the partial closure of stomata, which directly affects the net photosynthetic rate (Pn). Moreover, the changes in CO_2 concentration caused by the decrease of the stomatal conductance (Gs) further limit the photosynthetic function [6]. Under strong heat stress, the membrane structure of the thylakoid would be damaged, resulting in a series of changes and even damages caused to structures and functions in photosystem II (PSII) [5], which is considered to be the most thermo-sensitive component of the photosynthetic apparatus [7].

Recently, in vivo chlorophyll a fluorescence rise (OJIP) kinetics has been applied extensively as a rapid and non-invasive tool for elucidating the activity of PSII in higher plants [3,8]. A quantitative analysis of the OJIP curves, called the JIP test, is used to reveal the environmental effect on the structure, conformation, and function of the photosynthetic organisms [8,9]. A large number of studies have proved the extreme sensitivity of the OJIP transient to heat stress, such as the L step (reflecting the energetic connectivity of the PSII units), the K step (relating to the inactivation of the oxygen-evolving complex (OEC)), and the J step, combining the changes of the IP phase (reflecting the status of Q_A^- and electron transport from Q_A to Q_B) [8,10]. Moreover, the photosynthetic activity in a PSII reaction center complex could be reflected by the index PI_{ABS}, which is taken into consideration for three main functional steps, including light energy absorption, trapping of excitation energy, and conversion of the excitation energy to electron transport [3]. Therefore, research on the photosynthetic capacity of PSII based on OJIP curves has become an irreplaceable approach in heat-tolerance breeding in plants.

Moreover, proteomic approaches have provided important information for understanding the complex molecular mechanisms of plants in response to heat stress. Global protein expression profiles can be analyzed and compared using a two-dimensional gel-based protein separation method coupled with protein identification by mass spectrometry (MS). Among them, the expression and degradation of chloroplast proteins are regarded as necessary processes for normal growth and development in plants, as well as a common response mechanism to environmental changes. For example, the expressions of OEC and oxygen-evolving enhancer protein (OEE) can promote the absorption of light energy [11]. Magnesium chelatase plays an important role in maintaining the chlorophyll content and maximally absorbing light energy during chlorophyll synthesis [12]. Ferredoxin nicotinamide adenine dinucleotide phosphate (NADP) reductase can catalyze the synthesis of NADPH and thus plays an extremely important role in photosynthetic electron transport [13]. Moreover, a variety of enzymes involved in carbon assimilation, such as ribulose-1,5-bisphosphate carboxylase/oxygenase (rubisco), rubisco activase (RCA), phosphoribulose kinase, and transketolase (TK), have different abundances under heat stress, revealing the regulation of photosynthesis by chloroplast proteins [14,15].

Flowering crabapples (*Malus* spp.), members of Rosaceae, are a group of small landscape trees or shrubs with charming flowers, colorful fruits, and many tree shapes [16]. They originated in the temperate regions of northern China, and hundreds of varieties have been developed over the course of their more than 200-year history of cultivation, which are famous worldwide for their unique ornamental and economical values [17,18]. However, their cultivation in southern China, particularly within subtropical regions, is greatly limited because of the high-temperature environment in summer, which severely inhibits normal growth and development, especially photosynthesis [19]. *Malus*. 'Prairifire' is a spring-red leaf cultivar with excellent ornamental traits and strong thermal adaptability [20] and has passed the examination and approval of improved varieties of forest trees in Jiangsu Province (Su S-ETS-MP-007-2017). Therefore, *M.* 'Prairifire' is expected to be an ideal source for investigating the thermal adaption mechanism of crabapples. In the present study, chlorophyll a fluorescence combined with 2-dimensional electrophoresis (2-DE) analysis was used to study the response characteristics of photosynthesis to heat stress in leaves of *M.* 'Prairifire'. The results of this study will provide a basis for the mechanism of heat resistance and genetic breeding of flowering crabapples. In addition, these results will provide scientific reference for the selection of subtropical landscape plants.

2. Materials and Methods

2.1. Plant Materials and Growth Conditions

Samples of one-year-old *M.* 'Prairifire' seedlings were selected from the arboretum of the national repository of *Malus* spp. germplasm (Yangzhou City, Jiangsu Province, China) in November 2015 and were transferred separately into individual plastic pots (23 cm tall × 30 cm diameter; one plant per pot), filled with prepared soil. The soil used in this experiment was yellow-brown soil collected from the arboretum and was homogenized, air-dried, and sieved through a 4.0-mm sieve. The soil was neutral (pH = 6.68), mixed with suitable organic fertilizer, containing 6.79 g organic matter, 1.86 g nitrogen, 127 mg available phosphorus, 295 mg available potassium per kg soil, and less than 1 mg mercury per kg soil. A total of 40 sample pots were placed in the arboretum for culturing, and the fully expanded, healthy leaves from the third branch (from the top) of the seedlings were marked for future experiments.

2.2. Treatments

In May 2016, uniform and healthy potted samples were transferred to an artificial climate chest, with the light intensity set as 600 $\mu mol \cdot m^{-2} \cdot s^{-1}$, under a light/dark cycle of 12/12 h, and 70–75% relative humidity at a temperature of 26 ± 1 °C for controlled growth (CG). 500 mL water was added to per pot every 2 days until the end of all the measurements. In June 2016, potted samples grown in the artificial climate chest were randomly arranged for artificial heat shock treatment. The heat shock temperature was set as 38 °C during 6:00 am–6:00 pm, which referred to a local average temperature under extreme high temperature conditions; while that was 26 °C during 6:00 pm–6:00 am. The other settings were the same as the CG conditions. The photosynthetic gas exchange parameters and chlorophyll a fluorescence transient were measured under the CG conditions and heat shock treatments for 1 day, 2 days, 4 days, and 6 days. The time of per measurement was under 9:00 am–11:00 am. Moreover, fresh leaves under CG conditions and those treated with 38 °C heat shock for 48 h were collected, quickly placed in liquid nitrogen for freezing, and then saved under −80 °C conditions for 2-DE analysis.

2.3. Measurements of Photosynthetic Gas Exchange Parameters

The photosynthetic gas exchange parameters of leaves of M. 'Prairifire' were measured using an LI-6400 portable photosynthesis system (Li-Cor, USA), with a 6 cm^2 leaf chamber. The setting value of photosynthetically active radiation (PAR) was 600 $\mu mol\ m^{-2}\ s^{-1}$ with a red–blue LED light source (6400-02B). The gas flow rate was 500 $\mu mol\ s^{-1}$. The air temperature and relative humidity were consistent with the settings in the artificial climate chest. The measured parameters included the Pn, Gs, intercellular CO_2 concentration (Ci), and transpiration rate (Tr). Six marked leaves were measured, and the data were recorded when the rate of CO_2 uptake was stable. Finally, the average was calculated as the final result for each measurement time.2.4. Measurements of the chlorophyll a fluorescence transient

The chlorophyll a fluorescence transient for the leaves of M. 'Prairifire' was measured under CG conditions and on days 1, 2, 4, and 6 after the heat shock treatment. The leaves were dark-adapted for 30 min before the measurement; the OJIP curve was then determined using PEA-Senior (Hansatech, UK) and was analyzed using the JIP test, according to the methods of Strasser et al. [8]. Details of the introduced parameters are listed below: φP_O (F_V/F_M), maximum quantum yield of primary PSII photochemistry; Vj, the variable fluorescence at 2 ms; Sm, the normalized area (assumed to be proportional to the number of reduction and oxidation of one $Q_A{}^-$ molecule during the fast OJIP transient and therefore related to the number of electron carriers per electron transport chain); N, the times Q_A was reduced to $Q_A{}^-$ in the time span from t_0 to t_{Fmax}; ABS/RC, average absorbed photon flux per PSII reaction center; DI_O/RC, the specific energy fluxes per reaction center for dissipation; TR_O/RC, the specific energy fluxes per reaction center for trapping; ET_O/RC, the specific energy fluxes per reaction center for electron transport; φE_O, the probability that an absorbed photon will move an electron into the electron transport chain; PI_{ABS}, the performance index of PSII; Ψ_O, efficiency with which a trapped exciton can move an electron into the electron transport chain; and RC/CS_O, the number of active PSII reaction centers per excited cross section.

2.4. Extraction and Quantification of Proteins in Leaves of M. 'Prairifire'

1 g leaf samples were collected and ground in liquid nitrogen. The powder was transferred to a 50-mL centrifuge tube and dissolved by pre-cooled 10% TCA-acetone solution (containing 0.1% DTT and 1 mM PMSF) at −20 °C overnight. The supernatant was discarded by centrifugation at 15,000× g at 4 °C for 20 min, and a pre-cooled acetone solution (containing 0.1% DTT and 1 mM PMSF) was added to the precipitate at −20 °C for 2 h. Then, the supernatant was discarded by centrifugation at 15,000× g at 4 °C for 20 min, and the precipitate was placed in a freeze vacuum dryer for 30 min. The dried protein powder was stored in a −80 °C freezer for later use.

Protein powder (100 µg) was redissolved in 800 µL pyrolysis buffer (7 M urea, 2 M thiourea, 4% CHAPS, 65 mM DTT, 0.5% Bio-Rad Ampholyte, 1 mM PMSF). Then, the supernatant was retained by centrifugation at 15,000× g at 4 °C for 10 min. The quantification of proteins was performed using a Bradford Method Protein Concentration Assay Kit, according to the manufacturer's instructions.

2.5. 2-DE Analysis

The sample buffer that was loaded with 100 µg protein and mixed with 400 µl pyrolysis buffer was applied to IPG adhesive strips (24 cm; pH 4–7; non-linear) for isoelectric focus (IEF). Mineral oil was used to prevent exposure to the air. The IEF procedure was set to 50 V for 14 h, 100 V for 1 h, 500 V for 1 h, 1000 V for 1 h, and 8000 V for 4 h, and the total voltage time product was 12000 V/h. After the end of IEF, the IPG strip was preserved in a 10-mL SDS equilibrium solution 1 (containing 6 M urea, 2% SDS, 0.375 M Tris-HCl, 20% glycerin, 1% DTT) and then subjected to a shaker balance for 15 min. Subsequently, the strip was cleaned, SDS equilibrium solution 2 was added (additional 2.5% IAA to equilibrium solution 1 without 1% DTT), followed by the use of the shaker balance for 15 min. After equilibrium was achieved, the IPG strip was placed above the 12% uniform polyacrylamide gel, and 10-µL protein marker was added to the SDS-polyacrylamide gel. The second dimension was performed until the bromophenol blue reached the bottom edge of the gel. The temperature of the cold cycle was set to 16 °C. After protein fixation in 40% methanol and 5% phosphoric acid for 1 h, the gels were stained with Coomassie brilliant blue G-250 for 20 h. The gels were then washed by distilled water, scanned in the Ettan DIGE Imager (GE Healthcare, Buckinghamshire, UK), and converted to electronic files, which were then analyzed using PDQuest software (Bio-Rad, Hercules, CA, USA).

2.6. Protein Identification and Database Search

Significant differentially expressed protein (DEP) was identified with a fold change ≥ 2 and a p-value ≤ 0.05. The DEP spots observed by 2-DE analysis were cut from the gel and washed by distilled water. Then, these DEP spots were destained by 100 mmol/L NH_4HCO_3 and were washed by 50% acetonitrile for 5 min. Next, the obtained DEP spots were added to 100% acetonitrile for 5 min, then digested by 5 µL 50 mM NH_4HCO_3 containing 10 ng Trypsin at 37 °C for 16 h. Subsequently, 20 µL NH_4HCO_3 was used to cover the tube. Then, 5% TFA was added for 10 min to stop the reaction. The dissolved samples mixed with saturated matrix HCCA at 1:1 were loaded onto the target instrument, dried, and subjected to MALDI-TOF-TOF-MS detection. The obtained mass fingerprint data were searched in the NCBInr database, and the search engine was Mascot.

2.7. Protein Annotation and Interaction

The identified protein ID was converted to UniProt ID. The identified proteins were then mapped to the Gene Ontology (GO) database by protein UniProt ID. Unannotated proteins were annotated by using Inter ProScan software through the protein sequence alignment method. Then, all annotated proteins were classified into three categories: biological process, cellular component, and molecular function.

The pathway of the identified protein was then annotated by the Kyoto Encyclopedia of Genes and Genomes (KEGG) database. The KEGG online service tool KAAS was used to annotate the protein's KEGG description. Then, all annotated proteins were mapped to the KEGG pathway using the KEGG online service tool KEGG mapper. Moreover, wolfpsort, an updated version of PSORT/PSORT II, was used to predict the subcellular localization of all identified proteins.

All identified proteins were searched against the STRING database version 10.0 for protein-protein interactions (PPI). The interaction confidence was determined by a metric "confidence score" defined by STRING. The interaction network from STRING was visualized in Cytoscape.

2.8. Statistical Analysis

Statistical analysis was conducted using SPSS software version 19.0 and Microsoft Excel 2010. Multiple comparison analyses were performed using one-way analysis of variance (ANOVA) with Duncan's test ($p < 0.05$). Microsoft Excel 2010 and PhotoShop CS6 were used to draw the plots.

3. Results

3.1. The Response of Photosynthetic Gas Exchange Parameters of M. 'Prairifire' Leaves Exposed to High Temperature Stress

As Figure 1A shows, the value of Pn under CG treatment was 12.95 mol $CO_2 \cdot m^{-2} \cdot s^{-1}$, while 38 °C heat shock for 1 day caused Pn to drop by 72.14% ($p < 0.05$). Then, the Pn showed a trend of increasing first and then decreasing from day 2 to day 6 under heat shock treatment, and the Pn value on day 6 was significantly lower than that on day 4 ($p < 0.05$). The variation of Gs was consistent with that of Pn (Figure 1B). The lowest value of Gs was present on day 6 under heat shock, which was significantly different from that of Gs at other times ($p < 0.05$). The values of Ci showed a trend of decreasing first and then increasing (Figure 1C), and the lowest value of Ci was present on day 4 under heat shock. It was noteworthy that the Ci value on day 6 increased by 19.33% compared with that on day 4 ($p < 0.05$), showing an opposite trend to that of Pn and Gs. A significant decrease of 46.52% in Tr occurred on day 2 under heat shock treatment compared with that on day 1 ($p < 0.05$). Then, the Tr increased significantly ($p < 0.05$) and finally reached 2.39 mmol $m^{-2} \cdot s^{-1}$ on day 6 of the heat shock treatment (Figure 1D).

3.2. The Response of OJIP Curves of M. 'Prairifire' to Heat Stress

As Figure 2A shows, the fluorescence rise kinetics of control plant samples exhibited a typical O-J-I-P shape. The fluorescence rise kinetics in leaves exposed to heat stress for 1–6 days still kept a whole O-J-I-P polyphasic transient curve. The heat shock did not cause significant changes in J, and P step, nor did it show significant K step, indicating that the function of OEC, reduction of $Q_A{}^-$ and fluorescence yield were not significantly affected. However, the increase of JI and IP phase may increase the burden of electron transport chain. Moreover, a large amount of information about the donor side, the acceptor side, and the reaction center of PS II was also provided by the JIP test (Figure 2B). Among these variables, Fv/Fm showed a slight decrease under 38 °C heat shock ($p > 0.05$), while PI_{ABS} was significantly reduced under the same condition ($p < 0.05$). Sm and N showed a fluctuating trend with a large variation range, which was significantly different from CG ($p < 0.05$). φEo showed similar variation to Sm and N, but only on day 1 of the heat shock was there a significant difference ($p < 0.05$). Higher values of TR_O/RC and ET_O/RC were present on day 2 and 6, respectively, under heat shock, which were significantly different from CG ($p < 0.05$). The variation of DI_O/RC was basically consistent with TR_O/RC and ET_O/RC, and the values of DI_O/RC during the heat shock process were significantly higher than those of CG ($p < 0.05$). Moreover, Vj increased slightly at first and then decreased gradually, which was opposite to Ψ_O. RC/CS_O increased slightly on day 1 of heat shock, decreased significantly on day 2 ($p < 0.05$), and then gradually increased ($p < 0.05$). ABS/RC showed a rising trend under the heat shock process, and the values of ABS/RC during the heat shock process were significantly higher than those of CG ($p < 0.05$).

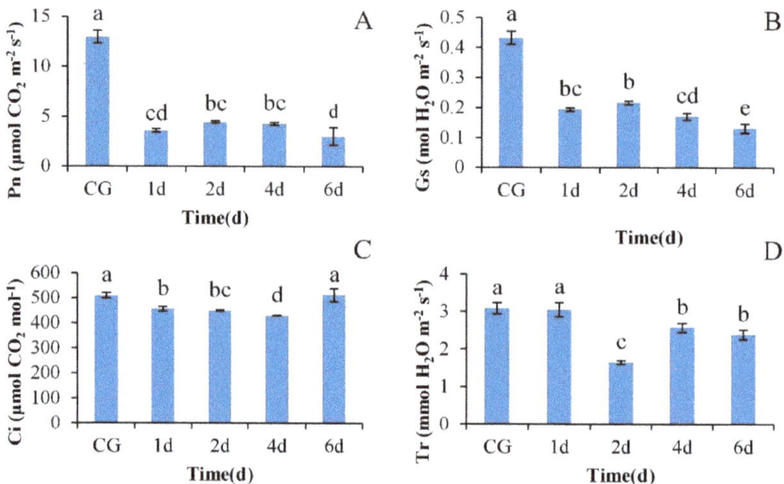

Figure 1. The changes of gas exchange parameters in Pn (**A**), Gs (**B**), Ci (**C**), Tr (**D**) in *M.* 'Prairifire' leaves exposed to 38 °C heat stress. Values are means ± SE ($n = 6$). Different lowercase letters indicate a significant difference at the 0.05 level between different treatments. CG: Plants were treated with CG conditions; 1d: Plants were treated with 38 °C heat shock for 1 day; 2d: Plants were treated with 38 °C heat shock for 2 days; 4d: Plants were treated with 38 °C heat shock for 4 days; 6d: Plants were treated with 38 °C heat shock for 6 days.

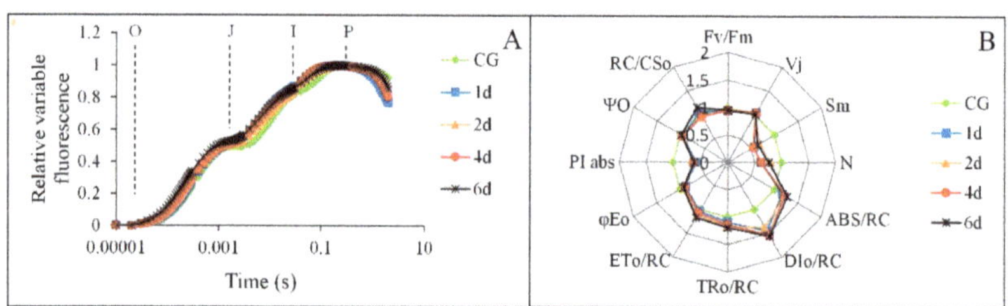

Figure 2. Effects of 38 °C heat stress on chlorophyll a fluorescence transient (**A**) and chlorophyll a fluorescence parameters (**B**) in *M.* 'Prairifire' leaves. Data are each the mean of 6 independent measurements. CG: Plants were treated with CG conditions; 1d: Plants were treated with 38 °C heat shock for 1 day; 2d: Plants were treated with 38 °C heat shock for 2 days; 4d: Plants were treated with 38 °C heat shock for 4 days; 6d: Plants were treated with 38 °C heat shock for 6 days.

3.3. Identification of DEPs in Leaves of M. 'Prairifire' between CG and Shock Treatment for 48 h

The representative maps of 2-DE in leaves of *M.* 'Prairifire' under CG and 48-h heat shock treatments are shown in Figs. 3A and 3B, respectively. The protein spots that showed large repetitive changes through automatic detection, matching, and manual editing were considered to be DEPs. A total of 38 reliable DEP spots were detected, including 19 that were down-regulated and 19 that were up-regulated (Figure 3C). All the DEP spots were extracted and identified by mass spectrometry. The obtained data were blasted with the sequence of known proteins in the NCBInr database, and 36 protein spots were identified successfully by excluding the same proteins. Therefore, compared with CG, 19 DEGs were down-regulated and 17 DEGs were up-regulated in leaves of *M.* 'Prairifire' under heat shock treatment for 48 h (Table S1).

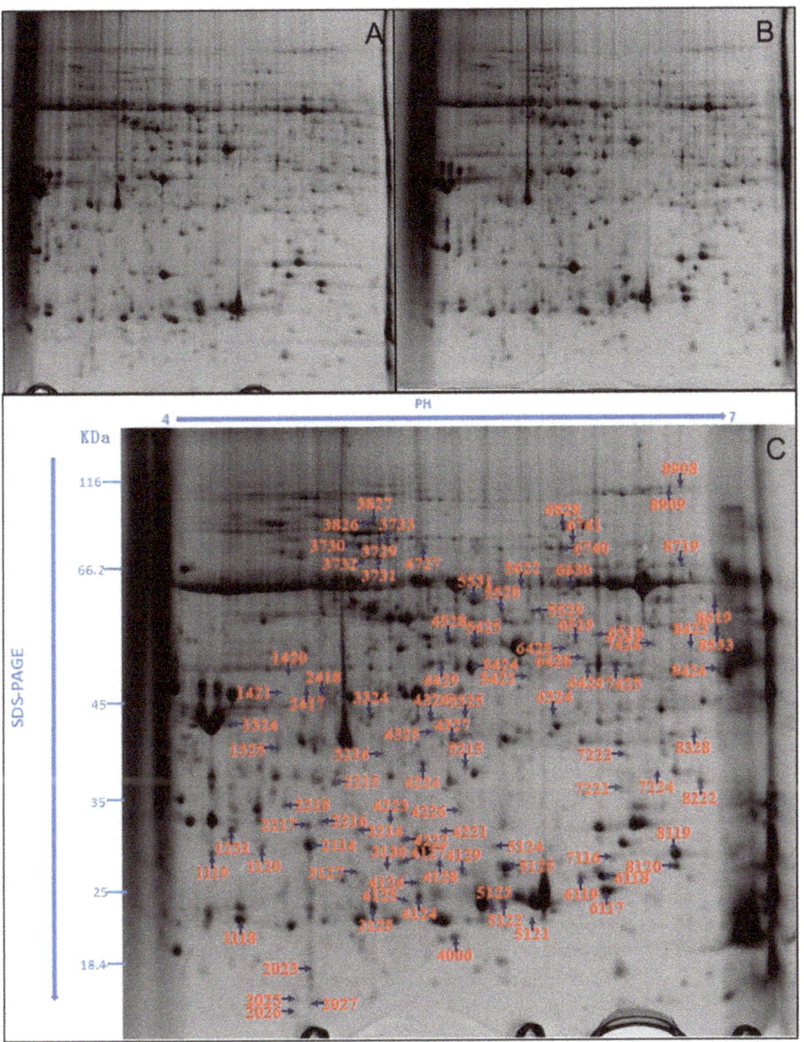

Figure 3. Representative 2-DE profiles of total proteins extracted from *M.* 'Prairifire' leaves treated with CG (**A**), and 38 °C heat shock for 48 h (**B**). (**C**) Representative 2-DE profiles of differential expressed proteins extracted from *M.* 'Prairifire' leaves between CG conditions and 38 °C heat shock for 48 h.

The GO terms representing all the DEPs in leaves of *M.* 'Prairifire' under heat shock treatment for 48 h were classified into three categories: cellular component, molecular function, and biological process (Figure 4A). In the biological process, classification, oxidative stress response, and temperature stimulus response accounted for a large proportion. The cytoplasm, chloroplasts, and plastids were the most important components in the classification of cell components. In the molecular functional classification, metal ion binding and cationic binding were the main functions.

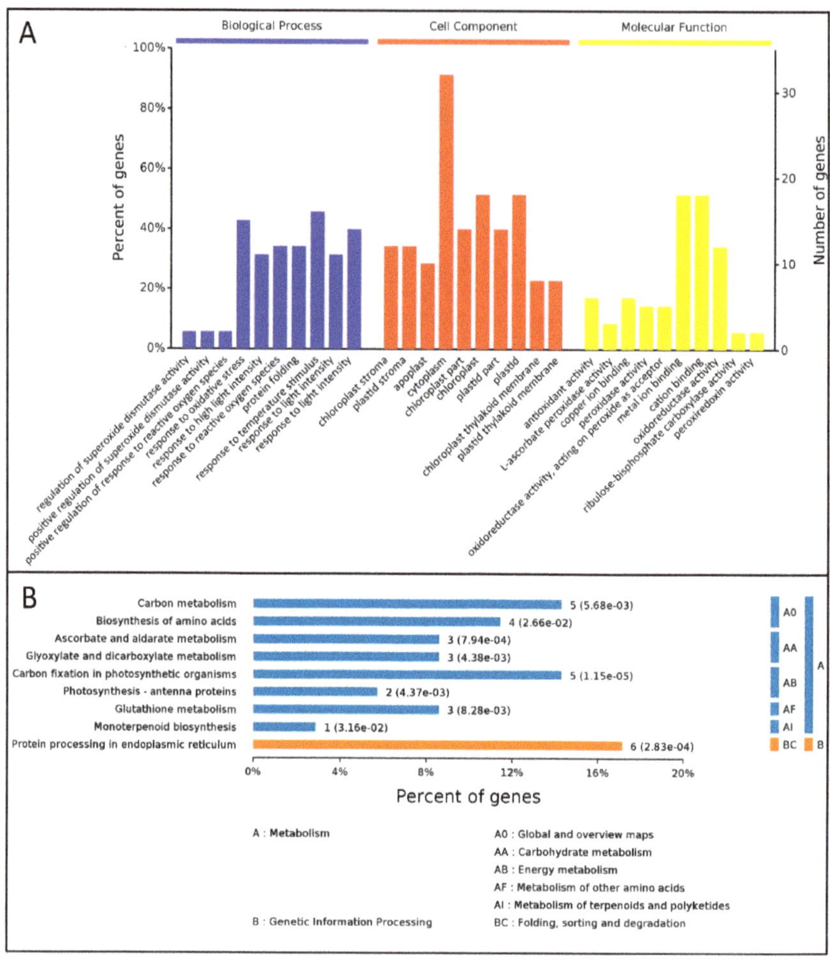

Figure 4. The function classification of GO (**A**) and KEGG pathway annotation (**B**) in *M.* 'Prairifire' leaves after 38 °C heat shock for 48 h.

Moreover, 36 DEPs were classified into various KEGG pathways, of which nine metabolic pathways were significantly enriched ($p < 0.05$) (Figure 4B). In these metabolic pathways, more proteins were enriched in the endoplasmic reticulum, followed by carbon metabolism and carbon fixation in photosynthetic organics.

3.4. PPI Network for the DEPs in Leaves of M. 'Prairifire' between CG and Heat Shock Treatment for 48 h

The PPI network for the DEPs in leaves of *M.* 'Prairifire' between CG and heat shock treatment for 48 h was mapped to understand the dynamic changes in metabolic pathways and to generate hypotheses about the relationship between DEPs (Figure 5). In the present study, the enriched pathways of the PPI network were mainly involved in the endoplasmic reticulum, carbon fixation in the photosynthetic apparatus, and photosynthetic antenna proteins. The predicted interactions of proteins mainly included heat shock proteins, oxidoreductase, and chloroplast-related proteins. Interestingly, there was a highly reliable protein interaction between 20 kDa chloroplastic chaperonin proteins (CPN20) and several chloroplast proteins, such as TK, ATP-dependent zinc metalloproteinase FTSH2 (FTSH2), and glyceraldehyde-3-phosphate dehydrogenase A (GADPH), and between CPN20 and several oxidoreductases as well as heat shock proteins (HSPs), including HSP17.6 and HSP18.1 (Table S2).

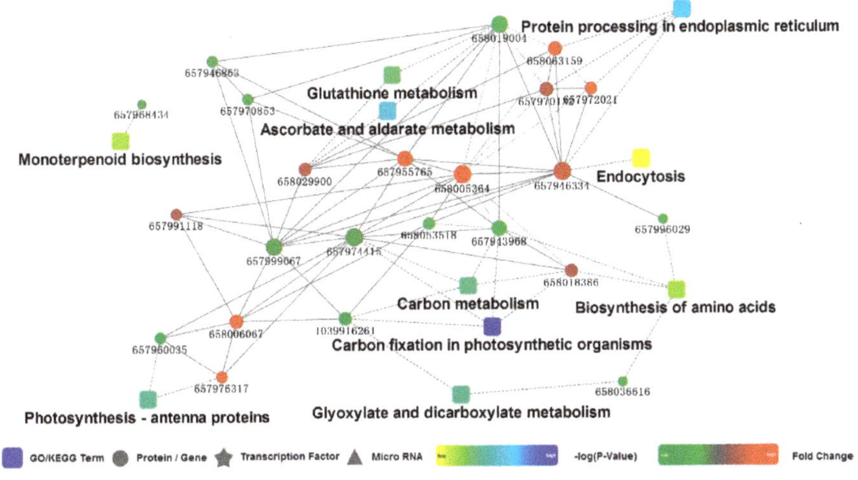

Figure 5. PPI speculation in *M.* 'Prairifire' leaves after 38 °C heat shock for 48 h.

3.5. Differential Expression of Photosynthesis-Related Proteins in Leaves of M. 'Prairifire' between CG and Heat Shock Treatment for 48 h

In the present study, a total of 14 photosynthesis-related proteins with different expressions under heat stress were obtained (Table 1). Most of these proteins were predicted to be located in the chloroplast. Among these proteins, the PS I chlorophyll a/b binding protein 3-1 (LHCA3), TK, OEE1, two CPN20 proteins, and a chloroplastic small HSP (HSP21) were up-regulated. Among these, LHCA3 and a CPN20 protein were more highly expressed under heat stress, i.e., 8.559 and 11.158 times higher, respectively, than in CG. The remaining proteins, including rubisco, ribulose diphosphate carboxylase long chain (rbcL), chlorophyll a-b binding protein of LHC II type 1 (LHCB1), 2-cysteine peroxidase BAS1 (2-Cys), FTSH2, GADPH, and thioredoxin H (Trx-H), were down-regulated. The differential expressions of the proteins mentioned above were involved in the light and dark reactions in the photosynthetic process, which indicated a unique regulation in response to heat stress.

Table 1. Differential expressed photosynthesis-related proteins in leaves of *M.* 'Prairifire' between CG and heat shock treatment for 48 h.

Number	Protein Name	gi Number	UniProt ID	Fold Change
1	PREDICTED: thioredoxin H-type	gi|657970853	P29448	0.445
2	PREDICTED: triosephosphate isomerase, cytosolic-like	gi|657943968	P48491	0.292
3	PREDICTED: 2-Cys peroxiredoxin BAS1, chloroplastic-like	gi|657999067	Q9C5R8	0.486
4	PREDICTED: ATP-dependent zinc metalloprotease FTSH 2, chloroplastic	gi|658053518	O80860	0.344
5	PREDICTED: chlorophyll a-b binding protein of LHCII type 1	gi|657960035	Q39142	0.143
6	PREDICTED: ribulose bisphosphate carboxylase large chain	gi|1039916261	O03042	0.059
7	ribulose-1,5-bisphosphate carboxylase/oxygenase large subunit, partial (plastid)	gi|817992125	O03042	0.243
8	PREDICTED: glyceraldehyde-3-phosphate dehydrogenase A, chloroplastic	gi|657974415	Q9LPW0	0.498
9	PREDICTED: photosystem I chlorophyll a/b-binding protein 3-1, chloroplastic-like	gi|657976317	Q9SY97	8.559
10	PREDICTED: small heat shock protein, chloroplastic	gi|657970132	P31170	2.356
11	PREDICTED: 20 kDa chaperonin, chloroplastic-like	gi|658005364	O65282	11.158
12	PREDICTED: 20 kDa chaperonin, chloroplastic-like	gi|657947332	O65282	2.415
13	PREDICTED: transketolase, chloroplastic, partial	gi|658018386	F4IW47	2
14	PREDICTED: oxygen-evolving enhancer protein 1, chloroplastic	gi|658006067	Q9S841	5.688

4. Discussion

Photosynthesis, one of the main means for plants to obtain energy, is very sensitive to heat stress [21]. Our results showed that heat stress induced a significant decrease in Pn in the leaves of *M.* 'Prairifire', and with the extension of heat shock duration, Pn was further inhibited. According to the variation of gas exchange parameters in the present study and the judgment basis proposed by Farquhar and Sharkey [22], stomatal limitation was considered to be an important factor that resulted in the decrease of Pn under heat stress, which indicated a sensitive stomatal behavior in response to heat stress. Combined with the changes of Tr during the heat shock process, the partial closure of stomata may provide a protective strategy for maintaining water homeostasis in leaves of *M.* 'Prairifire' in response to heat stress, that is, maintaining a strict transpiration rate to ensure nearly constant leaf water potential and relative water content [23]. However, after long-term exposure (6 days) to heat shock, the opposite trend of Ci and Pn or Gs implied an increased proportion of nonstomatal restriction, which indicated that the mesophyll cell activity of *M.* 'Prairifire' decreased gradually. To reveal the underlying mechanisms of photosynthesis in *M.* 'Prairifire' responses to heat stress, chlorophyll a fluorescence transient combined with 2-DE analysis were performed, and the findings are discussed below.

4.1. Electron Transport Chain and Related Proteins Involved in the Light Reaction

LHCs are major constituents of the antenna systems in higher plant photosystems. The peripheral antennas of PSII are composed of major trimeric and minor monomeric LHCII proteins [24], while four LHCA subunits are tightly bound to the PSI core complex, forming its outer antenna moiety called LHCI. In the present study, the expression of LHCB1 was downregulated in *M.* 'Prairifire' under heat stress, which showed an opposite trend to that of LHCA3, indicating that LHCII in leaves of *M.* 'Prairifire' was more susceptible than that of LHCI to heat stress. LHCII proteins are switched from efficient light-harvesting state to photoprotective state in response to the changes of external environmental conditions. The reversible state transitions could balance the distribution of excitation energy between PSII and PSI, and are required for the functionality of PSI. Therefore, the thermal sensitivity of LHCII in the leaves of *M.* 'Prairifire' may be a protective mechanism to prevent damage to PSI, which is arguably harder to recover from heat stress injury. Moreover, a significant decrease in Sm and N and a gradual decrease in φEo in leaves of *M.* 'Prairifire' under heat stress were measured by the JIP test, suggesting that the function of the acceptor side in PSII was inhibited under heat stress, which could be attributed to the loss of the PQ pool, and the restraint of electron transport with respect to Q_A, the primary quinone electron acceptor of PSII [8,25]. Therefore, the blocking of electron transport of PSII may become one of the main nonstomatal reasons for the decrease of photosynthetic ability in *M.* 'Prairifire' under heat stress.

However, from the JIP test, the performance of K steps under heat stress showed rare changes compared with that of CG, indicating functionally intact OEC in leaves of *M.* 'Prairifire' under heat stress. OEE is a key structural component of many different types of OECs and functions to stabilize the manganese cluster and to modulate the Ca^{2+} and Cl^- requirements for oxygen evolution [26]. Therefore, the upregulation of OEE1 in this study under heat stress could explain the activity of OEC under heat stress. Moreover, the level of Vj and Ψ_O showed little changes under heat treatment compared with the CG, indicating that the opening ratio of the active reaction center was less affected by the heat stress. With the extension of the heat shock process, ABS/RC, RC/CS$_O$, and TR$_O$/RC increased gradually, further indicating that heat stress increased the efficiency of energy utilization and consumption in the active reaction center [27,28]. These combined results showed an efficient reaction center in leaves of *M.* 'Prairifire' in response to heat stress. The improvement of heat dissipation and the efficiency of the active reaction center could explain the adaptive mechanism of the PSII reaction center under heat stress, combined with the changes of DI$_O$/RC and ET$_O$/RC [29]. Interestingly, the significant variations of PI$_{ABS}$ between CG and heat shock treatments were more sensitive than those of Fv/Fm, indicating that PI$_{ABS}$ could be selected as a key indicator reflecting the changes of the photosynthetic apparatus in *M.* 'Prairifire' under heat stress.

4.2. Proteins Involved in the Dark Reaction and Regulatory Mechanisms

Rubisco is a key functional protein involved in the dark reaction process of plants and can act as an oxygenase involved in catalyzing the first step of the plant photorespiration pathway. Moreover, it is also a carboxylase that mediates CO_2 assimilation [30]. In the present study, the levels of rubisco and rbcL under heat shock treatment were down-regulated, indicating that carbon assimilation in the leaves of *M.* 'Prairifire' was significantly inhibited due to heat stress. Moreover, the levels of GADPH and phosphotriose isomerase were down-regulated after 48 h of heat shock, further indicating that the Calvin cycle of *M.* 'Prairifire' leaves was inhibited, which may be the main reason contributing to the increased proportion of non-stomatal limitations under heat stress. Previous studies have reported oxidative damage caused to lipids, proteins, and even DNA by heat stress [31]. In the present study, the down-regulation of 2-cys and Trx-H in *M.* 'Prairifire' implied a risk of gradual accumulation of reactive oxygen species (ROS) in chloroplasts [31–33]. The down-regulation of FTSH2 further indicated a decrease of the ability to degrade photooxidatively damaged products of the D1 protein [34]. Therefore, one of the possible reasons for the down-regulation of photosynthetic regulatory protein levels under heat stress was attributed to the accumulation of ROS leading to oxidative stress.

However, TK was detected to be highly expressed after heat shock treatment, indicating that the regeneration ability of ribulose 1, 5 diphosphate were enhanced [35]. Besides of the regulation of carbon fixation in higher plants, TK is also reported to be involved in the response of higher plants to abiotic stress, thus enhancing their stress resistance [36–38]. Weber reported that increased TK activity could alter photosynthate allocation in favor of sucrose biosynthesis, and regulate the flux into phenylpropanoid metabolism [39], which conferred added heat resistance to photosynthetic apparatus. Therefore, the thermal stability of the photosynthetic metabolism in leaves of *M.* 'Prairifire' was probably due to the high level of TK, which is considered as a regulation of photosynthetic adaptability. Interestingly, according to the PPI results from STRING, CPN20 was predicted to have a highly reliable interactions with a series of photosynthesis-related proteins, such as TK and HSP21. Studies have proved the functions of chloroplastic HSPs that improve the thermotolerance of plants [40]. As a unique co-chaperonin in higher plant chloroplasts, CPN20 has also been reported to be involved in the folding of specific client proteins by forming chaperonin systems [41]. Therefore, the high level of photosynthetic proteins such as TK in the leaves of *M.* 'Prairifire' under heat stress may be results of interacting with CPN20. The expression and interaction verification of related proteins is expected to be the key to further reveal the heat resistance of *M.* 'Prairifire'.

5. Conclusions

The assessment of the results allows us to conclude that heat stress significantly inhibited Pn in leaves of *M.* 'Prairifire' through both stomatal and non-stomatal limitations. According to the changes of gas exchange parameters, a sensitive stomatal behavior in *M.* 'Prairifire' may be a protective strategy for maintaining water homeostasis under heat stress. Chlorophyll a fluorescence combined with 2-DE analysis was conducted to illustrate the response mechanism of photosynthetic apparatus to heat stress. The increased proportion of non-stomatal factors under heat stress could be derived from the sensitivity of LHCII to heat stress, the inhibition of electron transport in PSII, and the down-regulated expression of key enzymes in the Calvin cycle. It was noteworthy that, the photosynthetic apparatus of *M.* 'Prairifire' demonstrated a functionally intact OEC, which could be explained by higher levels of OEE1, and an efficient reaction center in PSII under heat stress. These combined results could responsible for strong photosynthetic adaptability in *M.* 'Prairifire' leaves. Moreover, the high level of TK under heat stress was regarded as a regulatory mechanism to provide some protection to the photosynthetic apparatus against heat stress. Interestingly, CPN20 was predicted from STRING to have a reliable interaction with TK, indicating a possible protective effect that improves the expression of photosynthetic proteins in chloroplasts under heat stress.

Supplementary Materials: The following are available online at http://www.mdpi.com/2223-7747/9/8/1040/s1, Table S1. The mass spectrometry identification of differential expressed proteins in leaves of *M.* 'Prairifire' between CK and heat shock treatment for 48 h; Table S2. PPI score.

Author Contributions: Conceptualization, T.W. and Y.X.; methodology, T.W. and S.L.; investigation, S.L., Y.M., and L.L.; data curation, T.W. and S.L.; writing—original draft preparation, T.W.; writing—review and editing, T.W.; supervision, Y.X. and W.Z.; project administration and funding acquisition, Y.X. and W.Z. All authors have read and agreed to the published version of the manuscript.

Funding: This research was supported by the Priority Academic Program Development of Jiangsu Higher Education Institutions (PAPD) and the Key Research and Development Program of Modern Agriculture in Jiangsu Province (BE2019389).

Acknowledgments: We thank LetPub (www.letpub.com) for its linguistic assistance during the preparation of this manuscript.

Conflicts of Interest: The authors declare no conflict of interest.

References

1. Wahid, A.; Gelani, S.; Ashraf, M.; Foolad, M.R. Heat tolerance in plants: An overview. *Environ. Exp. Bot.* **2007**, *61*, 199–223. [CrossRef]

2. Yuan, L.; Liu, S.; Zhu, S.; Chen, L.; Liu, F.; Zou, M.; Wang, C. Comparative response of two wucai (*Brassica campestris* L.) genotypes to heat stress on antioxidative system and cell ultrastructure in root. *Acta Physiol. Plant* **2016**, *38*, 223. [CrossRef]

3. Zhang, M.; Shan, Y.J.; Kochian, L.; Strasser, R.J.; Chen, G. Photochemical properties in flag leaves of a super-high-yielding hybrid rice and a traditional hybrid rice (*Oryza sativa* L.) probed by chlorophyll a fluorescence transient. *Photosynth. Res.* **2015**, *126*, 275–284. [CrossRef] [PubMed]

4. Bechtold, U.; Field, B. Molecular mechanisms controlling plant growth during abiotic stress. *J. Exp. Bot.* **2018**, *69*, 2753–2758. [CrossRef] [PubMed]

5. Chen, K.; Sun, X.; Amombo, E.; Zhu, Q.; Zhao, Z.; Chen, L.; Xu, Q.; Fu, J. High correlation between thermotolerance and photosystem II activity in tall fescue. *Photosynth. Res.* **2014**, *122*, 305–314. [CrossRef] [PubMed]

6. Dias, A.S.; Semedo, J.; Ramalho, J.C.; Lidon, F.C. Bread and durum wheat under heat stress: A comparative study on the photosynthetic performance. *J. Agron. Crop Sci.* **2011**, *197*, 50–56. [CrossRef]

7. Yan, K.; Chen, P.; Shao, H.; Zhang, L.; Xu, G. Effects of short-term high temperature on photosynthesis and photosystem II performance in sorghum. *J. Agron. Crop Sci.* **2011**, *197*, 400–408. [CrossRef]

8. Strasser, R.J.; Tsimilli-Michael, M.; Srivastava, A. Analysis of the chlorophyll a fluorescence transient. In *Chlorophyll a Fluorescence a Signature of Photosynthesis*; Papageorgiou, G., Govindjee, C., Eds.; Springer: Dordrecht, The Netherlands, 2004; Volume 19, pp. 321–362.

9. Chen, S.; Yang, J.; Zhang, M.; Strasser, R.J.; Qiang, S. Classification and characteristics of heat tolerance in *Ageratina adenophora* populations using fast chlorophyll a fluorescence rise OJIP. *Environ. Exp. Bot.* **2016**, *122*, 126–140. [CrossRef]

10. Chen, S.; Xu, X.; Dai, X.; Yang, C.; Qiang, S. Identification of tenuazonic acid as a novel type of natural photosystem II inhibitor binding in Q_B-site of *Chlamydomonas reinhardtii*. *BBA Bioenerg.* **2007**, *1767*, 306–318. [CrossRef]

11. Zhang, M.; Li, G.; Huang, W.; Bi, T.; Chen, G.; Tang, Z.; Su, W.; Sun, W. Proteomic study of *Carissa spinarum* in response to combined heat and drought stress. *Proteomics* **2010**, *10*, 3117–3129. [CrossRef]

12. Yang, Y.; Chen, J.; Liu, Q.; Ben, C.; Todd, C.D.; Shi, J.; Yang, Y.; Hu, X. Comparative proteomic analysis of the thermotolerant plant *Portulaca oleracea* acclimation to combined high temperature and humidity stress. *J. Proteome Res.* **2012**, *11*, 3605. [CrossRef] [PubMed]

13. Ahsan, N.; Donnart, T.; Nouri, M.Z.; Komatsu, S. Tissue-specific defense and thermo-adaptive mechanisms of soybean seedlings under heat stress revealed by proteomic approach. *J. Proteome Res.* **2010**, *9*, 4189–4204. [CrossRef] [PubMed]

14. Ferreira, S.; Hjernø, K.; Larsen, M.; Wingsle, G.; Larsen, P.; Fey, S.; Roepstorff, P.; Salomé Pais, M. Proteome profiling of *Populus euphratica* Oliv. upon heat stress. *Ann. Bot. Lond.* **2006**, *98*, 361–377. [CrossRef] [PubMed]

15. Rocco, M.; Arena, S.; Renzone, G.; Scippa, G.S.; Lomaglio, T.; Verrillo, F.; Scaloni, A.; Marra, M. Proteomic analysis of temperature stress-responsive proteins in *Arabidopsis thaliana* rosette leaves. *Mol. Biosyst.* **2013**, *9*, 1257–1267. [CrossRef] [PubMed]

16. Zhang, W.X.; Zhao, M.M.; Fan, J.J.; Zhou, T.; Chen, Y.X.; Cao, F.L. Study on relationship between pollen exine ornamentation pattern and germplasm evolution in flowering crabapple. *Sci. Rep. UK* **2017**, *7*, 39759. [CrossRef] [PubMed]

17. Liu, F.; Wang, M.; Wang, M. Phenolic compounds and antioxidant activities of flowers, leaves and fruits of five crabapple cultivars (*Malus Mill.* species). *Sci. Hortic Amst.* **2018**, *235*, 460–467. [CrossRef]

18. Zhang, J.; Liu, Y.; Bu, Y.F.; Zhang, X.; Yao, Y. Factor analysis of MYB gene expression and flavonoid affecting petal color in three crabapple cultivars. *Front. Plant Sci.* **2017**, *8*, 137. [CrossRef]

19. Liu, C.F.; Zhang, W.X.; Sun, L.; Ruan, J.Y.; Wang, T.; Cao, F.L. Effects of high temperature on photosynthesis and growth of crabapple seedling. *J. Nanjing Forest. Univ. (Nat. Sci. Ed.)* **2013**, *4*, 17–22.

20. Wan, H.; Zhang, J.; Song, T.; Tian, J.; Yao, Y. Promotion of flavonoid biosynthesis in leaves and calli of ornamental crabapple (*Malus* sp.) by high carbon to nitrogen ratios. *Front. Plant Sci.* **2015**, *6*, 673. [CrossRef]

21. Zou, M.; Yuan, L.; Zhu, S.; Liu, S.; Ge, J.; Wang, C. Effects of heat stress on photosynthetic characteristics and chloroplast ultrastructure of a heat-sensitive and heat-tolerant cultivar of wucai (*Brassica campestris* L.). *Acta Physiol. Plant.* **2017**, *39*, 30. [CrossRef]

22. Farquhar, G.D.; Sharkey, T.D. Stomatal conductance and photosynthesis. *Annu. Rev. Plant Physiol.* **1982**, *33*, 317–345. [CrossRef]

23. Sade, D.; Sade, N.; Shriki, O.; Lerner, S.; Gebremedhin, A.; Karavani, A. Water balance, hormone homeostasis, and sugar signaling are all involved in tomato resistance to tomato yellow leaf curl virus. *Plant Physiol.* **2014**, *4*, 1684–1697. [CrossRef] [PubMed]

24. Liu, G.T.; Ma, L.; Duan, W.; Wang, B.C.; Li, J.H.; Xu, H.G.; Yan, X.Q.; Yan, B.F.; Li, S.H.; Wang, L.J. Differential proteomic analysis of grapevine leaves by iTRAQ reveals responses to heat stress and subsequent recovery. *BMC Plant Biol.* **2014**, *14*, 110. [CrossRef] [PubMed]

25. Yan, K.; Chen, P.; Shao, H.; Shao, C.; Zhao, S.; Brestic, M. Dissection of photosynthetic electron transport process in sweet sorghum under heat stress. *PLoS ONE* **2013**, *8*, e62100. [CrossRef] [PubMed]

26. Roose, J.L.; Wegener, K.M.; Pakrasi, H.B. The extrinsic proteins of photosystem II. *Photosynth. Res.* **2007**, *92*, 369–387. [CrossRef] [PubMed]

27. Markou, G.; Depraetere, O.; Muylaert, K. Effect of ammonia on the photosynthetic activity of Arthrospira and Chlorella: A study on chlorophyll fluorescence and electron transport. *Algal Res.* **2016**, *16*, 449–457. [CrossRef]

28. Stirbet, A. On the relation between the Kautsky effect (chlorophyll a fluorescence induction) and photosystem II: Basics and applications of the OJIP fluorescence transient. *J. Photochem. Photobiol. B* **2011**, *104*, 236–257. [CrossRef] [PubMed]

29. Jedmowski, C.; Brüggemann, W. Imaging of fast chlorophyll fluorescence induction curve (OJIP) parameters, applied in a screening study with wild barley (*Hordeum spontaneum*) genotypes under heat stress. *J. Photochem. Photobiol. B* **2015**, *151*, 153–160. [CrossRef] [PubMed]

30. Han, F.; Chen, H.; Li, X.J.; Yang, M.F.; Liu, G.S.; Shen, S.H. A comparative proteomic analysis of rice seedlings under various high-temperature stresses. *BBA Proteins Proteom.* **2009**, *1794*, 1625–1634. [CrossRef]

31. Choudhury, F.K.; Rivero, R.M.; Blumwald, E.; Mittler, R. Reactive oxygen species, abiotic stress and stress combination. *Plant J.* **2017**, *90*, 856–867. [CrossRef]

32. Kocsy, G.; Kobrehel, K.; Szalai, G.; Duviau, M.; Buzás, Z.; Galiba, G. Abiotic stress-induced changes in glutathione and thioredoxin h, levels in maize. *Environ. Exp. Bot.* **2004**, *52*, 101–112. [CrossRef]

33. Salekdeh, G.H.; Siopongco, J.; Wade, L.J.; Ghareyazie, B.; Bennett, J. Proteomic analysis of rice leaves during drought stress and recovery. *Proteomics* **2002**, *2*, 1131–1145. [CrossRef]

34. Yoshioka, M.; Yamamoto, Y. Quality control of Photosystem II: Where and how does the degradation of the D1 protein by FtsH proteases start under light stress? -Facts and hypotheses. *J. Photochem. Photobiol. B* **2011**, *104*, 229–235. [CrossRef] [PubMed]

35. Lee, D.G.; Ahsan, N.; Lee, S.H.; Kang, K.Y.; Bahk, J.D.; Lee, I.J.; Lee, B.H. A proteomic approach in analyzing heat-responsive proteins in rice leaves. *Proteomics* **2007**, *7*, 3369–3383. [CrossRef] [PubMed]

36. Bernacchia, G.; Schwall, G.; Lottspeich, F.; Salamini, F.; Bartels, D. The transketolase gene family of the resurrection plant Craterostigma plantagineum: Differential expression during the rehydration phase. *EMBO J.* **1995**, *14*, 610. [CrossRef]

37. Bi, H.; Dong, X.; Wu, G.; Wang, M.; Ai, X. Decreased TK activity alters growth, yield and tolerance to low temperature and low light intensity in transgenic cucumber plants. *Plant Cell Rep.* **2015**, *34*, 345–354. [CrossRef]

38. Zhang, Y.; Du, H. Differential accumulation of proteins in leaves and roots associated with heat tolerance in two Kentucky bluegrass genotypes differing in heat tolerance. *Acta Physiol. Plant.* **2016**, *38*, 213. [CrossRef]

39. Weber, A.P.M. Synthesis, export and partitioning of the end products of photosynthesis. In *The Structure and Function of Plastids*; Wise, R.R., Hoober, J.K., Eds.; Springer: Dordrecht, The Netherlands, 2007; Volume 23, pp. 273–292.

40. Kim, K.H.; Alam, I.; Kim, Y.G.; Sharmin, S.A.; Lee, K.W.; Lee, S.H.; Lee, B.H. Overexpression of a chloroplast-localized small heat shock protein OsHSP26 confers enhanced tolerance against oxidative and heat stresses in tall fescue. *Biotechnol. Lett.* **2012**, *34*, 371–377. [CrossRef]

41. Tsai, Y.C.C.; Mueller-Cajar, O.; Saschenbrecker, S.; Hartl, F.U.; Hayer-Hartl, M. Chaperonin cofactors, Cpn10 and Cpn20, of green algae and plants function as hetero-oligomeric ring complexes. *J. Biol. Chem.* **2012**, *287*, 20471–20481. [CrossRef]

Article

Comparative Physiological and Proteomic Analysis Reveals Different Involvement of Proteins during Artificial Aging of Siberian Wildrye Seeds

Xiong Lei [1,2,†], Wenhui Liu [3,†], Junming Zhao [1], Minghong You [2], Chaohui Xiong [4], Yi Xiong [1], Yanli Xiong [1], Qingqing Yu [1], Shiqie Bai [2,*] and Xiao Ma [1,*]

[1] College of Animal science and Technology, Sichuan Agricultural University, Chengdu 611130, China; lxforage@126.com (X.L.); Junmingzhao163@163.com (J.Z.); xiongyi95@126.com (Y.X.); yanlimaster@126.com (Y.X.); yuqinggzu93@126.com (Q.Y.)
[2] Sichuan Academy of Grassland Science, Chengdu 611731, China; ymhturf@163.com
[3] Key Laboratory of Superior Forage Germplasm in the Qinghai-Tibetan Plateau/Qinghai Academy of Animal Science and Veterinary Medicine, Xining 810016, Qinghai, China; qhliuwenhui@163.com
[4] College of Environmental Sciences, Sichuan Agricultural University, Chengdu 611130, China; sophyx1@163.com
* Correspondence: baishiqie@126.com (S.B.); maroar@126.com (X.M.)
† These authors contributed equally to this work.

Received: 27 September 2020; Accepted: 13 October 2020; Published: 15 October 2020

Abstract: Seed aging has an important effect on the germplasm preservation and industrialized production of Siberian wildrye (*Elymus sibiricus*) in the Qinghai-Tibet Plateau. However, so far its underlying molecular mechanisms still largely remain unknown. To shed light on this topic, one-year stored seeds of *E. sibiricus* were exposed to artificial aging treatments (AAT), followed by seed vigor characteristics and physiological status monitoring. Then global proteomics analysis was undertaken by the tandem mass tags (TMT) technique, and the proteins were quantified with liquid chromatography-tandem mass spectrometry on three aging time points (0 h, 36 h and 72 h). Finally, we verified the expression of related proteins by parallel reaction monitoring (PRM). Our results demonstrated that the seed vigor decreased remarkably in response to artificial aging, but the relative ion-leakage and malondialdehyde content, superoxide anion and hydrogen peroxide showed the opposite situation. Proteomic results showed that a total of 4169 proteins were identified and quantified. Gene Ontology (GO) analysis and Kyoto Encyclopedia of Genes and Genomes (KEGG) analysis indicated that a series of key pathways including carbohydrate metabolism, lipid metabolism, and antioxidant activity were severely damaged by aging treatments. Numerous key proteins such as glyceraldehyde triphosphate glyceraldehyde dehydrogenase, succinate dehydrogenase, lipoxygenase, peroxidase, glutathione-s-transferase and late embryogenesis abundant proteins were significantly down-regulated. However, the up-regulation of the heat shock protein family has made a positive contribution to oxidative stress resistance in seeds. This study provides a useful catalog of the *E. sibiricus* proteomes with insights into the future genetic improvement of seed storability.

Keywords: *Elymus sibiricus*, seed aging; isobaric tandem mass tag labeling; reactive oxygen species; parallel reaction monitoring

---

1. Introduction

Seeds are vital organs for the survival and dispersion of plant species, as well as the fundamental materials for agricultural production and germplasm conservation. However, seeds enter an irreversible and inevitable process of deterioration and aging after natural maturity [1]. With increasing duration of storage, seed vigor gradually decreases until total loss of seed viability. This phenomenon could further result in dramatic loss in economic costs and genetic diversity [2]. Seed aging is the most common form of seed deterioration and is mainly characterized by disturbances in a variety of essential metabolic pathways and cellular biochemical processes including disruption of cellular membranes, mitochondrial dysfunction, and damage to key biological macromolecules such as proteins, lipids, nucleic acids, and carbohydrates [3,4]. Currently, the overproduction of reactive oxygen species (ROS; such as hydrogen peroxide (H_2O_2), superoxide anion (O_2^-) etc.) and their direct attack on biological macromolecules is recognized as one of dominant causes of seed aging [5–7]. As the by-product of unsaturated lipid peroxidation, malondialdehyde (MDA) is often used as an indicator of measurement of seed aging degree due to excessive accumulation during seed storage [8]. Alternatively, preventing the excessive production of ROS or scavenging the formed ROS were mainly exerted through the antioxidant defense system, which is composed of endogenous antioxidant substances (non-enzymatic system) and antioxidant enzyme systems such as peroxidase (POD) and superoxide dismutase (SOD) [9]. However, it is usually difficult to achieve a complete scavenging ROS due to high seed lipid content and unmanageable ambient storage conditions [3].

In recent years, proteomic techniques based on two-dimensional gel electrophoresis (2-DE) have been extensively employed to elucidate the mechanism of seed aging in some species, such as *Arabidopsis* [10], rice (*Oryza sativa*) [11] and oat (*Avena sativa*) [12]. These investigations declared that many differentially expressed proteins (DEPs) involved in metabolism, energy, protein synthesis, cellular defense and rescue during seed aging. For example, the protein L-isoaspartyl-O-methyltransferase (PIMT) functions to eliminate the excessive accumulation of L-isoaspartic acid and asparagine residues in seed proteins during the aging process, which reduces the sensitivity of seed aging [13]. Furthermore, the down-regulation of storage proteins leads to the inhibition of new protein synthesis and decline of seed vigor, whereas the up-regulation of heat shock proteins (HSPs) improves seed storability [6]. However, just a small number of proteins can be separated and identified by a conventional 2-DE technique because of its limited resolution. Compared to the 2-DE technique, mass spectrometry (MS)-based proteomics is a powerful tool for large-scale protein identification and quantification. Recent technological advances of liquid chromatography coupled to tandem mass spectrometry (LC-MS/MS) have permitted high-throughput detection and quantification of thousands of proteins in biological samples [14]. At present, two popular quantitative proteomic approaches have been developed including label-free quantification (LFQ) and an isobaric labeling strategy such as tandem mass tags (TMT) and isobaric tags for relative and absolute quantitation (iTRAQ) [15,16]. Additionally, the isobaric labeling approach has faster and more stable results than label-free shotgun method [17]. So far, TMT technology has been utilized to explore the molecular mechanism of seed aging in some cereal crops, such as oat [18] and wheat (*Triticum aestivum*) [17].

Siberian wildrye (*Elymus sibiricus*), as the model species of genus *Elymus* belonging to the Triticeae tribe of the Poaceae family, is one of the perennial native grass species with high nutritional value, high yield and resistance to cold and drought, which plays a vital role in forage production and restoring degraded alpine grasslands in the Tibetan Plateau [19,20]. However, the seed vigor of *E. sibiricus* is easily decreased significantly with the extension of storage time under normal local storage conditions [21]. In the eastern Tibetan Plateau, one of the major cultivation regions of Siberian wildrye, seed aging is often exacerbated because of relatively high humidity and frequent rainfalls during the seed harvest season. For instance, the percentage of seed germination at local storage in the fifth year dropped to less than 30% of the initial level (unpublished data). A few previous works about aging of *E. sibiricus* seeds mainly focused on the exploration of artificial accelerated aging conditions [22], improvement of antioxidant enzymes activity by exogenous ascorbic acid (AsA) [23] and reduced genetic integrity [24] of germplasm during seed aging. However, the types and dynamic alteration of possible DEPs involved in the seed deterioration process of *E. sibiricus* has not been extensively described. Here, we utilized the one year-stored *E. sibiricus* seeds, subjected them to artificial aging treatments on different time points, and examined their physiological, biochemical and proteomics status, which could provide basic data and technical theoretical support for the genetic improvement of seed storability and germplasm preservation of *E. sibiricus*.

2. Results

2.1. Seed Germination Characteristics and Physio-Biochemical Changes of Artificial Aging Treatments (AAT)

To learn more about phenotypic and physiological changes of *E. sibiricus* seed in response to artificial aging treatments (AAT), *E. sibiricus* seed samples with three biological replicates on 0 h, 36 h and 72 h were employed to germination characteristics and physio-biochemical testing (Figure 1). The result implied that the germination percentage of *E. sibiricus* seeds without AAT was dramatically decreased compared with that at 36 h and 72 h of ATT (Figure 1A). Furthermore, the germination potential of the *E. sibiricus* seeds also significantly decreased (Figure 1B). Moreover, the shoot length and root length (Figure 1C,I) also showed a progressive decline ($p < 0.05$). As one of major antioxidants during seed aging, enzymatic activities of CAT (catalase) also showed a significantly decreased trend ($p < 0.05$) along with the increase of the aging time (Figure 1D).

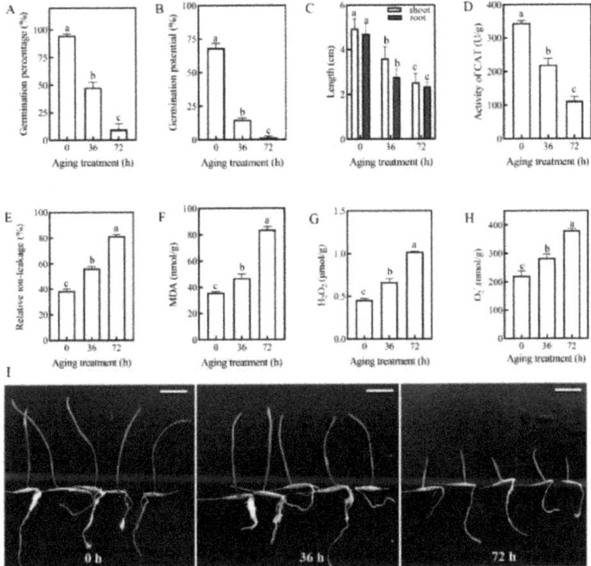

Figure 1. The seed germination characters and physio-biochemical status of *E. sibiricus* seeds with different aging treatments. (**A**) Germination percentage of *E. sibiricus* were calculated every day until 14 days; (**B**) Germination vigor in different aging time; (**C**) comparison of shoot length and root length in response to different aging treatments 14 days after germination; (**D**) the activity of catalase (CAT); (**E**) Relative ion-leakage (RIL); (**F**) malondialdehyde (MDA) content; (**G**) H_2O_2 content; (**H**) O_2^- content; (**I**) Phenotypes of *E. sibiricus* seeds applied by aging treatments (Bar = 1 cm). The different lowercase letters indicate significant differences at $p < 0.05$ and all treatments were conducted with three biological replicates.

As mentioned above, plant plasma membrane could be affected during artificial aging. To verify the integrity of the plasma membranes, the value of relative ion-leakage (RIL) and the MDA contents in *E. sibiricus* seeds with AAT on three time points (0 h, 36 h and 72 h) were measured. The results showed that both of RIL and MDA contents were increased with aging (Figure 1E,F). Furthermore, we also measured the contents of H_2O_2 and O_2^- which are considered the two major ROS in plants.

The results indicated that both H_2O_2 and O_2^- overproduced strongly with the extension of artificial aging (Figure 1G,H).

2.2. Changes in Protein Abundance after AAT

To obtain a global profile of the quantitative proteome, seeds by three biological replicates on each 0 h, 36 h and 72 h of AAT were employed with TMT labeling and quantified by LC-MS/MS. The result showed that a total of 5134 protein groups were identified at 95% confidence level, among which 4169 proteins were quantified. In order to classify DEPs, three comparison groups (36 h/0 h, 72 h/0 h, and 72 h/36 h) were constructed respectively. A total of 355 proteins (fold change ≥ 1.3 or ≤ 0.77, and $p < 0.05$) were differentially expressed in response to AAT (Table S1).

In detail, a total of 174 DEPs were identified in group 36 h/0 h consisting of 70 proteins up-regulated and 104 proteins down-regulated. In group 72 h/36 h, 16 proteins were up-regulated, and 60 were down-regulated. In group 72 h/0 h, a remarkably bigger number (276) of DEPs were identified with 47 proteins down-regulated and 229 proteins up-regulated (Figure 2A). Interestingly, in the overlapping region of the 3 comparison groups, we found only 30 DEPs including 3 proteins (A0A1D6C821, W5ECA4, W5FD68) were constantly up-regulated and 27 proteins were down-regulated in all of DEPs (Figure 2B).

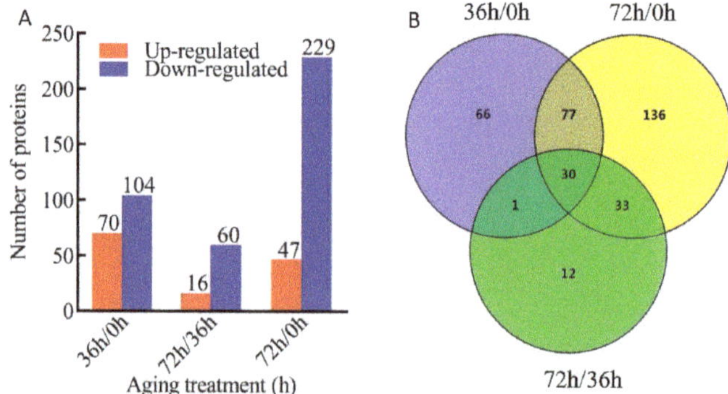

Figure 2. The profile of the differentially expressed proteins (DEPs). (**A**) Number of DEPs identified in different comparison groups with different artificial aging treatments (AAT); (**B**) Venn diagram illustrating the number of identified proteins in the 3 different comparison groups (36 h/0 h, 72 h/0 h and 72 h/36 h). The fold-change cutoff was set when proteins with quantitative ratios over 1.3-fold change was considered up-regulation while quantitative ratio less than 1/1.3-fold change was considered as down-regulation. There are three biological repeats for each treatment.

2.3. Gene Ontology (GO), Kyoto Encyclopedia of Genes and Genomes (KEGG) and Subcellular Location Analyses of Differentially Expressed Proteins (DEPs) in Response to Seed Aging

The result of Gene Ontology (GO) annotation (GO, http://www.ebi.ac.uk/GOA/, 12.19.2019) indicated that all identified DEPs were classified to 14 different biosynthetic processes in which the major categories were response to metabolic process, cellular process, single-organism process, and response to stimulus. It is worth noting that the DEPs involved in the developmental process were only detected in down-regulated DEPs in the comparison group 36 h/0 h (Figure 3A). Among cellular components, cell, membrane, organelle and macromolecular complex were the most abundant groups but the membrane-enclosed lumen was only detected in down-regulated DEPs in 72 h/36 h (Figure 3B). Additionally, in terms of molecular functions, the annotations such as binding, catalytic activity and antioxidant activity were overrepresented in DEPs (Figure 3A–C). Furthermore, the GO annotations such as antioxidant activity, electron carrier activity, transcription factor activity, and protein binding were only enriched in down-regulated DEPs identified in 36 h/0 h and 72 h/0 h comparison groups.

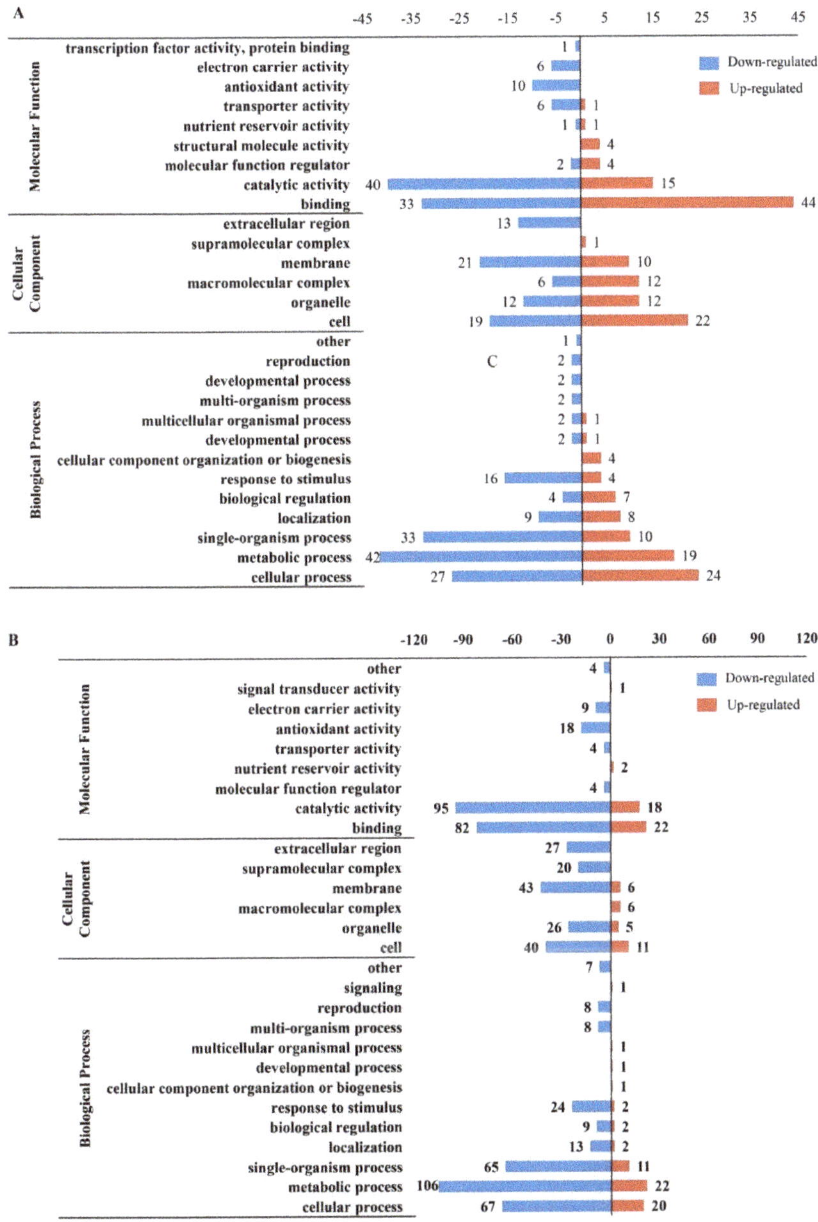

Figure 3. *Cont.*

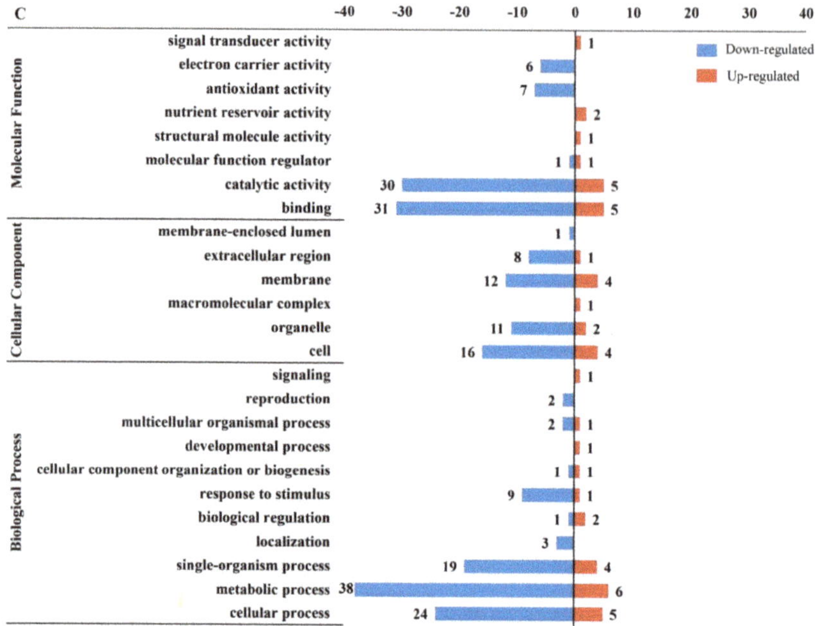

Figure 3. The functional distribution of DEPs in response to AAT by Gene Ontology (GO) level 2. The *x*-axis represents the number of enriched DEPs by GO annotation)and the note number mean protein number. (**A**) the DEPs of 36 h/0 h by GO; (**B**) the DEPs of 72 h/0 h by GO; (**C**) the DEPs of 72 h/36 h by GO.

The result of Kyoto Encyclopedia of Genes and Genomes (KEGG) analysis (KEGG, http://www.genome.jp/kegg/, 12.20.2019) showed that most up-regulated DEPs in 36 h/0 h enriched in pathways of phagosome, spliceosome, proteasome, RNA transport and protein processing in endoplasmic reticulum, whereas proteins involved in glyoxylate and dicarboxylate metabolism, phenylpropanoid biosynthesis, metabolic pathways, etc., were down-regulated. Most DEPs related to pyruvate metabolism, protein processing in endoplasmic reticulum, and carbon fixation in photosynthetic organisms increasing, whereas DEPs associated with glycerophospholipid metabolism, tyrosine metabolism, isoquinoline alkaloid biosynthesis decreased in 72 h/0 h. For the 72 h/36 h DEPs, the proteins involved in amino sugar and nucleotide sugar metabolism, metabolic pathways, etc., were up-regulated, whereas proteins in metabolic biosynthesis of phenylpropanoid, photosynthesis-antenna proteins, etc., were down-regulated (Figure 4A). Additionally, the subcellular localizations of the major DEPs were classified to be targeted to cytoplasm, chloroplasts, nucleus, vacuolar membranes and mitochondria (Figure 4B,C).

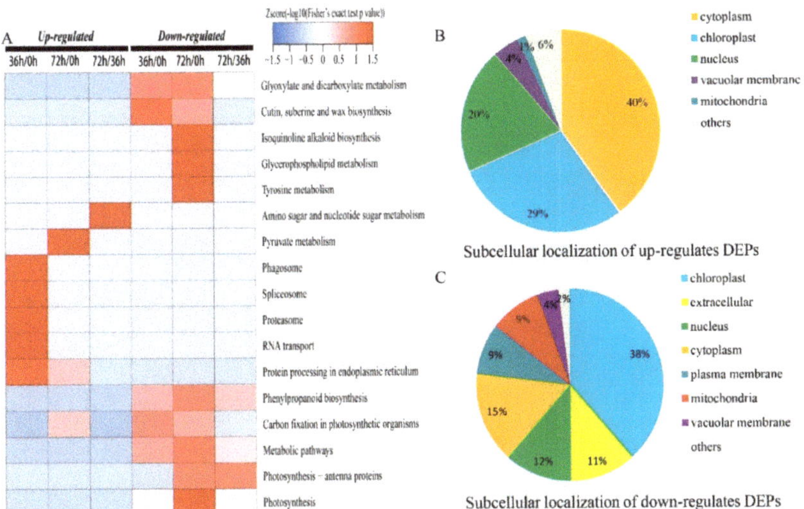

Figure 4. Kyoto Encyclopedia of Genes and Genomes (KEGG) enrichment analysis and subcellular localizations of DEPs. (**A**) the KEGG pathway; (**B**) the subcellular of up-regulated DEPs; (**C**) the subcellular of down-regulated DEPs.

2.4. Key Proteins among DEPs in Response to AAT

A number of important proteins associated with carbohydrate metabolism, lipid metabolism, antioxidant activities and stress response, dramatically changed in response to AAT. In the current study, 61 (44 down-regulated, 19 up-regulated) and 128 (106 down-regulated and 22 up-regulated) metabolic-related proteins in 36 h/0 h and 72 h/0 h were radically changed, respectively (Table S1). The DEPs of metabolic related-protein such as probable UDP-arabinose 4-epimerase 2 and Endo-beta-1,3-glucanase, which were involved in the starch and sucrose metabolism pathways, were both down-regulated remarkably. Furthermore, the down-regulation of the acyl CoA binding protein, like in the tricarboxylic acid cycle (TCA cycle), may cause insufficient energy supply during seed germination (Table 1). Furthermore, we also identified lots of antioxidant and stress response proteins in DEPs such as peroxidase, glutathione-s-transferase (GST), heat shock protein and late embryogenesis abundant protein (LEA), etc., which showed significant abundance alterations (fold change > 1.3, *p* value < 0.05) in response to seed aging (Table 1).

2.5. Candidate Functional Protein Selections and Parallel Reaction Monitoring (PRM) Validation

Among the DEPs in 36 h/0 h and 72 h/0 h, those having predicted annotations related to "metabolism", "protein binding", "oxidation-reduction process", "lipid transport" and "signaling pathways" are most likely to be responsive to AAT. Therefore, we annotated the DEPs that concentrated on the two comparisons i.e., 36 h/0 h and 72 h/0 h groups. In order to validate the TMT results, 20 DEPs were randomly selected for targeted parallel reaction monitoring (PRM) assay, of which 13 proteins had quantitative information and showed a good consistency with TMT results (r^2 = 0.898, *p* < 0.01), including 10 up-regulated and 3 down-regulated proteins in both 36 h/0 h and 72 h/0 h groups (Table 2).

Table 1. Fold changes of DEPs in key pathway in different groups.

Pathway	Accession No.	Protein Name	Log$_2$ (Fold Change)		
			36 h/0 h	72 h/0 h	72 h/36 h
glycolysis pathway	A0A1D5TTT1	Glyceraldehyde-3-phosphate dehydrogenase	−1.154c	−1.109	0.045
	A0A1D5YL69	Glyceraldehyde-3-phosphate dehydrogenase	−0.545	−1.092	−0.546
tricarboxylic acid cycle (TCA)	A0A1D6CTY7	Succinate dehydrogenase subunit 6	−0.369	−0.444	−0.075
	A0A1D6D5I2	Acyl-CoA-binding protein-like	−0.379	−0.307	0.072
	W5FCI5	Phosphoenolpyruvate carboxylase 2	0.516	0.561	0.045
starch and sucrose metabolism	W5B7S6	Probable UDP-arabinose 4-epimerase 2	−0.431	−0.484	−0.052
	Q1ERG1	Endo-beta-1,3-glucanase	−0.528	−1.234	−0.706
	A0A1D5UYF9	Glucan endo-1,3-beta-glucosidase 3-like isoform X1	−0.25	−0.387	−0.136
	W5EDL3	Fructose-bisphosphate aldolase	−0.345	−0.576	−0.23
	A0A1D5Z7K1	Predicted: beta-glucosidase BoGH3B-like	−0.318	−0.802	−0.484
	A0A1D6S663	Beta-glucosidase 31-like	−0.667	−0.76	−0.093
	A0A1D5ZSY5	Glucan endo-1,3-beta-glucosidase 4-like isoform X2	−0.165	−0.465	−0.299
	A0A1D5TQZ5	Serine/threonine-protein kinase	−0.666	−1.005	−0.339
	A0A1D6DGU9	Glucan endo-1,3-beta-glucosidase GII-like	−0.3	−0.589	−0.289
lipid metabolism	A0A1D6CGK6	Lipoxygenase	−0.536	−0.664	−0.128
	A0A1D6CX16	Non-specific lipid transfer protein GPI-anchored 1	−0.49	−0.697	−0.207
	A0A077RZ37	Glycerophosphodiester phosphodiesterase GDPD6-like	−0.175	−0.424	−0.248
	A0A0A0S1W1	12-oxo-phytodienoic acid reductase 2	0.01	−0.396	−0.405
	A0A1B5GFV1	Caleosin	−0.378	−0.828	−0.45
	A0A1B5GFV6	Caleosin	−0.548	−1.207	−0.659
	A0A1D6B2R1	Diacylglycerol kinase	−0.294	−0.406	−0.113

Table 1. *Cont.*

Pathway	Accession No.	Protein Name	Log$_2$ (Fold Change)		
			36 h/0 h	72 h/0 h	72 h/36 h
antioxidant activities	A0A1D5V3N8	Peroxidase	−0.353	−0.743	−0.391
	A0A1D5WDA8	Peroxidase	−0.946	−3.095	−2.148
	A0A1D6CQN7	Peroxidase	−0.269	−0.387	−0.118
	A0A1D6CV94	Peroxidase	−0.396	−0.717	−0.321
	W4ZYX8	Peroxidase	−0.448	−0.901	−0.453
	A0A1D5WT84	Peroxidase	−0.623	−0.896	−0.273
	W5FEC7	Peroxidase	−0.783	−0.923	−0.14
	W5G6B5	Peroxidase	−0.575	−0.997	−0.422
	A0A1D6CAM6	Probable glutathione S-transferase BZ2	−0.507	−0.855	−0.348
anti-stress protein	A0A024CKY0	LEA protein	−0.555	−0.683	−0.128
	A0A1D5ZWT3	11 kDa LEA protein-like	−0.487	−0.561	−0.073
	F4Y590	Heat shock protein 90	0.51	0.288	−0.222
	Q9SPH4	Heat shock protein 101	0.448	0.418	−0.03
	W5FD68	Chaperone protein ClpB1-like isoform X1	0.89	0.436	−0.454
	A0A1D6DGU9	Heat shock 70 kDa protein 14-like	0.407	0.155	−0.252
	A0A1D5Z4N3	Ribulose bisphosphate carboxylase small chain	−1.115	−1.428	−0.313

Note: Fold change > 1.3-fold or cut off less than 0.77-fold were considered up-regulated or down-regulated (*p* value < 0.05), respectively. Fold changes in Table 1 were carried out log$_2$ (fold change).

Table 2. Comparison between the isobaric tandem mass tag (TMT) labeling for relative quantitation and parallel reaction monitoring (PRM).

Protein Accession	Protein Description	36 h/0 h Ratio (TMT)	36 h/0 h Ratio (PRM)	72 h/0 h Ratio (TMT)	72 h/0 h Ratio (PRM)
A0A1D6B2D8	Carboxypeptidase	0.84	0.82	0.73	0.67
W5AC28	Glutathione-S-transferase DHAR2-like	0.69	0.64	0.74	0.81
Q8GVD3	Thioredoxin	0.74	0.67	0.79	0.72
A0A1D6RL87	26S proteasome non-ATPase Regulatory subunit 1 homolog A-like isoform X4	1.37	1.50	1.24	1.46
Q2PCD2	Non-specific lipid-transfer protein	0.81	0.87	0.75	0.72
A0A1D5WTN1	Predicted: transmembrane protein 214-B	0.89	0.77	0.76	0.73
A0A1D5Z9W3	Elongation factor 2-like	0.84	1.34	0.73	1.31
P11383	Ribulose bisphosphate carboxylase large chain	0.34	0.16	0.21	0.09
P12112	Adenosine triphosphate (ATP) synthase subunit alpha, chloroplastic	0.93	1.08	0.76	0.84
F4Y590	Heat shock protein 90	1.42	1.56	1.22	1.40
A0A1D5ZWT3	11 kDa late embryogenesis abundant protein-like	0.71	0.72	0.68	0.71
A0A1D6C2V8	Transmembrane protein 120 homolog	0.86	0.77	0.75	0.63
W5D591	Small ubiquitin-related modifier	0.79	0.79	0.67	0.65

3. Discussion

3.1. Changes of Carbohydrate Metabolism-Related DEPs during AAT

During long term storage of seeds, proteins related to nutrition storage, energy supply, and stress responses can be dramatically changed [1]. This change might in turn lead to a significant decrease in the germination rate, germination potential as well as shoot length and root length of artificially aged seeds (Figure 1A–C). [1] This study indicated that lots of down-regulated proteins in carbohydrate metabolism were mainly enriched in pathways of starch and sucrose metabolism, glycolysis and the TCA cycle, which led to impaired seed metabolism and energy supply, and eventually decreased seed germination and vigor.

It has been found that the seed vigor loss among wheat seed samples with different aging treatment was related to a significant reduction of sucrose (SU) and a slight enhancement of raffinose (Ra) as well as a subsequent increase in Ra/SU ratio [25]. In this study, we identified several down-regulated DEPs involving starch and sucrose metabolism pathways including UDP-arabinose 4-epimerase 2 (W5B7S6), endo-beta-1,3-glucanase (Q1ERG1), glucan endo-1,3-beta-glucosidase 3-like isoform X1 (A0A1D5UYF9), fructose-bisphosphate aldolase (W5EDL3), beta-glucosidase BoGH3B-like (A0A1D5Z7K1), beta-glucosidase 31-like (A0A1D6S663), glucan endo-1,3-beta-glucosidase 4-like isoform X2 (A0A1D5ZSY5), serine/threonine-protein kinase (A0A1D5TQZ5) and so on (Table 1). The down-regulation of these key proteins in starch and sucrose metabolism pathways might cause impairment of the energy supply chain during seed aging.

Glycolysis is the major biochemical process of the carbohydrate metabolism responsible for the conversion of glucose or glycogen to pyruvate or lactic acid in the cytoplasm, which produces a small amount of ATPs under anaerobic or hypoxic conditions [26]. Here, we observed a significant decrease in several isoforms of the glyceraldehyde triphosphate dehydrogenase (GAPDH, A0A1D5TTT1, A0A1D5YL69) of glycolysis at 36 h and 72 h compared to 0 h (Table 1). As is well known, the GAPDH transforms oxidatively phosphorylate glyceraldehyde 3-phosphate (G3P) into 1,3-bisphosphoglycerate (1,3-BPG) in the glycolysis pathway. The downregulation of GAPDH indicates an overall down-regulation of glycolysis during seed aging, and it could also be considered as an important indicator in the seed aging process of Siberian wildrye.

Moreover, numerous studies have shown that up-regulated proteins in the TCA cycle can provide energy for seeds germination [27]. In this study, lots of TCA cycle-related proteins including succinate dehydrogenase (SDH) subunit 6 (A0A1D6CTY7) in mitochondria, acyl-CoA-binding protein-like (A0A1D6D5I2) were significantly reduced, and phosphoenolpyruvate carboxylase 2 (W5FCI5) in the cytoplasm was up-regulated at 36 h and 72 h compared to 0 h (Table 1). It is reported that SDH, a marker enzyme for mitochondria, decreased in both cotyledons and embryonic axis tissues of aged sunflower seeds and leads to delayed growth of seedlings [28]. Furthermore, the SDH1 mutant reduced the activity of the electron transport chain in mitochondrial, which may cause rapid overproduction of ROS during seed germination and response to oxidative damage in Arabidopsis [29]. In consequence, down-regulated SDH enzyme during artificial aging could break redox balance between oxidative and reductive substances by ROS overproduction and then disrupted the respiratory capacity of mitochondrial, eventually resulting in poor germination vigor of aged seeds.

3.2. Changes of Lipid Metabolism-Related Proteins during AAT

Lipoxygenase (LOX) is an essential enzyme for fatty acid oxidation in plants. It not only catalyzes the production of fatty acid derivatives and ROS by phenolic glycerides, but is also strongly related to plant disease resistance and anti-injury responses [30,31]. In this study, the lipoxygenase (A0A1D6CGK6), non-specific lipid transfer protein GPI-anchored 1 (A0A1D6CX16) and diacylglycerol kinase (A0A1D6B2R1) abundance decreased continuously at 36 h and 72 h compared to 0 h. This indicated that seed aging could inhibit the process of enzymatic-lipid oxidation and reduced the defense against the harsh environment. However, the physio-biochemical measurement results indicated that both RIL and MDA increased with aging time, implying the plasma membrane was significantly destroyed (Figure 1E,F). This result also suggests that the overproduction of MDA, the final product of lipid peroxidation, was possibly produced by lipid peroxidation through non-enzymatic pathways instead of the enzymatic pathways [25,31].

3.3. Antioxidant Activities Responded to AAT

Generally, seeds under long-term storage are liable to suffer oxidative damage owing to the accumulation of ROS as well as the decrease of the antioxidant capacity in cells which accelerates the disturbances of the cellular redox homeostasis and a loss of germination ability [32]. The "redox homeostasis" is regulated by antioxidant activities species including the enzymatic (POD, SOD, CAT, etc.) and the non-enzymatic antioxidative systems such as glutathione (GSH), glutathione reductase (GR), and other antioxidants [33].

Broadly, peroxidases (PODs) are often heme-containing antioxidant enzymes, and they are divided into three superfamilies based on their structural and catalytic properties [34,35]. In particular, the three peroxidase superfamilies from plant source mainly consist of catalases (CAT), ascorbate peroxidase (APx), peroxidase (POX, EC 1.11.17, special indications of plant peroxidase) and glutathione peroxidase (GPx), which efficiently protect plants against damage by removing ROS in various biological processes [34]. Several studies have shown that the significant decrease activity of peroxidases is one of the landmark events of seed aging in rice [33], oat [36], and other crops. In the present study, we identified several down-regulated peroxidases such as A0A1D5V3N8, A0A1D5WDA8, A0A1D6CQN7, A0A1D6CV94, W4ZYX8, W5FEC7 and W5G6B5 (Table 1). Furthermore, we also found that the activity of CAT decreased dramatically in response to AAT (Figure 1D). Antioxidant enzymes such as CAT and POX are responsive to stresses when scavenging extra ROS to maintain the intracellular oxidative balance [37]. However, ROS overproduction increases lipid peroxidation and inhibits activities of antioxidant enzymes, triggering the inability of the cells to completely scavenge the radicals [38]. This may be part of the reason why the content of H_2O_2 and O_2^- had a positive trend with CAT activities over aging time of *E. sibiricus* (Figure 1D,G,H). Additionally, glutathione-s-transferase (GST) is a key enzyme that catalyzes the reaction of GSH and substrates and takes an essential role in primary and secondary metabolisms, stress metabolism, detoxification as well as seedling development [39,40]. It is worth noting that the overexpression of the GST gene such as *PpGST* in *Pyrus pyrifolia* and *GmGSTU4* in soybean significantly scavenged excess ROS and improved plant resistance to drought and salt [41]. Moreover, Cheng et al. [42] indicated that the dysfunction of GST, a member of the ascorbate-glutathione (AsA-GSH) antioxidant system, is attributed to the main cause of loss of seed vigor in artificially aged seed of oats. In this study, we found that the probable glutathione-s-transferase BZ2 (A0A1D6CAM6) showed a significant decrease at 36 h and 72 h compared to 0 h post AAT (Table 1). This result indicated that the reduction of GST may also cause excessive accumulation of ROS under oxidative stress, eventually resulting in decreased seed vigor and even seed death.

3.4. The Role of Stress-Related Proteins during AAT

LEA proteins are a group of key hydrophilic proteins during plant growth and seed development, which have been related to plant response to stresses including protecting cellular structures and improving cell thermal stability, and serving as a molecular chaperone protein to resist cell damages [43]. Wang et al. indicated that the overexpression of *OsLEA* enhanced the resistance to drought stress at seedling stage of rice [44]. Hundertmark et al. confirmed that the decrease of LEA proteins degraded the seed longevity in *Arabidopsis* seeds [45]. In this case, LEA protein (A0A024CKY0) and an isoform of 11 kDa LEA protein-like (A0A1D5ZWT3) were down-regulated at 36 h and 72 h compared to 0 h post AAT, implying that these LEAs might be associated with decreased seed vigor and germination in aged seeds of *E. sibiricus*.

Heat shock proteins (HSP), mainly located in the cytoplasm under normal physiological conditions, are widely distributed in animals and plants, including HSP 90, HSP 101, HSP 70, etc. [46]. When plants are subjected to high temperature stress, HSP is significantly up-regulated to reduce the excessive oxidative stress and maintain the correct spatial conformation of protein [47]. It is worth noting that the HSP 90 could regulate temperature-dependent seedling growth through stabilizing the auxin co-receptor F-box protein TIR1 in *Arabidopsis* [48]. Furthermore, the overexpression of HSP 101 improved the resistance to high temperatures in transgenic *Arabidopsis* seedlings [49]. In this study, the up-regulated HSP 90 (F4Y590), HSP 101 (Q9SPH4), chaperone protein ClpB1-like isoform X1 (W5FD68) and heat shock 70 kDa protein 14-like (A0A1D6DGU9) were observed at 36 h and 72 h compared to 0 h post AAT respectively (Table 1). These results demonstrated that the HSPs could improve the resistance of Siberian wildrye seeds under high temperature and humidity conditions, as confirmed by some similar investigations in *Arabidopsis* [47], oat [12] and other species [50].

Ribulose-1,5-bisphosphate carboxylase/oxygenase (RuBisCO), which localized to the stroma of chloroplast, is a principal enzyme for CO_2 fixation reaction in the Calvin cycle of plant photosynthesis [51]. Overexpression of RuBisCO increased the biomass productivity and the resistance to high temperature in seedlings of microalgae (*Nannochloropsis* spp.) and *Arabidopsis* [52]. However, when plants are subjected to stresses such as high temperature or tropospheric ozone (O_3) etc., the activities or content of RuBisCO could be significantly decreased [53]. Here, down-regulation of the RuBisCO small chain (A0A1D5Z4N3) was observed at 36 h and 72 h compared to 0 h post AAT (Table 1), implying AAT could restrain the RuBisCO activity and reduce the plant biomass and shoot length and root length (Figure 1C,I).

4. Materials and Methods

4.1. Plant Materials

The seeds of Siberian wildrye (*Elymus sibiricus* L. cv. 'Chuancao No.2') were harvested in Hongyuan county, Aba prefecture, Sichuan province of China in September 2018 (102.5441 E, 32.7734 W, 3500 m), which is located in the southeast of Qinghai-Tibetan Plateau. The original germination rate and moisture content was 95.01% ± 0.32 and 8.31% ± 0.25, respectively. The seeds with similar size and weight were selected and then stored at −20 °C.

4.2. Artificial Aging Treatment and Germination Test

AAT and the germination test were employed with the method described in the previous research with little modification [12]. Briefly, 150 g seeds which were packed in an aging chamber (LH-160, Zhejiang TOP Yunnong Technology Co., Ltd., Hangzhou, China) maintained at 45 ± 1 °C and 90 ± 1% relative humidity (RH), for 0 h, 36 h and 72 h. Each treatment had three replications containing 100 seeds, which were moistened to saturation with distilled water. Then the seeds were cultured in a germination chamber (GXZ, Ningbo, China) with a temperature of 25 °C and 70% relative humidity with 8/16 h light/darkness. Seed germination rate was calculated every day until 14 days and 10 seedlings of each replicate were randomly selected to measure the shoot length and root length after germinating 14 days. The calculation criterion of all the indexes was as follows. The germination rate was the proportion of seedlings to the number of seeds. The germination vigor was calculated based on the proportion of seedlings in the seventh day to the total number of seeds. Finally, the 0 h, 36 h and 72 h aged seeds, were stored at −80 °C for further analysis. All samples were three biological repeats.

4.3. Monitoring the Relative Ion-Leakage and Malondialdehyde Content

The value of RIL was measured as described previously [6]. The relative ion-leakage (%) = (A2 − A1)/(A3 − A1) × 100%. In the formula, A1 represented the initial conductivity. A2 was detected after the seed was shaken for 1 day at indoor temperature (DDS-307, Shanghai, China). A3 was the absolute conductivity of the seed after boiling in a water bath for 1 h.

The MDA content was measured using a TBA (thiobarbituric acid) method. Briefly, the seeds (0.1 g) were ground, homogenized in 10 mL 10% trichloroacetic acid (TCA), and centrifuged at 1500× *g* for 10 min. Then 2 mL supernatant was isovolumetric mixed with 0.6% TBA and boiled at 100 °C for 10 min. The absorbance values at 532 nm and 600 nm were used to calculate the MDA content. Each sample had three biological repeats.

4.4. Measurement of Catalase (CAT) Activity and Reactive Oxygen Species (ROS) Content

The activities of catalase (CAT) were examined using CAT-2-Y (Solarbio, Beijing, China). Briefly, the seeds were grinded on ice, mixed with 50 mmol PBS buffer (pH = 7.8), and then centrifuged at $2500\times g$ for 10 min. Finally, the absorbance value obtained from supernatants were used to calculate the activity of CAT.

The measurement of O_2^- and H_2O_2 content was based on Gong's method [54]. Briefly, the O_2^- reacts with MSDS (hydroxylamine hydrochloride) to generate NO^{2-}. Then a red azo compound is generated with the reaction of p-aminobenzenesulfonic acid and α-naphthylamine, which possesses an absorption peak at 530 nm. The yellow titanium peroxide composite generated by the reaction of H_2O_2 which has an absorption peak at 415 nm is used to calculate the H_2O_2 content. Each sample has three biological repeats.

4.5. Protein Extraction and Trypsin Digestion

The protein extraction and trypsin digestion were carried out as previous studies with appropriate modifications [17]. Briefly, aged seeds at 0 h, 36 h and 72 h were ground by liquid nitrogen and sonicated for three times on ice using a high intensity ultrasonic processor (Scientz, Zhejiang, China) in lysis buffer (8 mol/L urea, 2 mmol/L ethylenediaminetetraacetic acid (EDTA), 10 mmol/L dithiothreitol (DTT) and 1% Protease Inhibitor Cocktail), and then centrifuged at $20,000\times g$ at 4 °C for 10 min. Finally, the protein was reacted with pre-cooled 20% TCA for 2 h at 4 °C. After being cleaned completely with cold acetone, the protein was dissolved with 100 mmol/L TEAB buffer (pH 8.0). Then, the concentration of protein was determined with a 2-D Quant kit. For protein digestion, the protein solution was digested with 5 mmol/L DTT for 30 min at 56 °C and 11 mmol/L iodoacetamide (IAA) for 15 min. Finally, the diluted protein sample was digestion by 1:50 trypsin-to-protein mass ratio for the first digestion overnight and 1:100 trypsin-to-protein mass ratio for a second 4 h-digestion. Each sample has three biological replicates.

4.6. Tandem Mass Tags (TMT) Labeling, High-Performance Mass Chromatography (HPLC) Fractionation and Liquid Chromatography-Tandem Mass Spectrometry (LC-MS/MS) Analysis

Peptide reconstitution was performed in 0.5 mol/L TEAB using a 10-plex TMT kit (Thermo, Waltham, MA, USA). Follow by fractionated into fractions by high pH reverse-phase high-performance mass chromatography (HPLC) using Agilent 300Extend C18 column (5 μmol particles, 4.6 mm ID, 250 mm length). Finally, the peptides were dissolved in 0.1% formic acid (FA) for LC-MS/MS analysis by loading onto a constant flow rate of 700 nL/min on an EASY-nLC 1000 ultra-performance liquid chromatography (UPLC) system of reversed-phase analytical column (Thermo, Waltham, MA, USA).

For the LC-MS/MS analysis, the peptides were subjected to NSI source followed by tandem mass spectrometry (MS/MS) in Orbitrap Fusion TM (Thermo) coupled online to the UPLC. Peptides were selected for MS/MS using the normalized collision energy (NCE) setting as 35 while intact peptides and ion fragments were detected in the Orbitrap at a resolution of 60,000 and 15,000, respectively. A data-dependent procedure that alternated between one MS scan followed by 20 MS/MS scans was applied for the top 20 precursor ions above a threshold intensity greater than 5E3 in the MS survey scan with 30.0 s dynamic exclusion. The electrospray voltage applied was 2.0 kV. Automatic gain control (AGC) was used to prevent overfilling of the orbitrap; 5E4 ions were accumulated for generation of MS/MS spectra. For MS scans, the m/z scan range was 350 to 1550. Fixed first mass was set as 100 m/z.

4.7. MS/MS Database Search and Bioinformatics Analysis

The resulting MS/MS data were processed using MaxQuant with integrated Andromeda search engine (v.1.5.2.8). Tandem mass spectra were searched against Wheat (*Triticum aestivum*) database (www.uniprot.org/taxonomy/4565) concatenated with reverse decoy database. Trypsin/P was specified as cleavage enzyme allowing up to 2 missing cleavages. Mass error was set to 10 ppm (parts per million) for precursor ions and 0.02 Da for fragment ions. The thresholds of unique peptide were determined by false discovery rate (FDR) < 0.01. All the other parameters in MaxQuant were set to default values. For protein quantification, the protein ratios are calculated as the median of only unique peptides of the protein.

Proteins was considered as a DEP if its fold change was >1.3 (up-regulated protein) or <0.77 (down-regulated protein) and its p value < 0.05 (Student's t-test). The mass spectrometry proteomics data have been deposited to the ProteomeXchange Consortium via the Proteomics Identification Database (PRIDE) partner repository with the dataset identifier PXD021552.

For the bioinformatics analysis, the DEPs of GO annotation were derived from the UniProt-GOA database. Then DEPs were classified by Gene Ontology annotation based on three categories: biological process, cellular component and molecular function. For the KEGG pathway annotation, we first used the KEGG online service tool KEGG Automatic Annotation Server (KAAS) to annotate protein's KEGG database description. Then we mapped the annotation result on the KEGG pathway database using the KEGG online service tool KEGG mapper. Subcellular localization was conducted by wolfpsort (wolfpsort, http://wolfpsort.seq.cbrc.jp/, 12.20.2019) predication soft. Cluster membership was visualized by a heat map using the "heatmap.2" function from the "gplots" R-package.

4.8. Parallel Reaction Monitoring Validations

To verify the protein expression levels obtained by TMT analysis, 20 differentially abundant proteins (unique peptides≥ 2, fold change > 1.3 or < 0.77) were randomly chosen based on the TMT results and further quantified by the PRM assay according to Xu's method [55] at Jingjie PTM-Biolab Co., Ltd. (Hangzhou, China).

Briefly, peptides were prepared as described above for the TMT assay. These obtained peptide mixtures were subjected to Nitrogen soluble index (NSI) source followed by tandem mass spectrometry (MS/MS) in Q ExactiveTM Plus (Thermo) coupled online to the UPLC. A full MS was performed in the Orbitrap at a resolution of 70,000 (AGC target at 3E6; the maximum injection time at 50 ms and the m/z range was 350–1200), followed by 20 MS/MS scans on the Orbitrap at a resolution of 17,500 (AGC target was 1E5, and the maximum injection time was 100 ms). Mass window for precursor ion selection was 1.6 m/z. The isolation window for MS/MS was set at 2.0 m/z. The NCE was 27% with high energy collision dissociation (HCD). The FDR was set to 0.01 for the proteins and peptides. The resulting MS data were processed using Skyline (v.3.6) program as described before [56]. An analysis of three biological replicates was performed for each sample (0 h, 36 h and 72 h groups, respectively).

4.9. Statistical Analysis

SPSS19.0 software was used to test the variance of vitality parameters, and physiological and chemical indicators between different treatments. All results are shown as the mean ± standard deviation (SD) and the least significant difference (LSD) of the mean by the Duncan test ($p < 0.05$).

5. Conclusions

In summary, a model of AAT-mediated stress response in *E. sibiricus* was proposed (Figure 5). AAT induces ROS accumulation, osmotic pressure, and cell damage in *E. sibiricus* seeds, which significantly reduces the germination rate and seed vigor, and severely inhibits the embryos' development and seedling growth. Meanwhile, the proteomic analysis revealed changes in a variety of key proteins involved in carbohydrate metabolism, lipid metabolism, antioxidative systems and stress response induced by AAT, which in turn caused energy deficits and imbalance between ROS and the antioxidative defense system. These results provide basic information for exploring the molecular mechanisms that influence the seed aging of *E. sibiricus*. Nevertheless, a challenge for future work will be to elucidate the complex regulation network among those identified DEPs and their function in the context of artificial aging and natural aging.

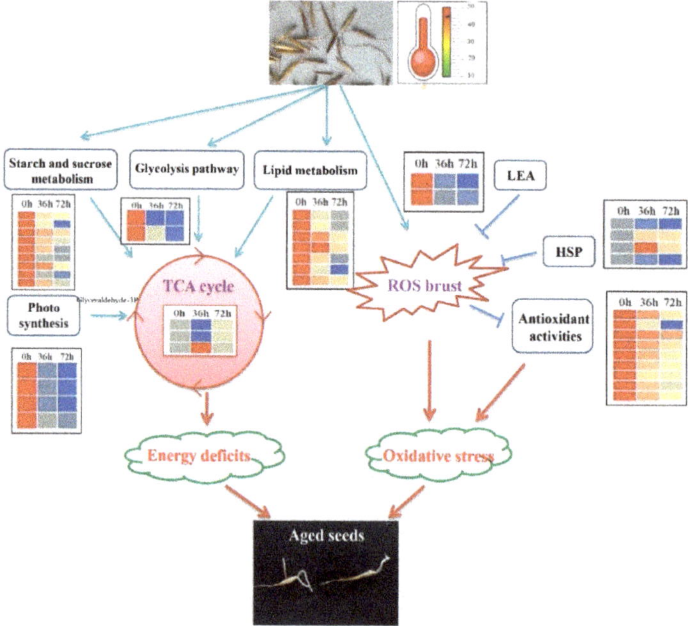

Figure 5. The overview of differentially modulated proteins during AAT. Reactive oxygen species, ROS; Late embryogenesis abundant proteins, LEA; Heat shock proteins, HSP; tricarboxylic acid cycle, TCA cycle.

Supplementary Materials: The following are available online at http://www.mdpi.com/2223-7747/9/10/1370/s1, Table S1: The DEPs of three groups of Siberian wildrye.

Author Contributions: Conceptualization, W.L., J.Z., M.Y., C.X., Y.X. (Yi Xiong), Y.X. (Yanli Xiong) and S.B.; Data curation, Y.X. (Yanli Xiong) and X.M.; Formal analysis, Y.X. (Yi Xiong) and Q.Y.; Funding acquisition, W.L., S.B. and X.M.; Investigation, J.Z., C.X. and Y.X. (Yanli Xiong); Methodology, Y.X. (Yi Xiong) and Q.Y.; Project administration, X.L., S.B. and X.M.; Resources, M.Y.; Software, M.Y.; Supervision, C.X. and Y.X. (Yanli Xiong); Visualization, J.Z.; Writing—review and editing, W.L., J.Z., M.Y., Y.X (Yanli Xiong)., Q.Y. and X.M. All authors have read and agreed to the published version of the manuscript.

Funding: This work was financially supported by the Key Research and Development Program of Sichuan Province; the Open subject of the Key Laboratory of Utilization of Excellent Forage Germplasm Resources in Qinghai-Tibet Plateau of Qinghai Province (2020-ZJ-Y03); National Forage Industry Technology System-Aba Comprehensive Experimental Station (CARS-34) and National Natural Science Foundation of China (3177131276).

Acknowledgments: We thank test support of laboratory staff in the Department of Grassland Science, Animal Science and Technology College, Sichuan Agricultural University.

Conflicts of Interest: The authors declare no conflict of interest.

Abbreviations

ABA	abscisic acid
CAT	catalase
GA3	gibberellin
GAPDH	glyceraldehyde triphosphate glyceraldehyde dehydrogenase
GO	Gene Ontology
GST	glutathione-S-transferase
H_2O_2	hydrogen peroxide
Hsps	heat shock proteins
KEGG	Kyoto Encyclopedia of Genes and Genomes
LC-MS/MS	liquid chromatography-tandem mass spectrometry
LEAs	late embryogenesis abundant proteins
LOX	lipoxygenase
MDA	malondialdehyde
O_2^-	superoxide anion
-OH	hydroxyl radicals
PRM	parallel reaction monitoring
RIL	relative ion-leakage
ROS	reactive oxygen species
RuBisCO	Ribulose-1,5-bisphosphate carboxylase/oxygenase
SDH	succinate dehydrogenase
SOD	superoxide dismutase
TMT	tandem mass tags

References

1. Zhou, W.; Chen, F.; Luo, X.; Dai, Y.; Yang, Y.; Zheng, C.; Yang, W.; Shu, K. A matter of life and death: Molecular, physiological, and environmental regulation of seed longevity. *Plant Cell Environ.* **2019**, *43*, 293–302. [CrossRef]
2. Dantas, A.F.; Fascineli, M.L.; José, S.C.B.R.; Pádua, J.G.; Grisolia, C.K. Loss of genetic integrity in artificially aged seed lots of rice (*Oryza sativa* L.) and common bean (*Phaseolus vulgaris* L.). *Mutat. Res. Genet. Toxicol. Environ.* **2019**, *846*, 403080. [CrossRef]
3. Liu, X.; Chen, Z.; Gao, Y.; Liu, Q.; Zhou, W.; Zhao, T.; Jiang, W.; Cui, X.; Cui, J.; Wang, Q. Combinative effects of *Azospirillum brasilense* inoculation and chemical priming on germination behavior and seedling growth in aged grass seeds. *PLoS ONE* **2019**, *14*, e0210453. [CrossRef]
4. Yan, H.; Jia, S.; Mao, P. Melatonin priming alleviates aging-induced germination inhibition by regulating β-oxidation, protein translation, and antioxidant metabolism in oat (*Avena sativa* L.) seeds. *Int. J. Mol. Sci.* **2020**, *21*, 1898. [CrossRef]
5. Parkhey, S.; Naithani, S.C.; Keshavkant, S. Protein metabolism during natural ageing in desiccating recalcitrant seeds of *Shorea robusta*. *Acta Physiol. Plant.* **2014**, *36*, 1649–1659. [CrossRef]
6. Yin, X.; He, D.; Gupta, R.; Yang, P. Physiological and proteomic analyses on artificially aged *Brassica napus* seed. *Front. Plant Sci.* **2015**, *6*, 112. [CrossRef]
7. Kurek, K.; Plitta-Michalak, B.; Ratajczak, E. Reactive oxygen species as potential drivers of the seed aging process. *Plants* **2019**, *8*, 174. [CrossRef]
8. Liu, J.; Wang, Q.; Karagić, D.; Liu, X.; Cui, J.; Gui, J.; Gu, M.; Gao, W. Effects of ultrasonication on increased germination and improved seedling growth of aged grass seeds of tall fescue and Russian wildrye. *Sci. Rep. UK* **2016**, *6*, 22403. [CrossRef]
9. Khan, F.; Hussain, S.; Tanveer, M.; Khan, S.; Hussain, H.A.; Iqbal, B.; Geng, M. Coordinated effects of lead toxicity and nutrient deprivation on growth, oxidative status, and elemental composition of primed and non-primed rice seedlings. *Environ. Sci. Pollut. Res.* **2018**, *25*, 21185–21194. [CrossRef]
10. Rajjou, L.; Lovigny, Y.; Groot, S.P.C.; Belghazi, M.; Job, C.; Job, D. Proteome-wide characterization of seed aging in Arabidopsis: A comparison between artificial and natural aging protocols. *Plant Physiol.* **2008**, *148*, 620–641. [CrossRef]

11. Gao, J.; Fu, H.; Zhou, X.; Chen, Z.; Luo, Y.; Cui, B.; Chen, G.; Liu, J. Comparative proteomic analysis of seed embryo proteins associated with seed storability in rice (*Oryza sativa* L) during natural aging. *Plant Physiol. Bioch.* **2016**, *103*, 31–44. [CrossRef] [PubMed]

12. Chen, L.; Chen, Q.; Kong, L.; Xia, F.; Yan, H.; Zhu, Y.; Mao, P. Proteomic and physiological analysis of the response of oat (*Avena sativa*) seeds to heat stress under different moisture conditions. *Front. Plant Sci.* **2016**, *7*, 896. [CrossRef] [PubMed]

13. Petla, B.P.; Kamble, N.U.; Kumar, M.; Verma, P.; Ghosh, S.; Singh, A.; Rao, V.; Salvi, P.; Kaur, H.; Saxena, S.C.; et al. Rice proteinl L-isoaspartyl methyltransferase isoforms differentially accumulate during seed maturation to restrict deleterious isoAsp and reactive oxygen species accumulation and are implicated in seed vigor and longevity. *New Phytol.* **2016**, *211*, 627–645. [CrossRef] [PubMed]

14. Kawashima, Y.; Watanabe, E.; Umeyama, T.; Nakajima, D.; Hattori, M.; Honda, K.; Ohara, O. Optimization of data-independent acquisition mass spectrometry for deep and highly sensitive proteomic analysis. *Int. J. Mol. Sci.* **2019**, *20*, 5932. [CrossRef]

15. Van Ulsen, P.; Kuhn, K.; Prinz, T.; Legner, H.; Schmid, P.; Baumann, C.; Tommassen, J. Identification of proteins of *Neisseria meningitidis* induced under iron-limiting conditions using the isobaric tandem mass tag (TMT) labeling approach. *Proteomics (Weinheim)* **2009**, *9*, 1771–1781. [CrossRef] [PubMed]

16. Li, Z.; Zhang, Y.; Xu, Y.; Zhang, X.; Peng, Y.; Ma, X.; Huang, L.; Yan, Y. Physiological and iTRAQ-based proteomic analyses reveal the function of spermidine on improving drought tolerance in white clover. *J. Proteome Res.* **2016**, *15*, 1563–1579. [CrossRef]

17. Lv, Y.; Zhang, S.; Wang, J.; Hu, Y. Quantitative proteomic analysis of wheat seeds during artificial ageing and priming using the isobaric tandem mass tag labeling. *PLoS ONE* **2016**, *11*, e0162851. [CrossRef]

18. Zhao, Z.; Liu, J.; Jia, R.; Bao, S.; Chen, X. Physiological and TMT-based proteomic analysis of oat early seedlings in response to alkali stress. *J. Proteom.* **2019**, *193*, 10–26. [CrossRef]

19. Ma, X.; Zhang, X.; Zhou, Y.; Bai, S.; Liu, W. Assessing genetic diversity of *Elymus sibiricus* (Poaceae: Triticeae) populations from Qinghai-Tibet Plateau by ISSR markers. *Biochem. Syst. Ecol.* **2008**, *36*, 514–522. [CrossRef]

20. Li, P.; Bai, S.; You, M.; Shen, Y. Effects of maturity stage and lactic acid bacteria on the fermentation quality and aerobic stability of Siberian wildrye silage. *Food Sci. Nutr.* **2016**, *4*, 664–670. [CrossRef]

21. Yan, H.; Mao, P.; Sun, Y.; Li, M. Impacts of ascorbic acid on germination, antioxidant enzymes and ultrastructure of embryo cells of aged *Elymus sibiricus* seeds with different moisture contents. *Int. J. Agric. Biol.* **2015**, *18*, 176–183. [CrossRef]

22. Yan, H.; Mao, P. Optimizing the accelerated ageing condition of Siberian wildrye seeds. *Seed* **2013**, *32*, 1–6.

23. Xia, F.; Cheng, H.; Chen, L.; Zhu, H.; Mao, P.; Wang, M. Influence of exogenous ascorbic acid and glutathione priming on mitochondrial structural and functional systems to alleviate aging damage in oat seeds. *BMC Plant Biol.* **2020**, *20*, 104. [CrossRef]

24. Huang, F.; Li, J.; Li, H.; Liu, L.; Shi, W.; Li, Z. Assessment of genetic integrity of Siberian wild rye (*Elymus sibiricus* L.) germplasm conserved ex situ and under accelerated aging using microsatellite markers. *Genet. Resour. Crop. Evol.* **2020**, *67*, 367–379. [CrossRef]

25. Lehner, A.; Mamadou, N.; Poels, P.; Côme, D.; Bailly, C.; Corbineau, F. Changes in soluble carbohydrates, lipid peroxidation and antioxidant enzyme activities in the embryo during ageing in wheat grains. *J. Cereal Sci.* **2008**, *47*, 555–565. [CrossRef]

26. Plaxton, W.C. The organization and regulation of plant glycolysis. *Annu. Rev. Plant Phys.* **1996**, *47*, 185–214. [CrossRef]

27. He, D.; Yang, P. Proteomics of rice seed germination. *Front. Plant Sci.* **2013**, *4*, 246. [CrossRef]

28. Kannababu, N.; Karivaratharaju, T.V. Effect of seed ageing on malate dehydrogenase and succinate dehydrogenase activity in seedling of sunflower. *Indian J. Plant Physiol.* **2000**, *5*, 397–399.

29. Jardim, M.D.; Caverzan, A.; Rauber, R.; Ferreira, E.D.S.; Margis, P.M.; Galina, A. Succinate dehydrogenase (mitochondrial complex II) is a source of reactive oxygen species in plants and regulates development and stress responses. *New Phytol.* **2015**, *208*, 776–789. [CrossRef] [PubMed]

30. Feussner, I.; Wasternack, C. The lipoxygenase pathway. *Annu. Rev. Plant Biol.* **2002**, *53*, 275–297. [CrossRef] [PubMed]

31. Oenel, A.; Fekete, A.; Krischke, M.; Faul, S.C.; Gresser, G.; Havaux, M.; Mueller, M.J.; Berger, S. Enzymatic and non-enzymatic mechanisms contribute to lipid oxidation during seed aging. *Plant Cell Physiol.* **2017**, *58*, 925–933. [CrossRef] [PubMed]

32. Mansouri-Far, C.; Goodarzian-Ghahfarokhi, M.; Saeidi, M.; Abdoli, M. Antioxidant enzyme activity and germination characteristics of different maize hybrid seeds during ageing. *Environ. Exp. Biol.* **2015**, *13*, 177–182.

33. Yin, G.; Xin, X.; Song, C.; Chen, X.; Zhang, J.; Wu, S.; Li, R.; Liu, X.; Lu, X. Activity levels and expression of antioxidant enzymes in the ascorbate-glutathione cycle in artificially aged rice seed. *Plant Physiol. Biochem.* **2014**, *80*, 1–9. [CrossRef] [PubMed]

34. Susumu, H.; Katsutomo, S.; Hiroyuki, I.; Yuko, O.; Hirokazu, M. A large family of class III plant peroxidases. *Plant Cell Physiol.* **2001**, *42*, 462–468.

35. Gill, R.S.; Gupta, A.K.; Taggar, G.K.; Taggar, M.S. Review article: Role of oxidative enzymes in plant defenses against insect herbivory. *Acta Phytopathol. Acad. Sci. Hung.* **2010**, *45*, 277–290. [CrossRef]

36. Xia, F.; Wang, X.; Li, M.; Mao, P. Mitochondrial structural and antioxidant system responses to aging in oat (*Avena sativa* L.) seeds with different moisture contents. *Plant Physiol. Biochem.* **2015**, *94*, 122–129. [CrossRef]

37. Jiang, M.; Zhang, J. Effect of abscisic acid on active oxygen species, antioxidative defence system and oxidative damage in leaves of maize seedlings. *Plant Cell Physiol.* **2001**, *42*, 1265–1273. [CrossRef]

38. Parkhey, S.; Naithani, S.C.; Keshavkant, S. ROS production and lipid catabolism in desiccating *Shorea robusta* seeds during aging. *Plant Physiol. Biochem.* **2012**, *57*, 261–267. [CrossRef]

39. Hernández Estévez, I.; Rodríguez Hernández, M. Plant Glutathione S-transferases: An overview. *Plant Genet. Resour. C* **2020**, *23*, 100233.

40. Liu, D.; Liu, Y.; Rao, J.; Wang, G.; Li, H.; Ge, F.; Chen, C. Overexpression of the glutathione S-transferase gene from *Pyrus pyrifolia* fruit improves tolerance to abiotic stress in transgenic tobacco plants. *Mol. Biol.* **2013**, *47*, 515–523. [CrossRef]

41. Kissoudis, C.; Kalloniati, C.; Flemetakis, E.; Madesis, P.; Labrou, N.E.; Tsaftaris, A.; Nianiou-Obeidat, I. Stress-inducible GmGSTU4 shapes transgenic tobacco plants metabolome towards increased salinity tolerance. *Acta Physiol. Plant.* **2015**, *37*, 128. [CrossRef]

42. Cheng, H.; Ma, X.; Jia, S.; Li, M.; Mao, P. Transcriptomic analysis reveals the changes of energy production and AsA-GSH cycle in oat embryos during seed ageing. *Plant Physiol. Biochem.* **2020**, *153*, 40–52. [CrossRef] [PubMed]

43. Rajjou, L.C.; Debeaujon, I. Seed longevity: Survival and maintenance of high germination ability of dry seeds. *Comptes Rendus Biol.* **2008**, *331*, 796–805. [CrossRef]

44. Wang, X.; Zhu, H.; Jin, G.; Liu, H.; Wu, W.; Zhu, J. Genome-scale identification and analysis of LEA genes in rice (*Oryza sativa* L.). *Plant Sci.* **2007**, *172*, 414–420. [CrossRef]

45. Hundertmark, M.; Buitink, J.; Leprince, O.; Hincha, D.K. The reduction of seed-specific dehydrins reduces seed longevity in *Arabidopsis thaliana*. *Seed Sci. Res.* **2011**, *21*, 165–173. [CrossRef]

46. Kalemba, E.A.M.; Pukacka, S.A. Possible roles of LEA proteins and HSPs in seed protection:a short review. *Biol. Lett.* **2007**, *44*, 3–16.

47. Jungkunz, I.; Link, K.; Vogel, F.; Voll, L.M.; Sonnewald, S.; Sonnewald, U. AtHsp70-15-deficient Arabidopsis plants are characterized by reduced growth, a constitutive cytosolic protein response and enhanced resistance to TuMV. *Plant J.* **2011**, *66*, 983–995. [CrossRef]

48. Wang, R.; Zhang, Y.; Kieffer, M.; Yu, H.; Kepinski, S.; Estelle, M. HSP90 regulates temperature-dependent seedling growth in Arabidopsis by stabilizing the auxin co-receptor F-box protein TIR1. *Nat. Commun.* **2016**, *7*, 10269. [CrossRef]

49. Queitsch, C.; Hong, S.W.; Vierling, E.; Lindquist, S. Heat shock protein 101 plays a crucial role in thermotolerance in Arabidopsis. *Plant Cell* **2000**, *12*, 479–492. [CrossRef]

50. Xu, M.; He, D.; Teng, H.; Chena, L.; Song, H.; Huang, Q. Physiological and proteomic analyses of coix seed aging during storage. *Food Chem.* **2018**, *260*, 82–89. [CrossRef]

51. Tabita, F.R.; Hanson, T.E.; Li, H.; Satagopan, S.; Chan, S. Function, structure, and evolution of the RubisCO-like proteins and their RubisCO homologs. *Microbiol. Mol. Biol. Rev.* **2008**, *71*, 576–599. [CrossRef]

52. Wei, L.; Wang, Q.; Xin, Y.; Lu, Y.; Xu, J. Enhancing photosynthetic biomass productivity of industrial oleaginous microalgae by overexpression of RuBisCO activase. *Algal Res.* **2017**, *27*, 366–375. [CrossRef]

53. Leitao, L.; Bethenod, O.; Biolley, J.P. The impact of ozone on juvenile maize (*Zea mays* L.) plant photosynthesis: Effects on vegetative biomass, pigmentation, and carboxylases (PEPc and Rubisco). *Plant Biol.* **2007**, *9*, 478–488. [CrossRef]

54. Gong, B.; Wen, D.; Vandenlangenberg, K.; Wei, M.; Yang, F.; Shi, Q.; Wang, X. Comparative effects of NaCl and NaHCO$_3$ stress on photosynthetic parameters, nutrient metabolism, and the antioxidant system in tomato leaves. *Sci. Hortic. Amst.* **2013**, *157*, 1–12. [CrossRef]

55. Xu, X.; Liu, T.; Yang, J.; Chen, L.; Liu, B.; Wang, L.; Jin, Q. The first whole-cell proteome and lysine-acetylome-based comparison between trichophyton rubrum conidial and mycelial stages. *J. Proteome Res.* **2018**, *17*, 1436–1451. [CrossRef]

56. Kim, H.J.; Lin, D.; Lee, H.J.; Li, M.; Liebler, D.C. Quantitative profiling of protein tyrosine kinases in human cancer cell lines by multiplexed parallel reaction monitoring assays. *Mol. Cell. Proteom.* **2016**, *15*, 682–691. [CrossRef] [PubMed]

Publisher's Note: MDPI stays neutral with regard to jurisdictional claims in published maps and institutional affiliations.

 *plants*

Review

Function and Mechanism of WRKY Transcription Factors in Abiotic Stress Responses of Plants

Weixing Li [†], **Siyu Pang** [†], **Zhaogeng Lu** and **Biao Jin** *

College of Horticulture and Plant Protection, Yangzhou University, Yangzhou 225009, China;
liwx@yzu.edu.cn (W.L.); pangsiyu_0212@163.com (S.P.); luzhaogeng@163.com (Z.L.)
* Correspondence: bjin@yzu.edu.cn
† Contributed equally to this work.

Received: 26 September 2020; Accepted: 4 November 2020; Published: 8 November 2020

Abstract: The WRKY gene family is a plant-specific transcription factor (TF) group, playing important roles in many different response pathways of diverse abiotic stresses (drought, saline, alkali, temperature, and ultraviolet radiation, and so forth). In recent years, many studies have explored the role and mechanism of WRKY family members from model plants to agricultural crops and other species. Abiotic stress adversely affects the growth and development of plants. Thus, a review of WRKY with stress responses is important to increase our understanding of abiotic stress responses in plants. Here, we summarize the structural characteristics and regulatory mechanism of WRKY transcription factors and their responses to abiotic stress. We also discuss current issues and future perspectives of WRKY transcription factor research.

Keywords: WRKY transcription factor; abiotic stress; gene structural characteristics; regulatory mechanism; drought; salinity; heat; cold; ultraviolet radiation

1. Introduction

As a fixed-growth organism, plants are exposed to a variety of environmental conditions and may encounter many abiotic stresses, for example, drought, waterlogging, heat, cold, salinity, and Ultraviolet-B (UV-B) radiation. To adapt and counteract the effects of such abiotic stresses, plants have evolved several molecular mechanisms involving signal transduction and gene expression [1,2]. Transcription factors (TFs) are important regulators involved in the process of signal transduction and gene expression regulation under environmental stresses. TFs can be combined with *cis*-acting elements to regulate the transcriptional efficiency of target genes by inhibiting or enhancing their transcription [3,4]. Accordingly, plants may show corresponding responses to external stresses via TFs regulating target genes. Although some TF families (MYB, bZIP, AP2/EREBP, NAC) are associated with adversity [2,5], WRKY is the most extensively studied TF family in plant stress responses.

The WRKY family is a unique TF superfamily of higher plants and algae, which play important roles in many life processes, particularly in response against biotic and abiotic stress [6,7]. In 1994, the SWEET POTATO FACTOR1 (*SPF1*) gene of the WRKY family was first found in *Impoea batatas* [8]. Later, *ABF1* and *ABF2* were found in wild *Avena sativa*, and showed regulatory roles in seed germination [9]. A previous study successively cloned *WRKY1*, *WRKY2*, and *WRKY3* from *Petroselinum crispum*, named the WRKY TF, and proved for the first time that WRKY protein can regulate plant responses to pathogens [10]. With an increase in available published genomes, many members of the WRKY TF family have been identified in various species, including 104 from *Populus* [11], 37 from *Physcomitrella patens* [12], 45 from *Hordeum vulgare* [13], 55 from *Cucumis sativus* [14], 74 from *Arabidopsis thaliana* [15], 83 from *Pinus monticola* [16], 81 from *Solanum lycopersicum* [17], and 102 from *Oryza sativa* [18]. WRKY TFs exist as gene families in plants, and the number of WRKY TFs varies among species. In plants exposed to abiotic stresses (salt, drought, temperature, and so forth), WRKY family members play important roles in diverse stress responses. In addition, these TFs affect the growth and development of plants [19,20]. Therefore, WRKY TFs have attracted broad attention. Although some reviews on WRKY TFs are available, in this review we focus on the structural characteristics and regulatory mechanisms of WRKY TFs and summarize recent progress in understanding the roles of WRKY TFs during exposure to abiotic stresses such as drought, temperature, salt, and UV radiation.

2. Structural Characteristics of WRKY TFs

The WRKY structure consists of two parts: the N-terminal DNA binding domain and the C-terminal zinc-finger structure [21]. The DNA binding domain sequence of WRKY is based on the heptapeptide WRKYGQK (Figure 1), but there are some differences, such as WRKYGQK, WRKYGKK, WRKYGMK, WSKYGQK, WKRYGQK, WVKYGQK, and WKKYGQK [17,22]. Zinc-finger structures mainly include C_2H_2 type and C_2HC type [23], whereas some exist in the form of $CX_{29}HXH$ and $CX_7CX_{24}HXC$ [17] (Figure 1). According to the number of WRKY domains and the structure of their zinc-finger motifs, WRKY can be divided into groups I, II, and III [23] (Figure 1). Group I usually contains two WRKY domains and one C_2H_2 zinc-finger structure. Those in group II and group III contain only one WRKY domain. The difference is that the zinc-finger structure in group II is C_2H_2 and that in group III is C_2HC [19,21,23] (Figure 1). According to the phylogenetic relationship of the amino acid sequence of the primary structure, group II can be further divided into subgroups a–e [7,23,24]. Evolutionary analyses have shown that the WRKY of group II is not generally a single source, mainly including five categories I, IIa + IIb, IIc, IId + IIe, and III [7,24]. In addition, some WRKY proteins contain a glutamate enrichment domain, a proline enrichment domain, and a leucine zipper structure [25].

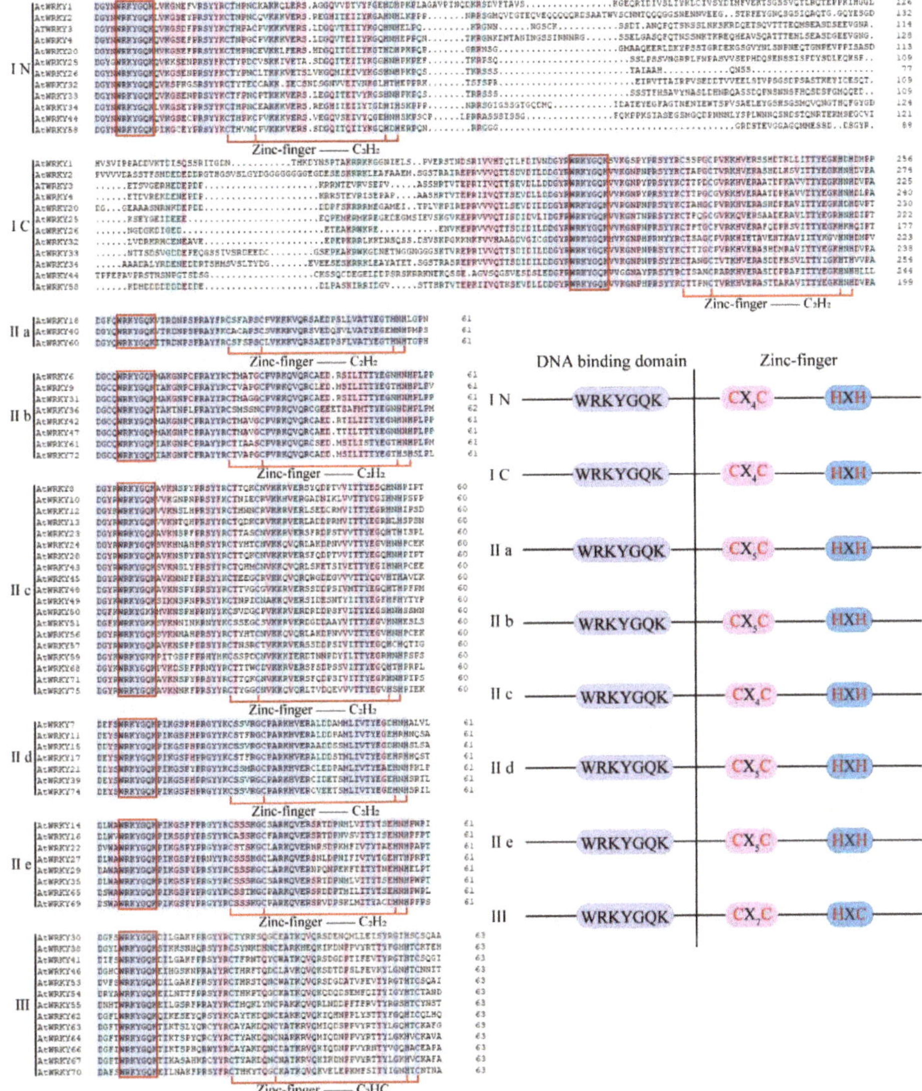

Figure 1. The domain of WRKY genes in *Arabidopsis thaliana*. The WRKY gene family is classified into the **I** (**I N** and **I C**), **IIa**, **IIb**, **IIc**, **IId**, **IIe**, and **III** subfamilies. The aligned conserved domains (DNA binding and zinc-finger structures) are highlighted (left panel) and simplified (right panel).

3. Regulatory Mechanism of WRKY TFs

WRKY family members have diverse regulatory mechanisms. Briefly, WRKY protein can be effectively combined with W-box elements to activate or inhibit the transcription of downstream target genes. Moreover, it can also bind other acting elements to form protein complexes, which enhances the activity of transcription binding [21].

WRKY TFs can effectively activate the expression of downstream genes by binding conserved W-box *cis*-acting elements in the downstream gene promoter region [21,26]. There are abundant W-box elements in the self-promoter of most WRKY TFs. Therefore, these WRKY TFs can bind with their own promoters to achieve self-regulation or cross-regulation networks by combining with other WRKY TFs [27]. For example, *CaWRKY6* of *Capsicum frutescens* can activate *CaWRKY40* and make the plant more tolerant to high temperature and humidity. *Glycine max GmWRKY27* not only inhibits the activity of downstream *GmNAC29* promoter by independent inhibition, but also cooperatively interacts with *GmMYB174* to inhibit the expression of *GmNAC29*, thereby increasing drought and salt stress resistances [28]. Moreover, chromatin immunoprecipitation (ChIP) studies have shown that when *Petroselinum crispum* is infected by pathogenic bacteria, *PcWRKY1* promoter can effectively bind to itself and the W-box of *PcWRKY3* promoter, and transcriptional activation can be achieved through self-negative feedback regulation and cross-regulation with other WRKY proteins [29]. In addition, WRKY TFs can interact with non-W-box elements. For example, *Oryza sativa OsWRKY13* can interact with PRE4 (TGCGCTT) elements [30]. *Hordeum vulgare HvWRKY46* and *Nicotiana tabacum NtWRKY12* can effectively combine with the sucrose response element SURE [31,32]. These results indicate that there are multiple binding modes between WRKY TFs and structural genes. Different binding patterns and preferences of binding sites allow for the regulation of downstream target genes, providing WRKY TFs with versatile functions in the plant transcriptional regulation network.

4. WRKY TF Involved in Abiotic Stress Responses

When plants sense stress, the corresponding signaling is activated and transferred to the cell interior. Reactive oxygen species (ROS) and Ca^{2+} ions are usually exchanged as the signal transduction in the cell. Protein kinases such as MPKs are subsequently activated to regulate the activities of related TFs. Consequently, the plant presents a stress response [31,32]. In response to abiotic stresses, some WRKY TFs can be rapidly differentially expressed, promoting signal transduction and regulating the expression of related genes [33]. The expression patterns and functional identifications of WRKYs in most studies are generally based on transcriptome analyses, real-time fluorescence quantitative PCR, gene chip analyses, and genetic transformation. Hence, WRKY genes can function effectively in most abiotic stress responses or tolerances in various plants (Table 1, Figure 2).

Table 1. WRKY transcription factors (TFs) involved in abiotic stress responses in plants.

No.	Gene	Species	Induced by Factors	Function	References
1	*AtWRKY25/26*	*Arabidopsis*	Heat	Tolerance to heat	[34]
2	*AtWRKY33*	*Arabidopsis*	NaCl, mannitol, H_2O_2	Tolerance to heat and NaCl, negative regulator in oxidative stress and abscisic acid (ABA)	[33]
3	*AtWRKY34*	*Arabidopsis*	Cold	Negative regulator in cold stress	[35]
4	*AtWRKY39*	*Arabidopsis*	Heat	Tolerance to heat	[36]
5	*AtWRKY53*	*Arabidopsis*	Drought, salt	Reduced drought resistance and H_2O_2, sensitive to salt	[37,38]
6	*AtWRKY57*	*Arabidopsis*	Drought	Tolerance to drought	[39]
7	*AtWRKY63*	*Arabidopsis*	ABA	Tolerance to drought, regulated ABA signaling	[40]
8	*AtWRKY54*	*Arabidopsis*	Heat	Response to heat stress	[41]
9	*POWRKY13*	*Populus tomentosa*	Heat	Response to heat stress	[42]
10	*GhWRKY21*	*Gossypium hirsutum*	Drought	Tolerance to drought	[43]
11	*GhWRKY25*	*Gossypium hirsutum*	Drought	Tolerance to salt, reduced drought resistance	[44]
12	*GhWRKY68*	*Gossypium hirsutum*	Salt, drought	Reduced salt tolerance and drought resistance, positive regulator in ABA signaling	[45]
13	*VvWRKY24*	*Vitis vinifera*	Cold	Upregulated expression at all stages of hypothermia	[46]

Table 1. *Cont.*

No.	Gene	Species	Induced by Factors	Function	References
14	*CaWRKY40*	*Capsicum annuum*	Heat	Tolerance to heat	[47]
15	*BdWRKY36*	*Brachypodium distachyon*	Drought	Tolerance to drought	[48]
16	*FcWRKY70*	*Fortunella crassifolia*	Salt	Tolerance to salt	[49]
17	*OsWRKY11*	*Oryza sativa*	Heat, drought	Tolerance to drought and heat	[50]
18	*OsWRKY72*	*Oryza sativa*	Drought, NaCl, ABA	Sensitive to salt, drought, sucrose, and ABA	[51]
19	*OsWRKY74*	*Oryza sativa*	Pi deprivation, cold	Tolerance to cold and Pi deprivation	[52]
20	*OsWRKY76*	*Oryza sativa*	Cold	Tolerance to cold	[53]
21	*OsWRKY89*	*Oryza sativa*	ABA, UV-B	Tolerance to UV	[54]
22	*GmWRKY13*	Soybean	Salt, drought	Sensitive to salt and mannitol, negative regulator in ABA signaling	[55]
23	*GmWRKY17*	Soybean	Salt	Reduced salt tolerance	[56]
24	*GmWRKY54*	Soybean	Salt, drought	Tolerance to salt and drought	[55]
25	*GmWRKY21*	*Glycine max*	NaCl, drought, cold	Tolerance to cold	[55]
26	*ZmWRKY17*	*Zea mays*	ABA, salt	Reduced salt tolerance	[57]
27	*TaWRKY2/19*	*Triticum aestivum*	NaCl, drought, ABA	Tolerance to salt and drought	[58]
28	*BcWRKY46*	*Brassica campestris*	NaCl, drought, cold	Tolerance to salt and drought	[59]
29	*BhWRKY1*	*Boea hygrometrica*	Dehydration, ABA	Tolerance to drought	[60]
30	*VpWRKY1*	*Vitis pseudoreticulata*	NaCl, ABA	Tolerance to salt	[61]
31	*VpWRKY2*	*Vitis pseudoreticulata*	Cold, NaCl, ABA	Tolerance to salt and cold	[61]
32	*VpWRKY3*	*Vitis pseudoreticulata*	Drought, ABA, salicylic acid (SA)	Tolerance to salt	[62]
33	*TcWRKY53*	*Thlaspi caerulescens*	Cold, PEG, NaCl	Negative regulator in osmotic stress	[63]
34	*NaWRKY3*	*Nicotiana attenuate*	Mechanical damage	Sensitive to mechanical damage	[64]
35	*JrWRKY2/7*	*Juglans regia*	Drought, cold	Tolerance to drought and cold	[65]
36	*SbWRKY30*	*Sorghum bicolor*	Salt, drought	Tolerance to salt and drought	[66]
37	*SbWRKY50*	*Sorghum bicolor*	Salt	Tolerance to salt	[67]
38	*IbWRKY47*	*Ipomoea batatas*	Salt	Tolerance to salt	[68]
39	*IbWRKY2*	*Ipomoea batatas*	Salt, drought	Tolerance to salt and drought	[69]
40	*MdWRKY30*	*Malus domestica*	Salt, osmotic stress	Tolerance to salt and osmotic stress	[70]
41	*MdWRKY100*	*Malus domestica*	Salt	Sensitive to salt	[71]
42	*SlWRKY81*	*Solanum lycopersicum*	Drought	Reduced drought tolerance	[72]
43	*GbWRKY1*	*Gossypium barbadense*	Salt	Tolerance to salt	[73]
44	*VbWRKY32*	*Verbena bonariensis*	Cold	Tolerance to cold	[74]
45	*PgWRKY33/62*	*Pennisetum glaucum*	Salt, drought	Tolerance to salt and drought	[75]
46	*PagWRKY75*	*Populus alba*	Drought	Negative regulator in salt and osmotic stress	[76]

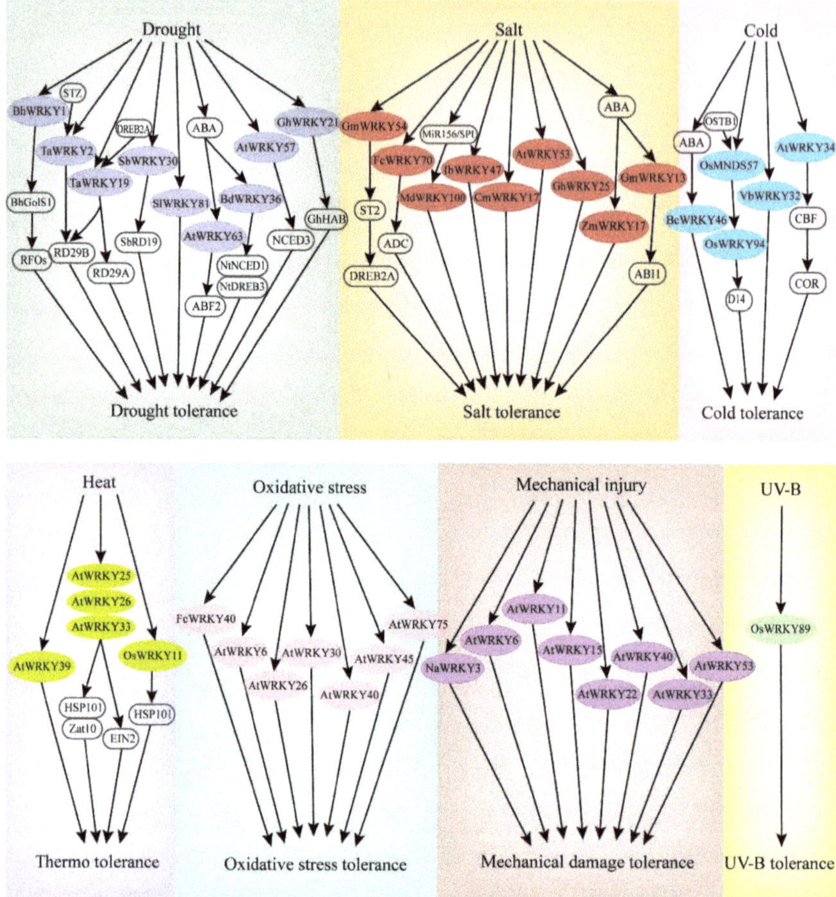

Figure 2. Some WRKY genes involved in the response pathways of major abiotic stresses (drought, salt, cold, heat, oxidative stress, mechanical injury, UV-B).

4.1. WRKY TFs and Drought Stress

Drought has a major impact on plant growth and development, resulting in a significant decrease in grain and other types of crop yield [77]. Under drought stress, drought-tolerant plants can accumulate oligosaccharides through sucrose metabolism to improve drought resistance. For example, when *Arabidopsis* is subjected to drought stress, the expression of *AtWRKY53* combined with the Qua-Quine Starch (QQS) promoter sequence is rapidly induced, hydrogen peroxide content is reduced, and the glucose metabolism pathway is significantly enhanced, thereby regulating stomatal opening and ultimately affecting drought tolerance [37]. In *Boea hygrometrica*, *BhWRKY1* effectively regulates the expression of the *BhGolS1* gene, and the overexpression of *BhGolS1* and *BhWRKY1* induces the accumulation of raffinose family oligosaccharides (RFOs) in transgenic *Nicotiana tabacum*, thus improving the ability of seedlings to resist drought [60].

WRKY protein can directly regulate the expression of drought-resistant genes. For example, in sorghum, *SbWRKY30* regulates the drought stress response gene *SbRD19* by binding with W-box elements of the *SbRD19* promoter, and protects plant cells from the damage of reactive oxygen species by improving ROS scavenging capability, enhancing drought tolerance [66]. *TaWRKY2* of wheat can bind to *STZ* and downstream drought-resistant gene *RD29B* promoter, with a positive regulatory effect on the expression of *RD29B* [58]. *DREB2A* regulates the expression of dehydration stress-related genes [78], while *TaWRKY19* can bind to *DREB2A* promoter, ultimately activating the expression of *RD29A*, *RD29B*, and *Cor6.6* in transgenic *Arabidopsis* plants [58]. Similarly, *Arabidopsis AtWRKY57* positively regulates the expression of *RD29A* and *NCED3* genes by binding their W-box elements in the promoter regions [39]. In addition, WRKY protein can act on other TFs to play regulatory roles in drought tolerance. For example, *TcWRKY53* of *Thlaspi arvense* significantly inhibits the expression of *NtERF5* and *NterEBp-1* of the AP2/ERF TF family, thus improving plant resistance to drought stress [63].

WRKY TFs also regulate plant tolerance through abscisic acid (ABA) and ROS-related signaling pathways. During drought stress, higher ABA levels were accumulated in plants, and leaf stomata were closed to reduce transpiration rate, thus regulating water balance in plants. ABA accumulation in cells, integrated with a variety of stress signals, regulates the expression of downstream genes, consequently sensing and responding to the adverse environment [40]. *Arabidopsis AtWRKY63* has a specific effect on ABA-mediated stomatal closure and other signal transduction pathways, thus affecting the drought response [40]. *GhWRKY21* regulates ABA-mediated cotton drought tolerance by promoting the expression of *GhHAB* [43]. Overexpression of *BdWRKY36* in tobacco reduces the accumulation of ROS, activated *NtLEA5*, *NtNCED1*, and *NtDREB3* in the ABA biosynthetic pathway, and significantly enhances the drought resistance of plants [48]. In *Solanum lycopersicum*, *SlWRKY81* increases the drought tolerance of plants by inhibiting the accumulation of H_2O_2, playing a negative regulation role of stomatal closure [72].

4.2. WRKY TFs and Salt Stress

Salt stress is an important abiotic stress affecting crop productivity, particularly in arid and semiarid regions. WRKY TFs play essential roles in regulating the response to salt stress. To date, a total of 47 WRKY genes have been found to be expressed under salt stress in the wheat genome [79]. *STZ* is a protein related to ZPT2, which acts as a transcriptional inhibitor to downregulate the deactivation of other transcription factors. *GmWRKY54* of *Glycine max* inhibits *STZ* expression and responds to salt stress by positively regulating the DREB2A-mediated pathway [55]. *FcWRKY70* promotes the upregulation of arginine decarboxylase (ADC) expression, which is heterologously expressed in tobacco, and the content of lemon putrescine is significantly increased, thus enhancing the salt tolerance of plants [49]. The *IbWRKY47* gene positively regulates stress resistance-related genes and significantly improves the salt tolerance of *Ipomoea batatas* [68]. MiR156/SPL modulates salt tolerance by upregulation of *Malus domestica* salt tolerance gene *MdWRKY100* [71]. In *Sorghum bicolor*, *SbWRKY50* could directly bind to the upstream promoter of *SOS1* and *HKT1* and participate in plant salt response by controlling ion homeostasis [67]. In addition, some *WRKY* genes function as negative regulation factors involved in salt stress resistance. *Arabidopsis* RPD3-like histone deacetylase HDA9 inhibits salt stress tolerance by regulating the DNA binding and transcriptional activity of *WRKY53* [38]. *Chrysanthemum CmWRKY17* overexpressed in *Arabidopsis* allows the plants to be more sensitive to salt stress. The expression level of stress resistance-related genes in transgenic *Arabidopsis* is lower than that in wild-type plants, indicating that *CmWRKY17* may be involved in negatively regulating the salt stress response in *Chrysanthemum* [80]. The expression of *GhWRKY68* is strongly induced in upland cotton and decreases salt tolerance [45]. In contrast, a high expression level of *GhWRKY25* enhances the salt tolerance of upland cotton, while transgenic tobacco shows a relatively weaker tolerance to drought stress [44], indicating that the regulatory effects of different WRKY TFs involved in drought response are different.

Plants can also respond to saline–alkali stress through ABA, H_2O_2, and other signal pathways. In *Glycine max*, the negative regulatory factor *ABI1* in the ABA pathway may be the downstream target gene of *GmWRKY13*. Genetic transformation experiments in *Arabidopsis* have shown that overexpression of *GmWRKY13* significantly increases the expression of *ABI1*, but plants show a low tolerance to salt stress [55]. Overexpression of *ZmWRKY17* has an inhibitory effect on the sensitivity of exogenous ABA treatment, resulting in a relatively lower tolerance to high levels of salinity [57]. Under salt-induced H_2O_2 and cytosolic Ca^{2+} stimulation, the activity of antioxidant enzymes increases, thus improving the tolerance to high-salinity environments [81]. ABA-induced WRKY gene expression is largely related to salt stress. Exogenous application of ABA and NaCl also induce *AtWRKY25* and *AtWRKY33* in *Arabidopsis* [33], *OsWRKY72* in rice [51], *GbWRKY1* in *Verbena bonariensis* [73], and *VpWRKY1/2* [61] and *VpWRKY3* [62] in grape.

4.3. WRKY TFs and Temperature Stress

Both low- and high-temperature stress can reduce crop yield and quality in plants. WRKY TFs play a role in the stress response through different signal transduction pathways. For example, in *Verbena bonariensis*, *VbWRKY32* as a positive regulator, upregulates the transcriptional level of cold response genes, which increases the antioxidant activity, maintains membrane stability, and enhances osmotic regulation ability, thereby improving the survival ability under cold stress [74]. The *BcWRKY46* gene of *Brassica campestris* is strongly induced by low temperature and ABA, activating related genes in the ABA signaling pathway to improve the low-temperature tolerance of plants [59]. *CBF* TFs regulate the expression of *COR*, and the overexpressed transgenic lines of *CBF1*, *CBF2*, and *CBF3* show stronger cold resistance [82]. *AtWRKY34* has a negative regulatory effect on the CBF-mediated cold response pathway; it is specifically expressed in mature pollen grains after exposure to low temperatures, resulting in resistance to low temperatures [35]. In addition, plants respond to temperature changes by coordinating organ development in an adverse environment. At low temperatures, rice MADS-Box TF *OsMADS57* and its interacting protein *OsTB1* synergistically activate the transcriptional regulation of *OsWRKY94*, preventing tillering by inhibiting transcription of the organ development gene *D14* [83].

Due to global climate change, high-temperature stress has attracted significant attention. There is evidence that, to a certain extent, high temperatures will lead to biochemical changes in plants [84]. Thermal stimulation can activate Ca^{2+} channels to maintain a higher intracellular Ca^{2+} concentration, thereby activating calmodulin protein expression and inducing thermal-shock protein transcriptional expression [85]. In *Arabidopsis*, *AtWRKY54* significantly responds to heat shock whereas basic leucine zipper factors (bZIPs) respond to prolonged warming [41]. Overexpression of *AtWRKY39* can make plants more heat-sensitive. *AtWRKY39* is highly homologous to *AtWRKY7*, and both of them can effectively bind calmodulin in plants, indicating a similar function [36]. In addition, *AtWRKY25*, *AtWRKY26*, and *AtWRKY33* can improve tolerance to high-temperature stress in transgenic *Arabidopsis* by regulating the *Hsp101* and *Zat10* genes [34]. Plants subjected to heat stress can also activate the oxidative stress response through ethylene [86]. Under high-temperature stress, the expressions of *AtWRKY25*, *AtWRKY26*, and *AtWRKY33* in *Arabidopsis* are induced by ethylene, the feedback factor *EIN2* is transcriptionally regulated, and the effective activation of ethylene signal transduction contribute to relatively stronger heat resistance. In *Oryza sativa*, *HSP101* promoter can activate the expression of the *OsWRKY11* gene. Under heat treatment, the leaves wilted more slowly and the green part of the plant was less damaged, which makes it more heat-resistant [50]. In addition, some noncoding RNAs, such as miR396, play a role in the response to heat stress by regulating its target *WRKY6* [87].

4.4. WRKY TFs and Other Abiotic Stresses

WRKY TFs are also involved in oxidative stress, mechanical damage, UV radiation, and other abiotic stresses (Figure 3). *FcWRKY40* overexpression can significantly enhance the resistance of transgenic tobacco to oxidative stress [88]. When *Arabidopsis* is treated with ROS, the expressions of *AtWRKY30*, *AtWRKY40*, *AtWRKY75*, *AtWRKY6*, *AtWRKY26*, and *AtWRKY45* are significantly upregulated [89]. After mechanical injury, the expression levels of *AtWRKY11*, *AtWRKY15*, *AtWRKY22*, *AtWRKY33*, *AtWRKY40*, *AtWRKY53* [90] and *AtWRKY6* [64] are upregulated. Similarly, *NaWRKY3* is strongly expressed in tobacco. By contrast, the sensitivity of transgenic plants is increased when *NaWRKY3* is knocked out [64]. In two previous studies, UV-B radiation treatment induced three WRKY genes in *Arabidopsis* and the *OsWRKY89* gene in rice, resulting in a thick waxy substance on the leaf surface and improved tolerance to heat [54,91].

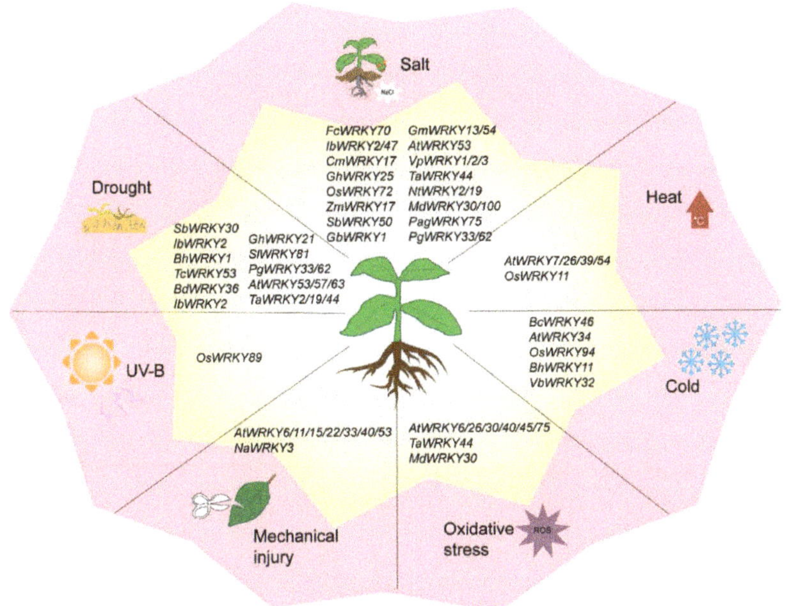

Figure 3. WRKY transcription factors in response to abiotic stresses.

In addition, a single WRKY TF can play multiple roles in different stress responses via various signal pathways and regulatory networks. For example, *TaWRKY44* expression in tobacco can improve resistance to drought, salt stress, and osmotic stress [92], while *PgWRKY62* and *PgWRKY33* in *Pennisetum glaucum* respond to salt and drought simultaneously [75]. *BhWRKY1* protein in *Boea hygrometrica* binds to the promoter of *BhGolS1* and is associated with both low-temperature resistance and drought tolerance [60]. *IbWRKY2* can interact with *IbVQ4*, and drought and salt treatment can induce the expression of *IbVQ4*, thus improving the tolerance of plants to drought and salt stress [69]. *MdWRKY30* overexpression enhances tolerance to salt and osmotic stress in transgenic apple callus through transcriptional regulation of stress-related genes [70]. *PagWRKY75* negatively regulates the tolerance of 84 K poplar (*Populus alba* × *P. glandulosa*) to salt and osmotic stress by reducing the scavenging capacity of ROS and the accumulation of proline, thus actively regulates the rate of leaf water loss [76].

5. Conclusions and Perspectives

As one of the largest TF families, WRKY plays an important and indispensable role in normal life activities of plants. Over the years, it has been shown that WRKY TFs not only participate in plant growth and development, but also show complex regulatory mechanisms and networks involved in external abiotic stresses. A large number of WRKYs have been functionally characterized in model plants, providing abundant functional references for other plants. Given that crops usually face various stresses and WRKYs play important roles in stress responses, further in-depth studies on WRKY genes in more crops are required. As increasing plant genomes have been sequenced, particularly of economically important crops, the genome-wide identification of WRKY genes will facilitate screening for stress resistance-related functional genes in plants. Moreover, previous studies of WRKY gene functions were largely dependent on transcriptomics and functional predictions, whereas more applications of genetic verification combined with new technologies are accelerating the research progress of WRKY's novel functions. In addition, characterization of the downstream genes regulated by WRKY TFs or WRKY TF self-regulation will help clarify the regulatory network of abiotic stress responses. Furthermore, noncoding RNAs and epigenetic modifications involved in the regulation of WRKY TFs should be explored in future studies. Ultimately, using WRKY TFs to screen for stress-resistant plant cultivars and improve plant stress resistance will significantly benefit agricultural crop yield and quality in the context of aggravated climate change.

Author Contributions: Conceptualization, W.L. and B.J.; methodology, B.J.; software, S.P.; formal analysis, S.P. and Z.L.; investigation, S.P. and Z.L.; resources, S.P.; writing—original draft preparation, S.P. and W.L.; writing—review and editing, B.J., S.P., Z.L. and W.L.; visualization, B.J. and S.P.; supervision, B.J. and W.L.; project administration, W.L.; funding acquisition, W.L. All authors have read and agreed to the published version of the manuscript.

Funding: This research was founded by the Natural Science Foundation of China (grant numbers 31971408, 31670695 and 31971686), and Forestry Technology Innovation and Extension of Jiangsu Province (grant number LYKJ [2019]48).

Conflicts of Interest: The authors declare no conflict of interest.

References

1. Yoon, Y.; Seo, D.H.; Shin, H.; Kim, H.J.; Kim, C.M.; Jang, G. The role of stress-responsive transcription factors in modulating abiotic stress tolerance in plants. *Agronomy* **2020**, *10*, 788. [CrossRef]
2. Ma, Q.; Xia, Z.; Cai, Z.; Li, L.; Cheng, Y.; Liu, J.; Nian, H. *GmWRKY16* enhances drought and salt tolerance through an ABA-mediated pathway in *Arabidopsis thaliana*. *Front. Plant Sci.* **2019**, *9*, 9. [CrossRef]
3. Liu, Y.; Yang, T.; Lin, Z.; Guo, B.; Xing, C.; Zhao, L.; Dong, H.; Gao, J.; Xie, Z.; Zhang, S.-L.; et al. A WRKY transcription factor *PbrWRKY53* from *Pyrus betulaefolia* is involved in drought tolerance and AsA accumulation. *Plant Biotechnol. J.* **2019**, *17*, 1770–1787. [CrossRef] [PubMed]
4. Shrestha, A.; Khan, A.; Dey, N. cis-trans Engineering: Advances and perspectives on customized transcriptional regulation in plants. *Mol. Plant* **2018**, *11*, 886–898. [CrossRef] [PubMed]
5. Yamasaki, K.; Kigawa, T.; Inoue, M.; Watanabe, S.; Tateno, M.; Seki, M.; Shinozaki, K.; Yokoyama, S. Structures and evolutionary origins of plant-specific transcription factor DNA-binding domains. *Plant Physiol. Biochem.* **2008**, *46*, 394–401. [CrossRef] [PubMed]
6. Zhang, Y.; Wang, L. The WRKY transcription factor superfamily: Its origin in eukaryotes and expansion in plants. *BMC Evol. Biol.* **2005**, *5*, 1. [CrossRef]
7. Pandey, S.P.; Somssich, I.E. The role of WRKY transcription factors in plant immunity. *Plant Physiol.* **2009**, *150*, 1648–1655. [CrossRef] [PubMed]
8. Ishiguro, S.; Nakamura, K. Characterization of a cDNA encoding a novel DNA-binding protein, *SPF1*, that recognizes SP8 sequences in the 5′ upstream regions of genes coding for sporamin and beta-amylase from sweet potato. *Mol Gen Genet* **1994**, *244*, 563–571. [CrossRef]
9. Rushton, P.J.; Macdonald, H.; Huttly, A.K.; Lazarus, C.M.; Hooley, R. Members of a new family of DNA-binding proteins bind to a conserved cis-element in the promoters of Amy2 genes. *Plant Mol. Biol.* **1995**, *29*, 691–702. [CrossRef]
10. Rushton, P.J.; Torres, J.T.; Parniske, M.; Wernert, P.; Hahlbrock, K.; Somssich, I.E. Interaction of elicitor-induced DNA-binding proteins with elicitor response elements in the promoters of parsley PR1 genes. *EMBO J.* **1996**, *15*, 5690–5700. [CrossRef] [PubMed]
11. He, H.; Dong, Q.; Shao, Y.; Jiang, H.; Zhu, S.; Cheng, B.; Xiang, Y. Genome-wide survey and characterization of the WRKY gene family in *Populus trichocarpa*. *Plant Cell Rep.* **2012**, *31*, 1199–1217. [CrossRef] [PubMed]
12. Rensing, S.A.; Lang, D.; Zimmer, A.D.; Terry, A.; Salamov, A.A.; Shapiro, H.; Nishiyama, T.; Perroud, P.-F.; Lindquist, E.; Kamisugi, Y.; et al. The *Physcomitrella* genome reveals evolutionary insights into the conquest of land by plants. *Science* **2007**, *319*, 64–69. [CrossRef] [PubMed]

13. Mangelsen, E.; Kilian, J.; Berendzen, K.W.; Kolukisaoglu, Ü.; Harter, K.; Jansson, C.; Wanke, D. Phylogenetic and comparative gene expression analysis of barley (*Hordeum vulgare*) WRKY transcription factor family reveals putatively retained functions between monocots and dicots. *BMC Genom.* **2008**, *9*, 194. [CrossRef]

14. Ling, J.; Jiang, W.; Zhang, Y.; Yu, H.; Mao, Z.; Gu, X.; Huang, S.; Xie, B. Genome-wide analysis of WRKY gene family in *Cucumis sativus*. *BMC Genom.* **2011**, *12*, 1–20. [CrossRef]

15. Berri, S.; Abbruscato, P.; Faivre-Rampant, O.; Brasileiro, A.C.M.; Fumasoni, I.; Satoh, K.; Kikuchi, S.; Mizzi, L.; Morandini, P.; Pè, M.E.; et al. Characterization of WRKY co-regulatory networks in rice and *Arabidopsis*. *BMC Plant Biol.* **2009**, *9*, 1–22. [CrossRef]

16. Liu, J.-J.; Ekramoddoullah, A.K. Identification and characterization of the WRKY transcription factor family in *Pinus monticola*. *Genome* **2009**, *52*, 77–88. [CrossRef] [PubMed]

17. Huang, S.; Gao, Y.; Liu, J.; Peng, X.; Niu, X.; Fei, Z.; Cao, S.; Liu, Y. Genome-wide analysis of WRKY transcription factors in *Solanum lycopersicum*. *Mol. Genet. Genom.* **2012**, *287*, 495–513. [CrossRef]

18. Ross, C.A.; Liu, Y.; Shen, Q.J. The WRKY gene family in rice (*Oryza sativa*). *J. Integr. Plant Biol.* **2007**, *49*, 827–842. [CrossRef]

19. Rushton, P.J.; Somssich, I.E.; Ringler, P.; Shen, Q.J. WRKY transcription factors. *Trends Plant Sci.* **2010**, *15*, 247–258. [CrossRef]

20. Ülker, B.; Somssich, E.I. WRKY transcription factors: From DNA binding towards biological function. *Curr. Opin. Plant Biol.* **2004**, *7*, 491–498. [CrossRef]

21. Phukan, U.J.; Jeena, G.S.; Shukla, R.K. WRKY transcription factors: Molecular regulation and stress responses in plants. *Front. Plant Sci.* **2016**, *7*, 760. [CrossRef] [PubMed]

22. Xie, Z.; Zhang, Z.-L.; Zou, X.; Huang, J.; Ruas, P.; Thompson, D.; Shen, Q.J. Annotations and functional analyses of the rice WRKY gene superfamily reveal positive and negative regulators of abscisic acid signaling in aleurone cells. *Plant Physiol.* **2004**, *137*, 176–189. [CrossRef]

23. Eulgem, T.; Rushton, P.J.; Robatzek, S.; Somssich, I.E. The WRKY superfamily of plant transcription factors. *Trends Plant Sci.* **2000**, *5*, 199–206. [CrossRef]

24. Rushton, P.J.; Bokowiec, M.T.; Han, S.; Zhang, H.; Brannock, J.F.; Chen, X.; Laudeman, T.W.; Timko, M.P. Tobacco transcription factors: Novel insights into transcriptional regulation in the *Solanaceae*. *Plant Physiol.* **2008**, *147*, 280–295. [CrossRef]

25. Chen, L.; Song, Y.; Li, S.; Zhang, L.; Zou, C.; Yu, D. The role of WRKY transcription factors in plant abiotic stresses. *Biochim. Biophys. Acta BBA Bioenerg.* **2012**, *1819*, 120–128. [CrossRef]

26. Ciolkowski, I.; Wanke, D.; Birkenbihl, R.P.; Somssich, I.E. Studies on DNA-binding selectivity of WRKY transcription factors lend structural clues into WRKY-domain function. *Plant Mol. Biol.* **2008**, *68*, 81–92. [CrossRef]

27. Ezentgraf, U.; Laun, T.; Miao, Y. The complex regulation of *WRKY53* during leaf senescence of *Arabidopsis thaliana*. *Eur. J. Cell Biol.* **2010**, *89*, 133–137. [CrossRef]

28. Cai, H.; Yang, S.; Yan, Y.; Xiao, Z.; Cheng, J.; Wu, J.; Qiu, A.; Lai, Y.; Mou, S.; Guan, D.; et al. *CaWRKY6* transcriptionally activates *CaWRKY40*, regulates Ralstonia solanacearum resistance, and confers high-temperature and high-humidity tolerance in pepper. *J. Exp. Bot.* **2015**, *66*, 3163–3174. [CrossRef]

29. Turck, F.; Zhou, A.; Somssich, I.E. Stimulus-dependent, promoter-specific binding of transcription factor WRKY1 to Its native promoter and the defense-related gene *PcPR1-1* in *Parsley*. *Plant Cell* **2004**, *16*, 2573–2585. [CrossRef]

30. Cai, T.; Flanagan, L.B.; Jassal, R.S.; Black, T.A. Modelling environmental controls on ecosystem photosynthesis and the carbon isotope composition of ecosystem-respired CO_2 in a coastal Douglas-fir forest. *Plant Cell Environ.* **2008**, *31*, 435–453. [CrossRef]

31. Grierson, C.; Du, J.-S.; Zabala, M.D.T.; Beggs, K.; Smith, C.; Holdsworth, M.; Bevan, M.W. Separate cis sequences and trans factors direct metabolic and developmental regulation of a potato tuber storage protein gene. *Plant J.* **1994**, *5*, 815–826. [CrossRef] [PubMed]

32. Sun, C.; Palmqvist, S.; Olsson, H.; Borén, M.; Ahlandsberg, S.; Jansson, C. A novel WRKY transcription factor, *SUSIBA2*, participates in sugar signaling in barley by binding to the sugar-responsive elements of the iso1 promoter. *Plant Cell* **2003**, *15*, 2076–2092. [CrossRef] [PubMed]

33. Jiang, Y.; Deyholos, M.K. Functional characterization of *Arabidopsis* NaCl-inducible *WRKY25* and *WRKY33* transcription factors in abiotic stresses. *Plant Mol. Biol.* **2008**, *69*, 91–105. [CrossRef]

34. Li, S.; Fu, Q.; Chen, L.; Huang, W.-D.; Yu, D. *Arabidopsis thaliana* WRKY25, WRKY26, and WRKY33 coordinate induction of plant thermotolerance. *Planta* **2011**, *233*, 1237–1252. [CrossRef]

35. Zou, C.; Jiang, W.; Yu, D. Male gametophyte-specific WRKY34 transcription factor mediates cold sensitivity of mature pollen in *Arabidopsis*. *J. Exp. Bot.* **2010**, *61*, 3901–3914. [CrossRef]

36. Park, C.Y.; Lee, J.H.; Yoo, J.H.; Moon, B.C.; Choi, M.S.; Kang, Y.H.; Lee, S.M.; Kim, H.S.; Kang, K.Y.; Chung, W.S.; et al. WRKY group IId transcription factors interact with calmodulin. *FEBS Lett.* **2005**, *579*, 1545–1550. [CrossRef] [PubMed]

37. Sun, Y.; Yu, D. Activated expression of *AtWRKY53* negatively regulates drought tolerance by mediating stomatal movement. *Plant Cell Rep.* **2015**, *34*, 1295–1306. [CrossRef] [PubMed]

38. Zheng, Y.; Ge, J.; Bao, C.; Chang, W.; Liu, J.; Shao, J.; Liu, X.; Su, L.; Pan, L.; Zhou, D.-X. Histone deacetylase *HDA9* and WRKY53 transcription factor are mutual antagonists in regulation of plant stress response. *Mol. Plant* **2020**, *13*, 598–611. [CrossRef]

39. Jiang, Y.; Liang, G.; Yu, D. Activated expression of WRKY57 confers drought tolerance in *Arabidopsis*. *Mol. Plant* **2012**, *5*, 1375–1388. [CrossRef]

40. Ren, X.; Chen, Z.; Liu, Y.; Zhang, H.; Zhang, M.; Liu, Q.; Hong, X.; Zhu, J.-K.; Gong, Z. ABO3, a WRKY transcription factor, mediates plant responses to abscisic acid and drought tolerance in *Arabidopsis*. *Plant J.* **2010**, *63*, 417–429. [CrossRef]

41. Wang, L.; Ma, K.-B.; Lu, Z.-G.; Ren, S.-X.; Jiang, H.-R.; Cui, J.-W.; Chen, G.; Teng, N.-J.; Lam, H.-M.; Jin, B. Differential physiological, transcriptomic and metabolomic responses of *Arabidopsis* leaves under prolonged warming and heat shock. *BMC Plant Biol.* **2020**, *20*, 1–15. [CrossRef]

42. Ren, S.; Ma, K.; Lu, Z.; Chen, G.; Cui, J.; Tong, P.; Wang, L.; Teng, N.; Jin, B. Transcriptomic and metabolomic analysis of the heat-stress response of *Populus tomentosa*. Carr. For. **2019**, *10*, 383. [CrossRef]

43. Wang, J.; Wang, L.; Yan, Y.; Zhang, S.; Li, H.; Gao, Z.; Wang, C.; Guo, X. GhWRKY21 regulates ABA-mediated drought tolerance by fine-tuning the expression of *GhHAB* in cotton. *Plant Cell Rep.* **2020**, *39*. [CrossRef]

44. Liu, X.; Song, Y.; Xing, F.; Wang, N.; Wen, F.; Zhu, C. GhWRKY25, a group I WRKY gene from cotton, confers differential tolerance to abiotic and biotic stresses in transgenic *Nicotiana benthamiana*. *Protoplasma* **2015**, *253*, 1265–1281. [CrossRef]

45. Jia, H.; Wang, C.; Wang, F.; Liu, S.; Li, G.; Guo, X. GhWRKY68 reduces resistance to salt and drought in transgenic *Nicotiana benthamiana*. *PLoS ONE* **2015**, *10*, e0120646. [CrossRef]

46. Wang, M.; Vannozzi, A.; Wang, G.; Liang, Y.-H.; Tornielli, G.B.; Zenoni, S.; Cavallini, E.; Pezzotti, M.; Cheng, Z.-M. Genome and transcriptome analysis of the grapevine (*Vitis vinifera* L.) WRKY gene family. *Hortic. Res.* **2014**, *1*, 14016. [CrossRef] [PubMed]

47. Dang, F.F.; Wang, Y.N.; Yu, L.; Eulgem, T.; Lai, Y.; Liu, Z.Q.; Wang, X.; Qiu, A.L.; Zhang, T.X.; Lin, J.; et al. *CaWRKY40*, a WRKY protein of pepper, plays an important role in the regulation of tolerance to heat stress and resistance to Ralstonia solanacearum infection. *Plant Cell Environ.* **2013**, *36*, 757–774. [CrossRef]

48. Sun, J.; Hu, W.; Zhou, R.; Wang, L.; Wang, X.; Wang, Q.; Feng, Z.-J.; Yu, H.; Qiu, D.; He, G.; et al. The *Brachypodium distachyon BdWRKY36* gene confers tolerance to drought stress in transgenic tobacco plants. *Plant Cell Rep.* **2014**, *34*, 23–35. [CrossRef]

49. Gong, X.; Zhang, J.; Hu, J.; Wang, W.; Wu, H.; Zhang, Q.; Liu, J.-H. FcWRKY70, a WRKY protein of *Fortunella crassifolia*, functions in drought tolerance and modulates putrescine synthesis by regulating arginine decarboxylase gene. *Plant Cell Environ.* **2015**, *38*, 2248–2262. [CrossRef]

50. Wu, X.; Shiroto, Y.; Kishitani, S.; Ito, Y.; Toriyama, K. Enhanced heat and drought tolerance in transgenic rice seedlings overexpressing *OsWRKY11* under the control of *HSP101* promoter. *Plant Cell Rep.* **2009**, *28*, 21–30. [CrossRef] [PubMed]

51. Song, Y.; Chen, L.; Zhang, L.; Yu, D. Overexpression of *OsWRKY72* gene interferes in the abscisic acid signal and auxin transport pathway of *Arabidopsis*. *J. Biosci.* **2010**, *35*, 459–471. [CrossRef]

52. Dai, X.; Wang, Y.; Zhang, W.-H. *OsWRKY74*, a WRKY transcription factor, modulates tolerance to phosphate starvation in rice. *J. Exp. Bot.* **2016**, *67*, 947–960. [CrossRef]

53. Yokotani, N.; Sato, Y.; Tanabe, S.; Chujo, T.; Shimizu, T.; Okada, K.; Yamane, H.; Shimono, M.; Sugano, S.; Takatsuji, H.; et al. WRKY76 is a rice transcriptional repressor playing opposite roles in blast disease resistance and cold stress tolerance. *J. Exp. Bot.* **2013**, *64*, 5085–5097. [CrossRef]

54. Wang, H.; Hao, J.; Chen, X.; Hao, Z.; Wang, X.; Lou, Y.; Peng, Y.; Guo, Z. Overexpression of rice WRKY89 enhances ultraviolet B tolerance and disease resistance in rice plants. *Plant Mol. Biol.* **2007**, *65*, 799–815. [CrossRef]

55. Zhou, Q.-Y.; Tian, A.-G.; Zou, H.-F.; Xie, Z.-M.; Lei, G.; Huang, J.; Wang, C.-M.; Wang, H.-W.; Zhang, J.-S.; Chen, S.-Y. Soybean WRKY-type transcription factor genes, *GmWRKY13*, *GmWRKY21*, and *GmWRKY54*, confer differential tolerance to abiotic stresses in transgenic *Arabidopsis* plants. *Plant Biotechnol. J.* **2008**, *6*, 486–503. [CrossRef]

56. Yan, H.; Jia, H.; Chen, X.; Hao, L.; An, H.; Guo, X. The cotton WRKY transcription factor *GhWRKY17* functions in drought and salt stress in transgenic *Nicotiana benthamiana* through ABA signaling and the modulation of reactive oxygen species production. *Plant Cell Physiol.* **2014**, *55*, 2060–2076. [CrossRef]

57. Cai, R.; Dai, W.; Zhang, C.; Wang, Y.; Wu, M.; Zhao, Y.; Ma, Q.; Xiang, Y.; Cheng, B. The maize WRKY transcription factor *ZmWRKY17* negatively regulates salt stress tolerance in transgenic *Arabidopsis* plants. *Planta* **2017**, *246*, 1215–1231. [CrossRef]

58. Niu, C.-F.; Wei, W.; Zhou, Q.-Y.; Tian, A.-G.; Hao, Y.-J.; Zhang, W.; Ma, B.; Lin, Q.; Zhang, Z.-B.; Zhang, J.-S.; et al. Wheat WRKY genes *TaWRKY2* and *TaWRKY19* regulate abiotic stress tolerance in transgenic Arabidopsis plants. *Plant Cell Environ.* **2012**, *35*, 1156–1170. [CrossRef]

59. Wang, F.; Hou, X.; Tang, J.; Wang, Z.; Wang, S.; Jiang, F.; Li, Y. A novel cold-inducible gene from Pak-choi (*Brassica campestris* ssp. chinensis), *BcWRKY46*, enhances the cold, salt and dehydration stress tolerance in transgenic tobacco. *Mol. Biol. Rep.* **2011**, *39*, 4553–4564. [CrossRef]

60. Wang, Z.; Zhu, Y.; Wang, L.; Liu, X.; Liu, Y.; Phillips, J.; Deng, X. A WRKY transcription factor participates in dehydration tolerance in *Boea hygrometrica* by binding to the W-box elements of the galactinol synthase (*BhGolS1*) promoter. *Planta* **2009**, *230*, 1155–1166. [CrossRef]

61. Li, H.; Xu, Y.; Xiao, Y.; Zhu, Z.; Xie, X.; Zhao, H.; Wang, Y. Expression and functional analysis of two genes encoding transcription factors, *VpWRKY1* and *VpWRKY2*, isolated from Chinese wild *Vitis pseudoreticulata*. *Planta* **2010**, *232*, 1325–1337. [CrossRef]

62. Zhu, Z.; Shi, J.; Cao, J.; He, M.; Wang, Y. *VpWRKY3*, a biotic and abiotic stress-related transcription factor from the Chinese wild *Vitis pseudoreticulata*. *Plant Cell Rep.* **2012**, *31*, 2109–2120. [CrossRef] [PubMed]

63. Wei, W.; Zhang, Y.; Han, L.; Guan, Z.; Chai, T. A novel WRKY transcriptional factor from *Thlaspi caerulescens* negatively regulates the osmotic stress tolerance of transgenic tobacco. *Plant Cell Rep.* **2008**, *27*, 795–803. [CrossRef]

64. Skibbe, M.; Qu, N.; Galis, I.; Baldwin, I.T. Induced plant defenses in the natural environment: *Nicotiana attenuata* WRKY3 and WRKY6 coordinate responses to herbivory. *Plant Cell* **2008**, *20*, 1984–2000. [CrossRef]

65. Yang, G.; Zhang, W.; Liu, Z.; Yi-Maer, A.Y.; Zhai, M.; Xu, Z. Both *JrWRKY2* and *JrWRKY7* of *Juglans regia* mediate responses to abiotic stresses and abscisic acid through formation of homodimers and interaction. *Plant Biol.* **2017**, *19*, 268–278. [CrossRef]

66. Yang, Z.; Chi, X.; Guo, F.; Jin, X.; Luo, H.; Hawar, A.; Chen, Y.; Feng, K.; Wang, B.; Qi, J.; et al. *SbWRKY30* enhances the drought tolerance of plants and regulates a drought stress-responsive gene, *SbRD19*, in sorghum. *J. Plant Physiol.* **2020**, *246*, 153142. [CrossRef]

67. Song, Y.; Li, J.; Sui, Y.; Han, G.; Zhang, Y.; Guo, S.; Sui, N. The sweet sorghum *SbWRKY50* is negatively involved in salt response by regulating ion homeostasis. *Plant Mol. Biol.* **2020**, *102*, 603–614. [CrossRef]

68. Qin, Z.; Hou, F.; Li, A.; Dong, S.; Wang, Q.; Zhang, L. Transcriptome-wide identification of WRKY transcription factor and their expression profiles under salt stress in sweetpotato (*Ipomoea batatas* L.). *Plant Biotechnol. Rep.* **2020**, *14*, 599–611. [CrossRef]

69. Zhu, H.; Zhou, Y.; Zhai, H.; He, S.; Zhao, N.; Liu, Q. A novel sweetpotato WRKY transcription factor, *IbWRKY2*, positively regulates drought and salt tolerance in transgenic. *Arabidopsis* **2020**, *10*, 506. [CrossRef]

70. Dong, Q.; Zheng, W.; Duan, D.; Huang, D.; Wang, Q.; Liu, C.; Li, C.; Gong, X.; Li, C.; Mao, K.; et al. *MdWRKY30*, a group IIa WRKY gene from apple, confers tolerance to salinity and osmotic stresses in transgenic apple callus and *Arabidopsis* seedlings. *Plant Sci.* **2020**, *299*, 110611. [CrossRef] [PubMed]

71. Ma, Y.; Xue, H.; Zhang, F.; Jiang, Q.; Yang, S.; Yue, P.; Wang, F.; Zhang, Y.; Li, L.; He, P.; et al. The miR156/SPL module regulates apple salt stress tolerance by activating *MdWRKY100* expression. *Plant Biotechnol. J.* **2020**, *18*. [CrossRef]

72. Ahammed, G.J.; Li, X.; Yang, Y.; Liu, C.; Zhou, G.; Wan, H.; Cheng, Y. Tomato WRKY81 acts as a negative regulator for drought tolerance by modulating guard cell H_2O_2–mediated stomatal closure. *Environ. Exp. Bot.* **2020**, *171*, 103960. [CrossRef]

73. Luo, X.; Li, C.; He, X.; Zhang, X.; Zhu, L.-F. ABA signaling is negatively regulated by GbWRKY1 through JAZ1 and ABI1 to affect salt and drought tolerance. *Plant Cell Rep.* **2019**, *39*, 181–194. [CrossRef]

74. Wang, M.-Q.; Huang, Q.-X.; Lin, P.; Zeng, Q.-H.; Li, Y.; Liu, Q.-L.; Zhang, L.; Pan, Y.-Z.; Jiang, B.-B.; Zhang, F. The overexpression of a transcription factor gene *VbWRKY32* enhances the cold tolerance in *Verbena bonariensis*. *Front. Plant Sci.* **2020**, *10*, 1746. [CrossRef]

75. Chanwala, J.; Satpati, S.; Dixit, A.; Parida, A.; Giri, M.K.; Dey, N. Genome-wide identification and expression analysis of WRKY transcription factors in pearl millet (*Pennisetum glaucum*) under dehydration and salinity stress. *BMC Genom.* **2020**, *21*, 1–16. [CrossRef]

76. Zhao, K.; Zhang, D.; Lv, K.; Zhang, X.; Cheng, Z.; Li, R.; Zhou, B.; Jiang, T. Functional characterization of poplar WRKY75 in salt and osmotic tolerance. *Plant Sci.* **2019**, *289*, 110259. [CrossRef]

77. Anjum, S.A.; Xie, X.-Y.; Wang, L.-C.; Saleem, M.F.; Man, C.; Lei, W. Morphological, physiological and biochemical responses of plants to drought stress. *Afr. J. Agr. Res.* **2011**, *6*, 2026–2032. [CrossRef]

78. Sakuma, Y.; Maruyama, K.; Osakabe, Y.; Qin, F.; Seki, M.; Shinozaki, K.; Yamaguchi-Shinozaki, K. Functional analysis of an *Arabidopsis* transcription factor, *DREB2A*, involved in drought-responsive gene expression. *Plant Cell* **2006**, *18*, 1292–1309. [CrossRef] [PubMed]

79. Hassan, S.; Lethin, J.; Blomberg, R.; Mousavi, H.; Aronsson, H. In silico based screening of WRKY genes for identifying functional genes regulated by WRKY under salt stress. *Comput. Biol. Chem.* **2019**, *83*, 107131. [CrossRef]

80. Linxiao, W.; Song, A.; Gao, C.; Wang, L.; Wang, Y.; Sun, J.; Jiang, J.; Chen, F.; Chen, S. *Chrysanthemum* WRKY gene *CmWRKY17* negatively regulates salt stress tolerance in transgenic chrysanthemum and *Arabidopsis* plants. *Plant Cell Rep.* **2015**, *34*, 1365–1378. [CrossRef]

81. Shen, Z.; Yao, J.; Sun, J.; Chang, L.; Wang, S.; Ding, M.; Qian, Z.; Zhang, H.; Zhao, N.; Sa, G.; et al. *Populus euphratica* HSF binds the promoter of *WRKY1* to enhance salt tolerance. *Plant Sci.* **2015**, *235*, 89–100. [CrossRef] [PubMed]

82. Jaglo-Ottosen, K.R. Arabidopsis CBF1 overexpression induces COR genes and enhances freezing tolerance. *Science* **1998**, *280*, 104–106. [CrossRef]

83. Chen, L.; Zhao, Y.; Xu, S.; Zhang, Z.; Xu, Y.; Zhang, J.; Chong, K. *OsMADS57* together with *OsTB1* coordinates transcription of its target *OsWRKY94* and *D14* to switch its organogenesis to defense for cold adaptation in rice. *New Phytol.* **2018**, *218*, 219–231. [CrossRef]

84. Jin, B.; Wang, L.; Wang, J.; Jiang, K.-Z.; Wang, Y.; Jiang, X.X.; Ni, C.-Y.; Wang, Y.; Teng, N.-J. The effect of experimental warming on leaf functional traits, leaf structure and leaf biochemistry in *Arabidopsis thaliana*. *BMC Plant Biol.* **2011**, *11*, 35. [CrossRef] [PubMed]

85. Liu, H.-T.; Li, G.-L.; Chang, H.; Sun, D.-Y.; Zhou, R.-G.; Li, B. Calmodulin-binding protein phosphatase PP7 is involved in thermotolerance in *Arabidopsis*. *Plant Cell Environ.* **2007**, *30*, 156–164. [CrossRef] [PubMed]

86. Larkindale, J.; Huang, B. Effects of abscisic acid, salicylic acid, ethylene and hydrogen peroxide in thermotolerance and recovery for creeping bentgrass. *Plant Growth Regul.* **2005**, *47*, 17–28. [CrossRef]

87. Zhao, J.; He, Q.; Chen, G.; Wang, L.; Jin, B. Regulation of non-coding RNAs in heat stress responses of plants. *Front. Plant Sci.* **2016**, *7*, 1213. [CrossRef]

88. Gong, X.-Q.; Hu, J.-B.; Liu, J.-H. Cloning and characterization of *FcWRKY40*, A WRKY transcription factor from *Fortunella crassifolia* linked to oxidative stress tolerance. *Plant Cell Tissue Organ Cult. PCTOC* **2014**, *119*, 197–210. [CrossRef]

89. Cheong, Y.H.; Chang, H.-S.; Gupta, R.; Wang, X.; Zhu, T.; Luan, S. Transcriptional profiling reveals novel interactions between wounding, pathogen, abiotic stress, and hormonal responses in *Arabidopsis*. *Plant Physiol.* **2002**, *129*, 661–677. [CrossRef]

90. Robatzek, S.; Somssich, I.E. A new member of the *Arabidopsis* WRKY transcription factor family, *AtWRKY6*, is associated with both senescence- and defence-related processes. *Plant J.* **2001**, *28*, 123–133. [CrossRef]

91. Kilian, J.; Whitehead, D.; Horak, J.; Wanke, D.; Weinl, S.; Batistic, O.; D'Angelo, C.; Bornberg-Bauer, E.; Kudla, J.; Harter, K. The AtGenExpress global stress expression data set: Protocols, evaluation and model data analysis of UV-B light, drought and cold stress responses. *Plant J.* **2007**, *50*, 347–363. [CrossRef]

92. Han, Y.; Zhang, X.; Wang, Y.; Ming, F. The suppression of WRKY44 by GIGANTEA-miR172 pathway is involved in drought response of *Arabidopsis thaliana*. *PLoS ONE* **2013**, *8*, e73541. [CrossRef] [PubMed]

Publisher's Note: MDPI stays neutral with regard to jurisdictional claims in published maps and institutional affiliations.

Article

Transcriptome Analysis of High-NUE (T29) and Low-NUE (T13) Genotypes Identified Different Responsive Patterns Involved in Nitrogen Stress in Ramie (*Boehmeria nivea* (L.) Gaudich)

Longtao Tan [1,2], Gang Gao [1], Chunming Yu [1], Aiguo Zhu [1], Ping Chen [1], Kunmei Chen [1], Jikang Chen [1,*] and Heping Xiong [1,*]

[1]   Institute of Bast Fiber Crops, Chinese Academy of Agricultural Sciences, Changsha 410205, China; tanlongtao@isa.ac.cn (L.T.); gaogang@caas.cn (G.G.); yuchunming@caas.cn (C.Y.); zhuaiguo@caas.cn (A.Z.); Chenping02@caas.cn (P.C.); Chenkunmei@caas.cn (K.C.)

[2]   Institute of Subtropical Agriculture, The Chinese Academy of Sciences, Changsha 410125, China

*   Correspondence: chenjikang@caas.cn (J.C.); ramiexhp@vip.163.com (H.X.); Tel.: +86-0731-88998518 (J.C.); +86-0731-88998599 (H.X.)

Received: 15 May 2020; Accepted: 16 June 2020; Published: 19 June 2020

Abstract: Nitrogen-use efficiency (NUE) has significant impacts on plant growth and development. NUE in plants differs substantially in physiological resilience to nitrogen stress; however, the molecular mechanisms underlying enhanced resilience of high-NUE plants to nitrogen deficiency remains unclear. We compared transcriptome-wide gene expression between high-NUE and low-NUE ramie (*Boehmeria nivea* (L.) Gaudich) genotypes under nitrogen (N)-deficient and normal conditions to identify the transcriptomic expression patterns that contribute to ramie resilience to nitrogen deficiency. Two ramie genotypes with contrasting NUE were used in the study, including T29 (NUE = 46.01%) and T13 (NUE = 15.81%). Our results showed that high-NUE genotypes had higher gene expression under the control condition across 94 genes, including frontloaded genes such as GDSL esterase and lipase, gibberellin, UDP-glycosyltransferase, and omega-6 fatty acid desaturase. Seventeen stress-tolerance genes showed lower expression levels and varied little in response to N-deficiency stress in high-NUE genotypes. In contrast, 170 genes were upregulated under N deficiency in high-NUE genotypes but downregulated in low-NUE genotypes compared with the controls. Furthermore, we identified the potential key genes that enable ramie to maintain physiological resilience under N-deficiency stress, and categorized these genes into three groups based on the transcriptome and their expression patterns. The transcriptomic and clustering analysis of these nitrogen-utilization-related genes could provide insight to better understand the mechanism of linking among the three gene classes that enhance resilience in high-NUE ramie genotypes.

Keywords: ramie (*Boehmeria nivea* (L.) Gaudich); transcriptome; nitrogen deficiency; resilience; nitrogen-use efficiency

---

1. Introduction

Increasing population and consumption are placing unprecedented demands on agriculture and natural resources [1]. One of the greatest challenges of the 21st century is meeting the world's growing need for agricultural products while simultaneously reducing agriculture's environmental harm. Improvements in nitrogen-use efficiency (NUE) in crop production are critical for addressing the challenge [2]. Therefore, the development and cultivation of new varieties containing genetic traits associated with abiotic stress tolerance will be essential in order to sustainably grow high-yielding crops under increasingly stressful environmental conditions [3].

Fiber extracted from ramie (*Boehmeria nivea* L.) stem is the longest and one of the strongest natural fine textile fibers [4]. In recent years, the rapidly growing application of ramie for addressing needs of feed [5,6], phytoremediation [7], and mushroom production [8] has made the crop a research hotspot, and the commercial cultivation of this crop has increased in countries such as China, Brazil, and the Philippines [9]. Nitrogen is the primary nutrient required for optimal growth and fiber yield [10–12]. Despite the agronomic importance of ramie, our previous studies showed that nitrogen fertilizer was applied excessively in ramie fields and ramie presented a very low nitrogen agronomy efficiency (NAE, 23.2–27.8%) in traditional farming systems. In addition, ramie was mainly planted in Yangtze River Valley in China, which exhibits a rapid alternation from droughts to floods. Hence, decreasing the human and environmental costs and risks associated with nitrogen loss and pollution in ramie fields is critical.

It is imperative to understand the molecular mechanisms of stress tolerance for breeding and cultivation of high-NUE genotypes [13,14]. Some genetic variability in N uptake has been thoroughly investigated [15,16] and genome-wide responses to low-nitrogen (N) stress have been described in many plants [17–21]. There have also been a few studies discussing N utilization and metabolism in ramie [22], but none considered N-stress simultaneously with genotypes. Previous studies mainly focused on differentially expressed genes (DEGs) that are more sensitive to nitrogen rates and had higher plasticity during genetic modification, especially those genes expressed much more highly in high-NUE genotypes than in low-NUE genotypes. However, without comparing the transcriptional differences between low-N-tolerant and low-N-sensitive genotypes, separating stress-tolerance genes from stress-responsive genes was impossible [23]. Additionally, the express pattern of constitutive primed genes in higher-tolerance genotypes cannot be detected by a single type of experiment due to the variety specificity. Therefore, we hypothesized that classification of express patterns of ramie DEGs by testing genotypes with contrasting NUE would increase our understanding of the N-utilization mechanism and enable us to focus on targeted genes.

In the present study, we used *de novo* transcriptome assembly and digital gene expression (DGE) profiling to measure gene expression differences between high- and low-NUE genotypes under nitrogen-deficient (0 mmol·L^{-1}) and normal conditions (10 mmol·L^{-1}). The objectives were to investigate the effects of N deficiency on the gene expression of two contrasting ramie varieties differing in NUE, and the functional classification of DEGs was conducted to explore the resilience mechanism of ramie.

2. Materials and Methods

2.1. Plant Material and N Treatment

Two ramie varieties, H2000-03 (T29) and Ceheng Jiama (T13), were identified to have distinct NUE in our previous studies (Supplementary Information S1). The NUE of T29 and T13 was 46.01% and 15.81%, respectively. Cuttings of these varieties, 13.5 ± 1.5 cm in length, were collected and rooted in a hydroponic apparatus with water only in a plant growth chamber in May 2015. During the period of ramie culture, the relative humidity was 60%, temperature was 25 °C, and the photoperiod was 16 h/8 h (light/dark). On the seventh day, ramie plants were divided into two groups and provisioned with 15 L modified Hoagland solution [24] (Table 1). Thirty plants of each variety cultured with 10 mmol·L^{-1} N formed the control group (T29_C and T13_C), and others cultured with 0 mmol·L^{-1} N formed the treatment group (T29_T and T13_T).

Table 1. Components of macro- and micro-elements in nutrient solution.

Salts		Concentration (mg·L^{-1})	
		N0	**N10**
Macro-elements	Ca(NO$_3$)$_2$ 4(H$_2$O)	-	944.6
	KNO$_3$	-	202.2
	KCl	-	223.65
	K$_2$SO$_4$	435.65	-
	KH$_2$PO$_4$	136.07	136.07
	MgSO$_4$ 7H$_2$O	492.96	492.96
	CaCl2	554.9	110.98
Micro-elements	H$_3$BO$_3$	2.86	2.86
	MnSO$_4$·H$_2$O	1.55	1.55
	ZnSO$_4$·7H$_2$O	0.22	0.22
	CuSO$_4$·5H$_2$O	0.08	0.08
	H$_2$MoO$_4$·4H$_2$O	0.09	0.09
	FeNa·EDTA	13.00	13.00

Samples from roots, stems, and leaves were collected after 0.5, 1, 3, 5, and 7 days of culturing. Each sample was derived from three plants and two biological replicates. All samples of all treatments were mixed equally and sequenced to construct the transcriptome library. Subsequently, all samples of each treatment were mixed for gene expression analysis. Both treatments of each genotype were replicated twice for Illumina sequencing and DGE analysis. After 45 days of N treatment, shoots and roots were harvested separately for biomass testing.

2.2. RNA Isolation and Library Preparation for DGE Sequencing

The total RNA was extracted using an E.Z.N.A.® Plant RNA Kit (OMEGA Bio-Tek, Norcross, GA, USA) according to the manufacturer's protocol. The RNA-seq and assembly were performed at Novogene Bioinformatics Technology CO., LTD., Beijing, China (http://www.novogene.cn) using the Illumina platform. Briefly, 3 µg total RNA per sample was used and treated with DNase I. The RNA quality and integrity were detected by the Agilent Bioanalyzer 2100 system. RNA-seq libraries were constructed using NEBNext® Ultra™ RNA Library Prep Kit for Illumina® (NEB, Ipswich, MA, USA) according to the manufacturer's protocol. Library quality was assessed using an Agilent Bioanalyzer 2100 (Agilent Technologies, Santa Clara, CA, USA). Whole RNA sequencing datasets were submitted to NCBI.

2.3. Quality Control and Quantification of Gene Expression Levels

Raw reads of FASTQ format were first processed through in-house perl scripts. In this step, clean reads were obtained by removing reads containing adaptors, ploy-N residues, and low-quality reads from the raw data. At the same time, Q20, Q30, GC content, and sequence duplication level of clean data were calculated. All downstream analyses were based on clean, high-quality reads.

Gene expression levels were estimated using RNA-seq by RSEM (RNA-Seq by Expectation Maximization) [25]; clean data of each sample were mapped back onto the assembled transcriptome and a read count for each gene was obtained from the mapping results. Transcriptome assembly was accomplished based on the left.fq and right.fq using Trinity [26] with min_kmer_cov set to 2, and all other parameters at default settings.

2.4. Gene Annotation

Gene function was annotated based on the following databases: NCBI Non-redundant Protein Sequences (NR); Swiss-Prot; Gene Ontology (GO); NCBI Nucleotide Sequences (NT); Clusters of Orthologous Groups of proteins (COGs); Protein family (Pfam); and the KEGG Ortholog (KO).

2.5. Differential Expression Analysis

The differentially expressed genes of the two treatments were analyzed using the DESeq package in R 1.10.1 (R Foundation for Statistical Computing, Vienna, Austria). DESeq performs statistical analyses to determine differential expression in DGE data using a model based on the negative binomial distribution. The resulting *p* values were adjusted using the Benjamini–Hochberg procedure [27] to control the false discovery rate. Genes with an adjusted *p* value < 0.05 were considered differentially expressed.

2.6. Pathway Enrichment Analysis of DEGs

Gene ontology enrichment analysis of differentially expressed genes (DEGs) was done using the GOseq packages in R based on Wallenius' noncentral hypergeometric distribution [28], which adjusts for gene length bias in DEGs.

KEGG is a database resource for understanding high-level functions and utilities in biological systems [29]. In this study, KOBAS [30] software was used to test the statistical enrichment of differentially expressed genes in KEGG pathways.

2.7. Quantitative Real-Time PCR (qRT-PCR) Analysis

Gene quantification was performed using a two-step reaction process: reverse transcription (RT) and PCR. Each RT reaction consisted of 0.5 µg RNA, 2 µL of PrimerScript Buffer, 0.5 µL of oligo dT, 0.5 µL of random hexamers, and 0.5 µL of PrimerScript RT Enzyme Mix I (TaKaRa BioInc., Shiga, Japan) in a total volume of 10 µL. Reactions were performed in an Applied Biosystems® GeneAmp® PCR System 9700 (Thermo Fisher Scientific Inc., Waltham, MA, USA) for 15 min at 37 °C, followed by heat inactivation of RT for 5 s at 85 °C. The 10 µL RT reaction mix was then diluted (1:10) in nuclease-free water and stored at −20 °C.

Real-time PCR was performed using a LightCycler® 480 Real-Time PCR Instrument (Roche, Swiss) with a 10 µL PCR reaction mixture that included 1 µL of cDNA, 5 µL of 2 × LightCycler® 480 SYBR Green I Master (Roche, Swiss), 0.2 µL of forward primer, 0.2 µL of reverse primer, and 3.6 µL of nuclease-free water. Reactions were incubated in a 384 well optical plate (Roche, Swiss) at 95 °C for 10 min, followed by 40 cycles at 95 °C for 10 s, and 60 °C for 30 s. Each sample was run in triplicate. At the end of the PCR cycles, a melting-curve analysis was performed to validate the expected PCR products. Primer sequences were designed in the laboratory and synthesized based on mRNA sequences obtained from the NCBI database.

Messenger RNA expression levels were normalized for 15 genes and were calculated using the $2^{-\Delta\Delta Ct}$ method [31]. Briefly, melting-curve analysis of the amplified products was performed at the end of each PCR run to confirm that only one PCR product was amplified and detected. The Ct values of target genes were thus normalized with the 18S rRNA reference gene, while a mathematical model was used to determine the relative expression ratio. The real-time PCR-primers for validated gene expression in the results of DGEs and housekeeping genes are shown in the Supplementary Information S2.

3. Results

3.1. The Performance of Ramie Genotypes under Different Nitrogen Treatments

Both the genotypes showed a remarkable decrease in shoot and root biomass after 45 days of treatment under N-deficit stress (Figure 1). Thus, nitrogen deficiency inhibited the growth of ramie regardless of the NUE performance. The shoot biomass of T29 and T13 was decreased by 20% and 59%, respectively. A similar pattern was observed in root biomass. Therefore, we hypothesized that the low-NUE ramie genotype were more sensitive to N deficiency.

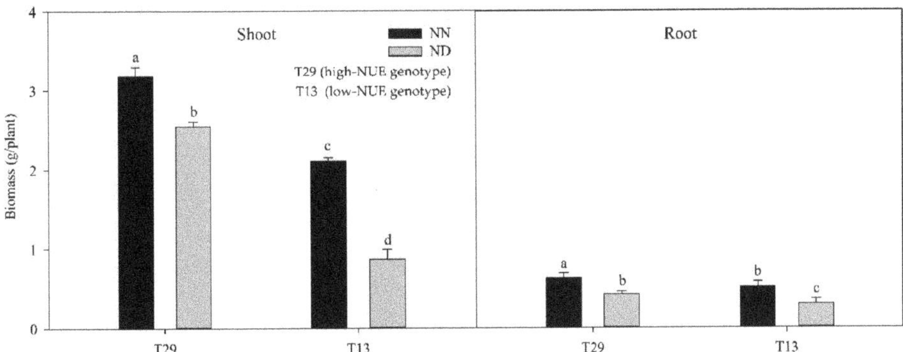

Figure 1. Plant growth performance of ramie genotypes under nitrogen deficiency. ND: N-deficient conditions, NN: normal-N conditions. The error bar represents the standard error. The different letters in the chart indicate significant differences at $p < 0.05$ between the genotypes according to SNK test.

3.2. De Novo Transcriptome Sequencing in High-NUE and Low-NUE Ramie Genotypes

Mixed samples of four control plants (T29_C1, T29_C2, T13_C1, and T13_C2) and four N-deficit-treated plants (T29_T1, T29_T2, T13_T1, and T13_T2) were used to construct libraries of total RNA. A total of 55.1 million clean sequence reads (96.58% of all raw reads) were obtained at an error rate below 1%. Clean reads were spliced into transcripts using Trinity software, resulting in a total of 80,493 transcripts with an average length of 919 bp. From these transcripts, 61,424 non-redundant unigenes were yielded. Seven public databases were used for validation and annotation of the assembled unigenes. The results indicated that 50.3%, 24.0%, 19.3%, 38.3%, 37.4%, 39.0%, and 23.2% of the unigenes showed similarity to known proteins in the NR, NT, KO, Swiss-Prot, PFAM, GO, and KOG databases, respectively (Table 2). Together, 34,251 (55.8%) unigenes showed similarity to known proteins in at least one of these seven databases.

Table 2. Functional annotation of the ramie transcriptome in the seven public databases searched.

Database	Number of Unigenes	Matching Proportion
NCBI Non-redundant Protein Sequences (NR)	30,919	50.3%
NCBI Nucleotide Sequences (NT)	14,730	24.0%
KEGG Ortholog (KO)	11,856	19.3%
Swiss-Prot	23,502	38.3%
Protein family (Pfam)	22,973	37.4%
Gene Ontology (GO)	23,968	39.0%
Clusters of Orthologous Groups of proteins (COGs)	14,230	23.2%
Total	34,251	55.8%

3.3. Global Analysis of Differential Gene Expression

All sequencing reads were realigned to the unigenes to determine the expression levels of the 61,424 unigenes assembled *de novo*. In total, the DGE libraries generated between 10.1 and 12.2 million raw reads. After removing insignificant reads, the total number of clean reads ranged from 9.93 to 11.89 million. Clean reads per library were mapped to the reference transcriptome database using RSEM software, and represented between 88.31% and 90.20% of the overall total. The FPKM values of all unigenes were compared between replicates. Significant correlations between replicates were observed, which suggested that the abundances of these unigene transcripts in the replicate libraries were similar (Figure 2).

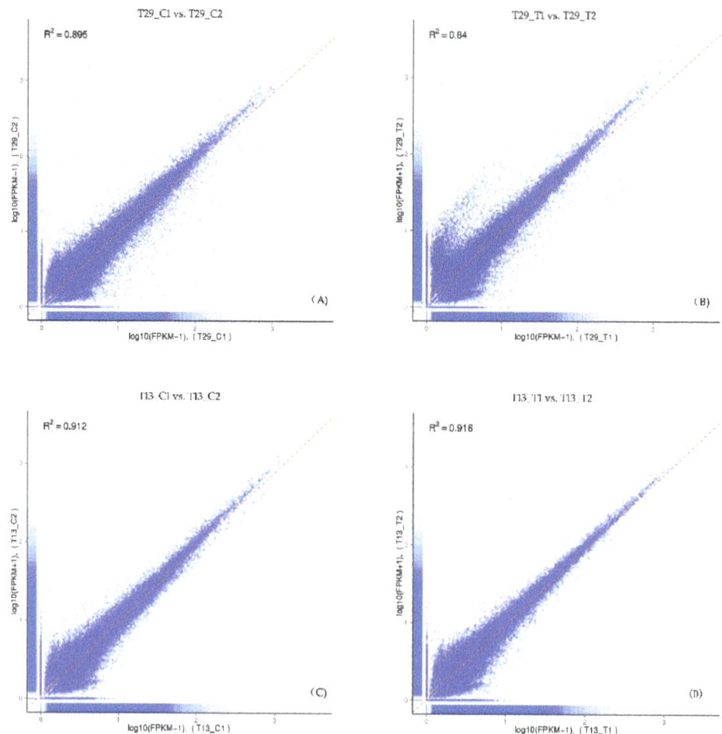

Figure 2. The correlation between the expression levels for each gene for the two biological replicates. Abscissa indicates log10(RPKM + 1) of Biological Replicate 1, Ordinate indicates log10(RPKM + 1) of Biological Replicate 2. T29 is the high-NUE genotype and T13 is the low-NUE genotype. Subfigures (**A**), (**B**), (**C**) and (**D**) show the correlation between the two biological replicates of T29_C (high-NUE genotype T29 under normal N treatment), T29_T (T29 under N-deficit treatment), T13_C (low-NUE genotype T13 under normal N treatment) and T13_T (T13 under N-deficit treatment), respectively.

The numbers of DEGs that were up- and downregulated were determined in the comparisons of the two treatments and genotypes (19, 312, 953, and 950 for all; Figure 3). Among these four comparisons, 1699 unigenes were identified, comprising 1265 unique NR database matches. These genes had an average fold-change of 22.35 (range: 1.49–4465.36) for 739 upregulated and −30.72 (range: −1.4−−7426.18) for 960 downregulated unigenes (Supplementary Information S3). The most highly upregulated gene was a hypothetical protein, POPTR_0008s16330g, belonging to the PHD finger protein family (c57966_g1), which are regarded as key factors in chromatin and transcription regulation [32]. The most downregulated gene was UPL1, a zinc finger protein with E3 ubiquitin ligase activity (c33163_g1). The classifications of the 1265 unigenes included kinase (57 unigenes), transcription factor (31 unigenes), cytochrome P450 (30 unigenes), photosystem (30 unigenes), disease resistance protein (27 unigenes), UDP-glycosyltransferase (14 unigenes), glutathione *S*-transferase (12 unigenes), and GDSL esterase and lipase (10 unigenes), among others.

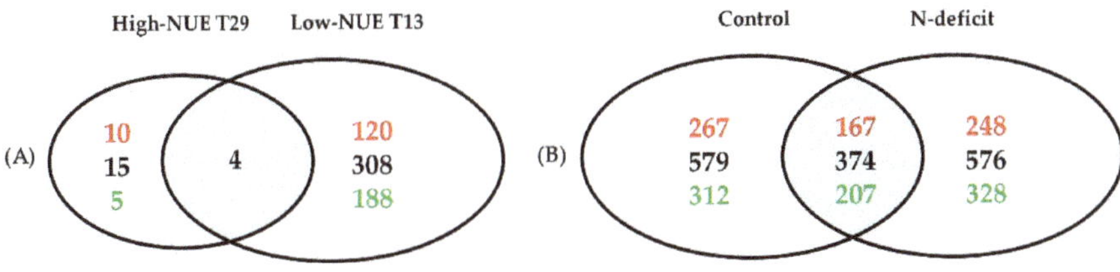

Figure 3. Venn diagram of DEGs in different treatments. The sum of the numbers in each large circle represents the total number of differentially expressed genes between combinations; the overlapping part of the circles colored with cyan represents common DEGs between combinations. Black numbers represent the total DEG number. (**A**) DEGs responding to N deficit compared to control, in which red numbers represent upregulated DEGs and green numbers represent downregulated DEGs. (**B**) DEGs responding to different genotypes, in which red numbers represent DEG numbers in high-NUE T29 and green numbers represent DEG numbers in low-NUE T13.

3.4. Genotype-Specific Gene Expression within Treatments

The two genotypes differed significantly in gene expression under the normal-N condition across 953 genes (434 higher in T29 (mean fold-change: 26.65); and 519 higher in T13 (mean fold-change: −53.69)) There were 30 GO categories between T29_C and T13_C related to catalytic activity (351 unigenes) and binding-related activity (130 unigenes) (Supplementary Information S3 and S4). The main significantly enriched categories were lysosome, glutathione metabolism (seven unigenes), retinol metabolism (six unigenes), metabolism of xenobiotics by cytochrome P450 (six unigenes), phenylpropanoid biosynthesis (six unigenes), and carbohydrate digestion and absorption (three unigenes). Under N-deficit conditions, high-NUE genotype T29 and low-NUE genotype T13 differed significantly in gene expression across 950 genes (415 higher in T29 (mean fold-change: 27.63); and 535 higher in T13 (mean fold-change: −25.11)) of these genes, 374 were commonly and differentially expressed between T29_C and T13_C, and T29_T and T13_T. There were 24 GO categories between T29_T and T13_T related to single organism metabolic process (207 unigenes) and oxidoreductase activity (117 unigenes). The main significantly enriched categories were carbon metabolism (20 unigenes), biosynthesis of amino acids (16 unigenes), phenylpropanoid biosynthesis (12 unigenes), and phenylalanine metabolism (9 unigenes).

3.5. Variety-Specific Responses to N-Deficit Conditions

Low-NUE genotype T13 (121 upregulated and 191 downregulated genes) showed exceedingly different transcriptome response to N deficit than high-NUE genotype T29 (13 upregulated and 6 downregulated genes). In T13, 308 genes reacted significantly to N deficit that did not in T29.

Not all 308 genes showed the same direction of change in the two varieties. A downregulated gene was abandoned due to lack of significant change in T29. A total of 121 upregulated genes showed a greater increase, and 187 downregulated genes showed a greater decrease in T13 (χ^2 test; $p < 0.01$).

In total, 94 of 120 genes in the high-NUE genotype T29_T, displayed reduced upregulation compared with higher expression in T29_C and T13_C (i.e., each gene's ratio in T29_C and T13_C expression levels) and lower expression levels in T29_T and T13_T, i.e., each gene's ratio in T29_T and T13_T fold-change values in the N-deficit treatment (Figure 4). The relationship showed that genes reacted less sensitively during N-deficit conditions in T29, and may have exhibited a higher level of gene expression before the onset of N-deficit. Another potential explanation is that many upregulated genes in T13 were already expressed at higher levels in T29, accounting for our results.

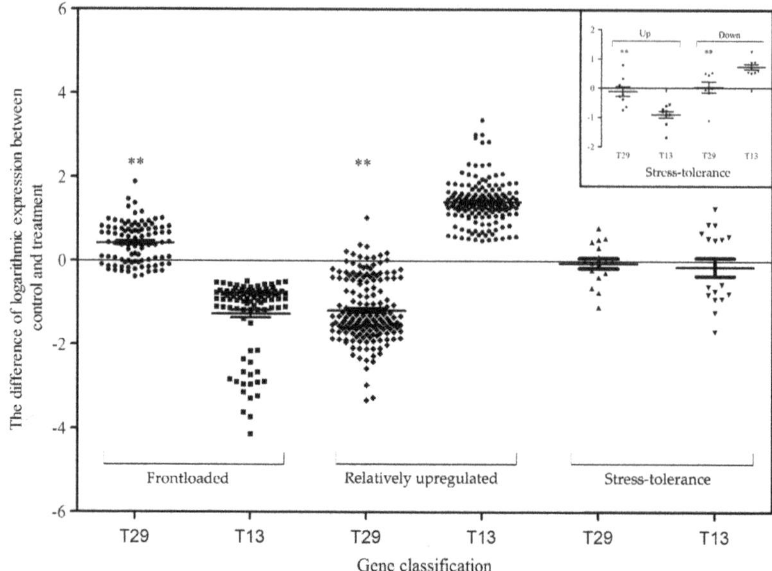

Figure 4. Scatter plot of expression of genes between high-NUE T29 and low-NUE T13, unique to the T13_T vs. T13_C. The difference of logarithmic expression between control and treatment is shown on the vertical axis ln(T29_C/T29_T) and ln(T13_C/T13_T); the types of differentially expression genes are shown on the horizontal axis. The small figure shows the up- and downregulated details of stress-tolerance genes in T13. ** indicates significant difference between the two genotypes at $p < 0.01$, vertical error bars represent mean $\pm$ SD.

Nine of 120 upregulated genes and 8 of 187 downregulated genes (Figure 5) responded less sensitively to N-deficit in T29 compared with T13_T, and showed equal or even lower expression in the T29_C compared with T13_C. The results showed that the high-NUE genotype T29 had a reduced reaction during N deficiency.

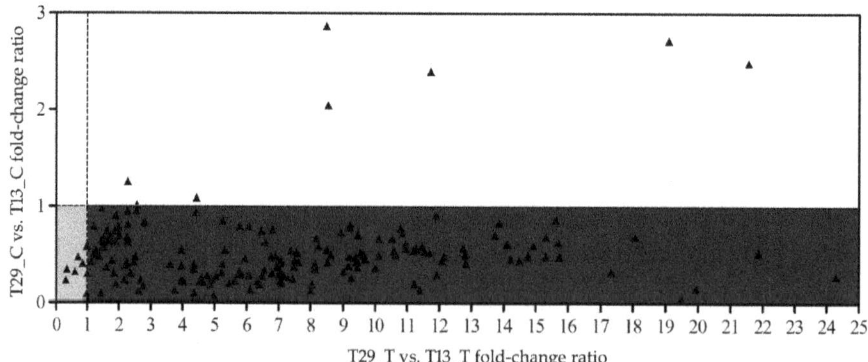

Figure 5. Scatter plot comparing the relative ratio of treatment-to-control fold-changes in expression between high-NUE T29 and low-NUE T13 across the 187 downregulated differently expressed genes unique to the T13_T vs. T13_C. y < 1 T29 control showed lower expression relative to T13 control, x > 1 T29 treatment showed higher expression relative to T13 treatment. Numbers greater than 25 are not shown. The gray portion of the graph represents the stress-tolerance genes and the dark portion represents relatively upregulated genes.

A total of 94 genes had higher constitutive expression levels under the control treatment but had a lower response to N deficiency in the high-NUE T29. These genes included GDSL esterase and lipase, zipper protein, NADH dehydrogenase, and UDP-glycosyltransferase. A total of 17 genes had lower constitutive expression levels and a lower reactivity to N deficiency in T29, including auxin-induced protein, 1-aminocyclopropane-1-carboxylate oxidase-1-like protein, and inter-alpha-trypsin inhibitor heavy chain H3. There were 170 genes that were relatively upregulated in T29 but downregulated in T13, including wall-associated receptor kinase, serine carboxypeptidase, homeobox-leucine zipper protein, and ferulate 5-hydroxylase. These genes represent potential candidates contributing to N-deficit-stress resistance. Overall, there were three markedly different expression patterns of these genes. Transcripts of these three gene categories were annotated with 15 GO classification terms, including single organism metabolic process (85 unigenes), oxidoreduction (63 unigenes), binding-related (19 unigenes), and dioxygenase activity (14 unigenes).

In contrast, eight genes with reduced upregulation in T13_T showed higher expression in T13_C and T29_C but lower expression in T13_T and T29_T (Figure 6). Two genes showed equal or lower expression in T13_C compared with T29_C, which showed a reduced reaction during N deficiency in the low-NUE genotype. Four genes were downregulated in T29 that responded as relatively upregulated in N-deficient stress in T13. Five of eight transcripts were annotated with one GO classification term, including oxidoreductase activity, acting on paired donors, with incorporation or reduction of molecular oxygen. DEGs were significantly enriched in alpha-linolenic acid metabolism.

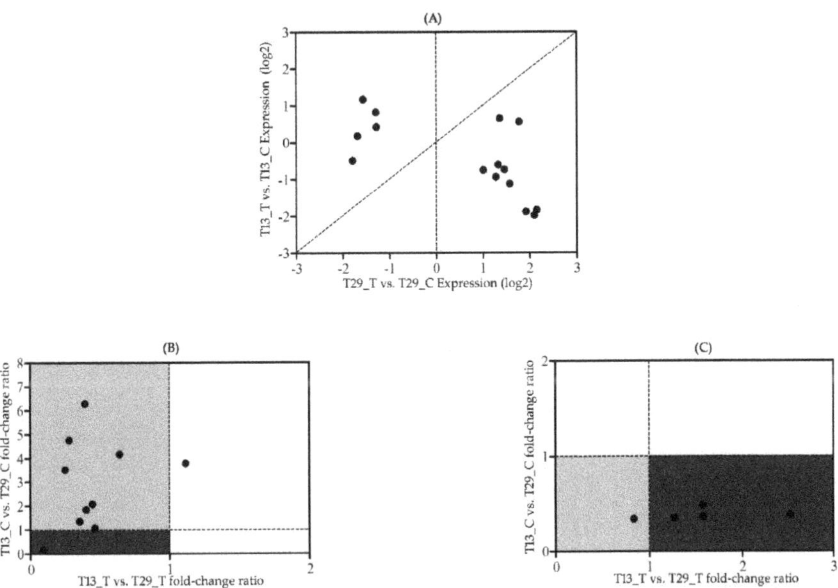

Figure 6. Scatter plot of expression between low-NUE T13 and high-NUE T29 across the 15 differently expressed genes unique to T29_T vs. T29_C. (**A**) Scatter plot of the log2 fold-changes in expression of 15 genes that were unique to T29_T vs. T29_C in response to N-deficit stress in T29 and T13. (**B**) Comparison of the relative ratio of treatment-to-control fold-changes in expression of the 10 upregulated DEGs unique to T29_T vs. T29_C between T29 and T13. (**C**) Comparison of the relative ratio of treatment-to-control fold-changes in expression of the five downregulated DEGs unique to T29_T vs. T29_C between T29 and T13. Each filled dot represents an individual gene; the dotted-line is a 1:1 line; the gray and dark portions of the (**B**) graph represent the frontloaded and stress-tolerance genes, respectively; the gray and dark portions of the (**C**) graph represent stress-tolerance or relatively upregulated genes, respectively.

There were four genes that displayed opposite expression patterns under N deficiency in T29 and T13. In T29, three genes were upregulated and one gene was downregulated, and T13 displayed the opposite pattern. The role of these genes in N-deficiency stress is unclear.

3.6. Quantitative Real-Time PCR

To validate the reliability of Illumina sequencing technology, 15 genes were randomly selected from both of the genotypes under N-deficit treatment for quantitative RT-PCR assays. The results showed that all 15 of these genes had different expression levels in the treatment and control, and the trend of expression changes based on qRT-PCR was the same as that detected by DGE analysis (Supplementary Information S5).

4. Discussion

4.1. Feasibility of Gene Expression Pattern in Ramie NUE

Among the essential plant nutrients, N plays the most important role in improving the agricultural production of ramie [33]. Hence, it is important to maximize nitrogen-use efficiency. As nitrogen-saving cultivation becomes a trend, such as ISSM (Integrated Soil-crop System Management) [34], strategies combining high yield and low nitrogen rate together should be applied in ramie farming. In the present study, significant differences were found in whole plant biomass, above-ground biomass, and shoot/root ratio between the two varieties under both normal and N-deficit conditions. Additionally, the genome-wide expression profile of molecular response to N deficiency in different ramie varieties was characterized for the first time. A total of 170 N-stress-responsive genes were identified. The results suggest that the NUE of ramie could be improved significantly by genetic strategies, and ramie cultivation could be operated more economically by using varieties with higher NUE under a lower nitrogen rate.

4.2. Functional Classification of N-Stress-Responsive Genes in Ramie

A number of studies have found that the basic ramie stress response involves a wide array of molecular processes which cause differences in physiological resilience [35–37]. Similarly, in the present study, genes related to numerous physiological functions were identified to be responsive to nitrogen deficiency. However, none of the previous studies involved varieties with contrast traits. Numerous genes showed altered expression levels under N-deficit conditions, and many of the genes that showed striking different responses to N deficiency in low-NUE genotypes showed reduced response in the high-NUE genotypes (Supplementary Information S6). That is, we found that low-NUE and high-NUE ramies showed different patterns of gene expression under N deficiency. Based on the transcriptome and the differential expression pattern, we categorized the potential key genes related to nitrogen utilization of ramie: (1) frontloaded genes, which already exhibiting a higher gene expression level in high-NUE genotypes compared with low-NUE genotypes under control conditions, but showed a reduced response to N-deficit in a single-genotype-experiment; (2) relatively upregulated genes, which displayed relatively high expression levels under N-deficit stress in high-NUE genotypes, but were downregulated in low-NUE genotypes; and (3) stress-tolerance genes, which showed lower expression levels and were not sensitive to N-deficiency stress in high-NUE genotypes compared to low-NUE genotypes. These gene categories may be useful for exploring the potential mechanisms of high NUE, and to explain why high-NUE genotypes can grow and yield well under N-deficiency stress. We discuss the possible mechanisms of these genes on nitrogen-utilization regulation in the following paragraphs.

4.3. Frontloaded Genes

Frontloaded genes can be defined as the genes already upregulated under control conditions in tolerant populations before responding to stress. Barshis et al. [38] proposed that constitutive preloading enables an individual coral (*Acropora hyacinthus*) to maintain physiological resilience during frequently encountered environmental stresses, an idea that has strong parallels in model systems such as yeast (*Saccharomyces cerevisiae*). A total 94 potentially frontloaded genes were found in T29 in our study. Among these genes identified, some might be critical in response to N deficiency, such as GDSL esterase and lipase [39–41], gibberellin 20-oxidase and 3-beta-dioxygenase [42–45], UDP-glycosyltransferase [46,47], and ω-6 fatty acid desaturase [32,48]. These results suggest that these high-NUE ramie genotypes with more physiological resilience follow a similar pattern of gene preloading.

Systemic acquired resistance refers to a distinct signal transduction pathway implicated in the ability of plants to defend themselves against biotic and abiotic stresses [49]. Biotic and abiotic stresses prime subsequent changes in tolerance in many plants. In eggplant, seedlings grew well under cold stress through enhanced antioxidant enzyme activity and related gene expression if they received a salicylic acid pre-treatment at a concentration of 0.3% [50]. In maize and cucumber, a pre-treatment with H_2O_2 and paraquat reduced salt-induced oxidative damage by increasing the antioxidative mechanisms [51,52]. These results are similar to the increased adaptability to N deficit seen in the high-NUE genotype T29. Frontloaded genes may reduce stress through faster protein mobilization and produce results similar to those acquired with pre-treatments.

4.4. Relatively Upregulated Genes

Numerous genes showed relatively high expression levels under N deficiency in the high-NUE genotype and were downregulated in the low-NUE genotype, including wall-associated receptor kinase (WAK), serine carboxypeptidase (SCP), and homeobox-leucine zipper protein (HD-Zip). Previous studies have suggested that some WAK members play an important role in responses to aluminum, cell elongation, and plant development [53–55]. Serine carboxypeptidases comprise a large family of protein-hydrolyzing enzymes that play roles in multiple cellular processes. *OsBISCPL1* and *GS5* encode a putative SCP, and *OsBISCPL1*-overexpressing plants showed an increased tolerance to oxidative stress and upregulated expression of oxidative-stress-related genes. Higher expression levels of *GS5* can increase grain size and yield in rice [56,57]. Homeodomain-leucine zipper proteins are transcription factors unique to plants. These proteins are generally involved in responses related to abiotic stress, abscisic acid (ABA), blue light, de-etiolation, and embryogenesis [58]. WAK, SCP, and HD-Zip are involved in reducing damage due to various stresses. A number of differentially expressed genes followed a similar pattern of relative upregulation in the high-NUE genotype. We hypothesized that these genes respond to nitrogen stress, polarize expression among varieties, and then lead to differentiation in production performance.

4.5. Stress-Tolerance Genes

Some genes responded less sensitively to N deficiency in high-NUE genotypes, and showed equal or even lower level of gene expression in T29_C compared with T13_C. In our dataset, 5.52% of the 308 reduced-reaction genes fell into this category, of which, 9 were upregulated and 8 were downregulated. In these stress-tolerance genes, one was a match to the inter-alpha-trypsin inhibitor heavy chain H3 precursor (ITIH3; c32816_g1 and c34225_g1; Supplementary Information S7), and the other was a BLAST match to 1-aminocyclopropane-1-carboxylate oxidase-1-like protein (ACC; c21936_g1 and c25703_g1). ACC synthase is a rate-limiting oxidase in the regulation of the ethylene biosynthesis. As a plant hormone, ethylene plays a significant role in plant maturation and senescence [59,60]. Saving energy and transferring substances for essential life activities are important to survival in adverse environments [61]. The present study showed that reduced expression changes in T29 might delay senescence, which would enable the plants to be resilient than T13 under N-deficiency stress.

5. Conclusions

Ramie plants respond to the environment in a complex fashion, and the magnitude of changes in gene expression was different between genotypes under N-deficiency stress. Many studies on the effect of low-N conditions on gene expression in one or two genotypes have been reported. However, our results went a step further to reveal the resilience-regulation mechanism in ramie by using two genotypes contrasting in NUE. The study detected many DEGs responding to nitrogen deficit, and we categorized these genes into three main groups based on the transcriptome and their expression patterns. A total 94 genes showed a higher expression level in high-NUE genotype T29 before exposure to N-deficient conditions compared to the low-NUE genotype T13; these are defined as frontloaded genes, which are important in reacting to N-deficiency stress in high-NUE genotypes. A total of 17 genes responded less sensitively to N deficiency in the high-NUE genotype T29, and showed equal or even low expression levels compared to the low-NUE genotype T13 under normal N conditions. Defined as stress-tolerance genes, these might enable the plant to have a more resilient adaptability to N deficiency. A total of 170 genes responded with relatively high expression levels under N deficiency in the high-NUE genotype, and were downregulated in the low-NUE genotype; defined as relatively upregulated genes, these were major drivers for NUE differentiation in ramie. There three categories of genes express patterns were used to construct a multi-dimensional molecular mechanism of ramie in N-stress response. These concepts could provide further insight into the mechanism linking these three gene classes and the enhanced resilience of high-NUE genotypes. We identified only eight frontloaded genes, two stress-tolerance genes, and four relatively upregulated genes in the low-NUE genotype. This result explains why high-NUE genotypes have increased yield and growth under N-deficiency stress than low-NUE genotypes. In addition, the three gene categories presented a hypothesis regarding the way ramie responds to N-deficiency stress, which might serve to explain other differences in plants under biotic and abiotic stresses. Different patterns of gene expression under different abiotic conditions (N deficiency) should be taken into consideration for ramie molecular breeding.

Supplementary Materials: The following are available online at http://www.mdpi.com/2223-7747/9/6/767/s1, S1 NUE of ramie genotypes used in the screening experiment. S2 Real-time PCR-primers for validated gene expression in results of DGEs and housekeeping genes. S3 Results for the 1699 differentially expressed genes (DEGs) between control and treatments. S4 Summary of the part differentially expressed genes grouped by functional category. S5 The DGEs validated by qRT-PCR. S6 Category of DGEs in T29 showed different responding characteristics under nitrogen deficit condition. S7 308 unigenes that were unique to the T13 DEG set.

Author Contributions: Conceptualization, L.T., J.C.; Data curation, L.T., J.C.; validation, P.C.; formal analysis, L.T., G.G.; investigation, L.T., J.C.; resources, H.X.; writing—original draft preparation, L.T., J.C.; writing—review and editing, G.G., K.C., A.Z.; visualization, L.T., J.C.; supervision, H.X.; project administration, C.Y., H.X.; funding acquisition, J.C., H.X. All authors have read and agreed to the published version of the manuscript.

Funding: This research was funded by National Natural Science Foundation of China (31701477), China Agriculture Research System for Bast and Leaf Fiber Crops (CARS-16), and Central Public-interest Scientific Institution Basal Research Fund (No.1610242020001).

Conflicts of Interest: The authors declare no conflict of interest.

References

1. Foley, J.A.; Ramankutty, N.; Brauman, K.A.; Cassidy, E.S.; Gerber, J.S.; Johnston, M.; Mueller, N.D.; O. Connell, C.; Ray, D.K.; West, P.C.; et al. Solutions for a cultivated planet. *Nature* **2011**, *478*, 337–342. [CrossRef] [PubMed]

2. Zhang, X.; Davidson, E.A.; Mauzerall, D.L.; Searchinger, T.D.; Dumas, P.; Shen, Y. Managing nitrogen for sustainable development. *Nature* **2015**, *528*, 51–59. [CrossRef]

3. Adbelrahman, M.; Burritt, D.J.; Tran, L.P. The use of metabolomic quantitative trait locus mapping and osmotic adjustment traits for the improvement of crop yields under environmental stresses. *Semin. Cell Dev. Biol.* **2017**. [CrossRef]

4. Mitra, S.; Saha, S.; Guha, B.; Chakrabarti, K.; Satya, P.; Sharma, A.K.; Gawande, S.P.; Kumar, M.; Saha, M. Ramie: The Strongest Bast Fibre of Nature. *Tech. Bull.* **2013**, *8*, 1–38. [CrossRef]

5. Angelini, L.G.; Tavarini, S. Ramie [*Boehmeria nivea* (L.) Gaud.] as a potential new fibre crop for the Mediterranean region: Growth, crop yield and fibre quality in a long-term field experiment in Central Italy. *Ind. Crop. Prod.* **2013**, *51*, 138–144. [CrossRef]

6. Kipriotis, E.; Heping, X.; Vafeiadakis, T.; Kiprioti, M.; Alexopoulou, E. Ramie and kenaf as feed crops. *Ind. Crop. Prod.* **2015**, *68*, 126–130. [CrossRef]

7. Luan, M.; Jian, J.; Chen, P.; Chen, J.; Chen, J.; Gao, Q.; Gao, G.; Zhou, J.; Chen, K.; Guang, X.; et al. Draft genome sequence of ramie, *Boehmeria nivea* (L.) Gaudich. *Mol. Ecol. Resour.* **2018**. [CrossRef] [PubMed]

8. Xie, C.; Gong, W.; Yan, L.; Zhu, Z.; Hu, Z.; Peng, Y. Biodegradation of ramie stalk by Flammulina velutipes: Mushroom production and substrate utilization. *AMB Express* **2017**, *7*, 171. [CrossRef]

9. Jose, S.; Tajna, S.; Ghosh, P. Ramie Fibre Processing and Value Addition. *Asian J. Text.* **2017**, *7*, 1–9. [CrossRef]

10. Angelini, L.G.; Lazzeri, A.; Levita, G.; Fontanelli, D.; Bozzi, C. Ramie (*Boehmeria nivea* (L.) Gaud.) and Spanish Broom (*Spartium junceum* L.) fibres for composite materials: Agronomical aspects, morphology and mechanical properties. *Ind. Crop. Prod.* **2000**, *11*, 145–161. [CrossRef]

11. Özgür, T.; Emre, I.; Aykut, T.F.; Hamdi, A.; Önder, C. Impact of Different Nitrogen and Potassium Application on Yield and Fiber Quality of Ramie (*Boehmeria nivea*). *Int. J. Agri. Biol.* **2010**, *12*, 369–372.

12. Subandi, M. The Effect of Fertilizers on the Growth and the Yield of Ramie (*Boehmeria nivea* L. Gaud). *Asian Econ. Soc. Soc.* **2012**, *2*, 126.

13. Raun, W.R.; Johnson, G.V. Improving nitrogen use efficiency for cereal production. *Agron. J.* **1999**, *91*, 357–363. [CrossRef]

14. Stanton, M.L.; Roy, B.A.; Thiede, D.A. Evolution in stressful environments. I. Phenotypic variability, phenotypic selection, and response to selection in five distinct environmental stresses. *Evolution* **2000**, *54*, 93–111. [CrossRef]

15. Huang, C.; Wei, G.; Luo, Z.; Xu, J.; Zhao, S.; Wang, L.; Jie, Y. Effects of nitrogen on ramie (*Boehmeria nivea*) hybrid and its parents grown under field conditions. *J. Agr. Sci. (Tor.)* **2014**, *6*, 230–243. [CrossRef]

16. Cabangbang, R.P. Fiber yield and agronomic characters of ramie as affected by plant density and fertilizer level. *Philipp. J. Crop. Sci.* **1978**, *3*, 78–89.

17. Wang, R.; Guegler, K.; LaBrie, S.T.; Crawford, N.M. Genomic analysis of a nutrient response in Arabidopsis reveals diverse expression patterns and novel metabolic and potential regulatory genes induced by nitrate. *Plant Cell* **2000**, *12*, 1491–1509. [CrossRef] [PubMed]

18. Lian, X.; Wang, S.; Zhang, J.; Feng, Q.; Zhang, L.; Fan, D.; Li, X.; Yuan, D.; Han, B.; Zhang, Q. Expression profiles of 10,422 genes at early stage of low nitrogen stress in rice assayed using a cDNA microarray. *Plant Mol. Biol.* **2006**, *60*, 617–631. [CrossRef]

19. Bi, Y.M.; Kant, S.; Clark, J.; Gidda, S.; Ming, F.; XU, J.Y.; Rochon, A.; Shelp, B.J.; Hao, L.X.; Zhao, R.; et al. Increased nitrogen-use efficiency in transgenic rice plants over-expressing a nitrogen-responsive early nodulin gene identified from rice expression profiling. *Plant Cell Environ.* **2009**, *32*, 1749–1760. [CrossRef]

20. Hao, Q.N.; Zhou, X.A.; Sha, A.H.; Wang, C.; Zhou, R.; Chen, S.L. Identification of genes associated with nitrogen-use efficiency by genome-wide transcriptional analysis of two soybean genotypes. *BMC Genom.* **2011**, *12*, 525. [CrossRef]

21. Chen, R.; Tian, M.; Wu, X.; Huang, Y. Differential global gene expression changes in response to low nitrogen stress in two maize inbred lines with contrasting low nitrogen tolerance. *Genes Genom.* **2011**, *33*, 491–497. [CrossRef]

22. Deng, G.; Liu, L.J.; Zhong, X.Y.; Lao, C.Y.; Wang, H.Y.; Wang, B.; Zhu, C.; Shah, F.; Peng, D.X. Comparative proteome analysis of the response of ramie under N, P and K deficiency. *Planta* **2014**, *239*, 1175–1186. [CrossRef] [PubMed]

23. Gelli, M.; Duo, Y.; Konda, A.R.; Zhang, C.; Holding, D.; Dweikat, I. Identification of differentially expressed genes between sorghum genotypes with contrasting nitrogen stress tolerance by genome-wide transcriptional profiling. *BMC Genom.* **2014**, *15*, 179. [CrossRef]

24. Chandna, R.; Ahmad, A. Nitrogen stress-induced alterations in the leaf proteome of two wheat varieties grown at different nitrogen levels. *Physiol. Mol. Biol. Plants* **2015**, *21*, 19–33. [CrossRef]

25. Li, B.; Dewey, C.N. RSEM: Accurate transcript quantification from RNA-Seq data with or without a reference genome. *BMC Bioinform.* **2011**, *12*, 323. [CrossRef]

26. Grabherr, M.G.; Haas, B.J.; Yassour, M.; Levin, J.Z.; Thompson, D.A.; Amit, I.; Adiconis, X.; Fan, L.; Raychowdhury, R.; Zeng, Q.; et al. Full-length transcriptome assembly from RNA-Seq data without a reference genome. *Nat. Biotechnol.* **2011**, *29*, 644–652. [CrossRef]

27. Benjamini, Y.; Hochberg, Y. Controlling the False Discovery Rate: A Practical and Powerful Approach to Multiple Testing. *J. R. Stat. Soc. Ser. B* **1995**, *57*, 289–300. [CrossRef]

28. Young, M.D.; Wakefield, M.J.; Smyth, G.K.; Oshlack, A. Gene ontology analysis for RNA-seq: Accounting for selection bias. *Genome Biol.* **2010**, *11*, R14. [CrossRef]

29. Kanehisa, M.; Araki, M.; Goto, S.; Hattori, M.; Hirakawa, M.; Itoh, M.; Katayama, T.; Kawashima, S.; Okuda, S.; Tokimatsu, T.; et al. KEGG for linking genomes to life and the environment. *Nucleic Acids Res.* **2007**, *36*, D480–D484. [CrossRef]

30. Mao, X.; Cai, T.; Olyarchuk, J.G.; Wei, L. Automated genome annotation and pathway identification using the KEGG Orthology (KO) as a controlled vocabulary. *Bioinformatics* **2005**, *21*, 3787–3793. [CrossRef]

31. Livak, K.J.; Schmittgen, T.D. Analysis of Relative Gene Expression Data Using Real-Time Quantitative PCR and the $2^{-\Delta\Delta C_T}$ Method. *Methods* **2001**, *25*, 402–408. [CrossRef]

32. Bienz, M. The PHD finger, a nuclear protein-interaction domain. *Trends Biochem. Sci.* **2006**, *31*, 35–40. [CrossRef] [PubMed]

33. Ullah, S.; Liu, L.; Anwar, S.; Tuo, X.; Khan, S.; Wang, B.; Peng, D. Effects of Fertilization on Ramie (*Boehmeria nivea* L.) Growth, Yield and Fiber Quality. *Sustainability* **2016**, *8*, 887. [CrossRef]

34. Cui, Z.; Zhang, H.; Chen, X.; Zhang, C.; Ma, W.; Huang, C.; Zhang, W.; Mi, G.; Miao, Y.; Li, X.; et al. Pursuing sustainable productivity with millions of smallholder farmers. *Nature* **2018**, *555*, 363–366. [CrossRef] [PubMed]

35. Gao, G.; Xiong, H.; Chen, K.; Chen, J.; Chen, P.; Yu, C.; Zhu, A. Gene expression profiling of ramie roots during hydroponic induction and adaption to aquatic environment. *Genom. Data* **2017**, *14*, 32–35. [CrossRef]

36. Huang, C.; Zhou, J.; Jie, Y.; Xing, H.; Zhong, Y.; She, W.; Wei, G.; Yu, W.; Ma, Y. A ramie (*Boehmeria nivea*) bZIP transcription factor BnbZIP3 positively regulates drought, salinity and heavy metal tolerance. *Mol. Breed.* **2016**, *36*. [CrossRef]

37. Zhu, S.; Zheng, X.; Dai, Q.; Tang, S.; Liu, T. Identification of quantitative trait loci for flowering time traits in ramie (*Boehmeria nivea* L. Gaud). *Euphytica* **2016**, *210*, 367–374. [CrossRef]

38. Barshis, D.J.; Ladner, J.T.; Oliver, T.A.; Seneca, F.O.; Traylor-Knowles, N.; Palumbi, S.R. Genomic basis for coral resilience to climate change. *Proc. Natl. Acad. Sci. USA* **2013**, *110*, 1387–1392. [CrossRef]

39. Naranjo, M.A.; Forment, J.; Roldan, M.; Serrano, R.; Vicente, O. Overexpression of Arabidopsis thaliana LTL1, a salt-induced gene encoding a GDSL-motif lipase, increases salt tolerance in yeast and transgenic plants. *Plant Cell Environ.* **2006**, *29*, 1890–1900. [CrossRef]

40. Hong, J.K.; Choi, H.W.; Hwang, I.S.; Kim, D.S.; Kim, N.H.; Choi, D.S.; Kim, Y.J.; Hwang, B.K. Function of a novel GDSL-type pepper lipase gene, CaGLIP1, in disease susceptibility and abiotic stress tolerance. *Planta* **2008**, *227*, 539–558. [CrossRef]

41. Zhu, S.; Tang, S.; Tang, Q.; Liu, T. Genome-wide transcriptional changes of ramie (*Boehmeria nivea* L. Gaud) in response to root-lesion nematode infection. *Gene* **2014**, *552*, 67–74. [CrossRef]

42. Huang, S.; Raman, A.S.; Ream, J.E.; Fujiwara, H.; Cerny, R.E.; Brown, S.M. Overexpression of 20-oxidase confers a gibberellin-overproduction phenotype in Arabidopsis. *Plant Physiol.* **1998**, *118*, 773–781. [CrossRef] [PubMed]

43. Coles, J.P.; Phillips, A.L.; Croker, S.J.; Garcia-Lepe, R.; Lewis, M.J.; Hedden, P. Modification of gibberellin production and plant development in Arabidopsis by sense and antisense expression of gibberellin 20-oxidase genes. *Plant J.* **1999**, *17*, 547–556.

44. Sakamoto, T. An Overview of Gibberellin Metabolism Enzyme Genes and Their Related Mutants in Rice. *Plant Physiol.* **2004**, *134*, 1642–1653. [CrossRef]

45. Fagoaga, C.; Tadeo, F.R.; Iglesias, D.J.; Huerta, L.; Lliso, I.; Vidal, A.M.; Talon, M.; Navarro, L.; Garcia-Martinez, J.L.; Pena, L. Engineering of gibberellin levels in citrus by sense and antisense overexpression of a GA 20-oxidase gene modifies plant architecture. *J. Exp. Bot.* **2007**, *58*, 1407–1420. [CrossRef]

46. Langenbach, C.; Campe, R.; Schaffrath, U.; Goellner, K.; Conrath, U. UDP-glucosyltransferase UGT84A2/BRT1 is required for Arabidopsis nonhost resistance to the Asian soybean rust pathogen Phakopsora pachyrhizi. *New Phytol.* **2013**, *198*, 536–545. [CrossRef]

47. Tognetti, V.B.; Van Aken, O.; Morreel, K.; Vandenbroucke, K.; van de Cotte, B.; De Clercq, I.; Chiwocha, S.; Fenske, R.; Prinsen, E.; Boerjan, W.; et al. Perturbation of Indole-3-Butyric Acid Homeostasis by the UDP-Glucosyltransferase UGT74E2 Modulates Arabidopsis Architecture and Water Stress Tolerance. *Plant Cell* **2010**, *22*, 2660–2679. [CrossRef]

48. Zhang, J.; Liu, H.; Sun, J.; Li, B.; Zhu, Q.; Chen, S.; Zhang, H. Arabidopsis fatty acid desaturase FAD2 is required for salt tolerance during seed germination and early seedling growth. *PLoS ONE* **2012**, *7*, e30355. [CrossRef]

49. Durrant, W.E.; Dong, X. Systemic Acquired Resistance. *Annu. Rev. Phytopathol.* **2004**, *42*, 185–209. [CrossRef]

50. Chen, S.; Zimei, L.; Cui, J.; Jiangang, D.; Xia, X.; Liu, D.; Yu, J. Alleviation of chilling-induced oxidative damage by salicylic acid pretreatment and related gene expression in eggplant seedlings. *Plant Growth Regul.* **2011**, *65*, 101–108. [CrossRef]

51. de Azevedo Neto, A.D.; Prisco, J.T.; Enéas-Filho, J.; Rolim Medeiros, J.; Gomes-Filho, E. Hydrogen peroxide pre-treatment induces salt-stress acclimation in maize plants. *J. Plant Physiol.* **2005**, *162*, 1114–1122. [CrossRef] [PubMed]

52. Lin, S.; Liu, Z.; Xu, P.; Fang, Y.; Bai, J. Paraquat pre-treatment increases activities of antioxidant enzymes and reduces lipid peroxidation in salt-stressed cucumber leaves. *Acta Physiol. Plant* **2011**, *33*, 295–304. [CrossRef]

53. Wagner, T.A.; Kohorn, B.D. Wall-Associated kinases are expressed throughout plant development and are required for cell expansion. *Plant Cell* **2001**, *13*, 303–318. [CrossRef]

54. Lally, D.; Ingmire, P.; Tong, H.Y.; He, Z.H. Antisense expression of a cell wall-Associated protein kinase, WAK4, inhibits cell elongation and alters morphology. *Plant Cell* **2001**, *13*, 1317–1331.

55. Hou, X. Involvement of a Cell Wall-Associated Kinase, WAKL4, in Arabidopsis Mineral Responses. *Plant Physiol.* **2005**, *139*, 1704–1716. [CrossRef]

56. Liu, H.; Wang, X.; Zhang, H.; Yang, Y.; Ge, X.; Song, F. A rice serine carboxypeptidase-like gene OsBISCPL1 is involved in regulation of defense responses against biotic and oxidative stress. *Gene* **2008**, *420*, 57–65. [CrossRef]

57. Li, Y.; Fan, C.; Xing, Y.; Jiang, Y.; Luo, L.; Sun, L.; Shao, D.; Xu, C.; Li, X.; Xiao, J.; et al. Natural variation in GS5 plays an important role in regulating grain size and yield in rice. *Nat. Genet.* **2011**, *43*, 1266–1269. [CrossRef]

58. Elhiti, M.; Stasolla, C. Structure and function of homodomain-leucine zipper (HD-Zip) proteins. *Plant Signal. Behav.* **2009**, *4*, 86–88. [CrossRef] [PubMed]

59. Starrett, D.A.; Laties, G.G. Involvement of Wound and Climacteric Ethylene in Ripening Avocado Discs. *Plant Physiol.* **1991**, *97*, 720–729. [CrossRef]

60. Yang, S.F.; Hoffman, N.E. Ethylene biosynthesis and its regulation in higher plants. *Annu. Rev. Plant Phys.* **1984**, *35*, 155–189. [CrossRef]

61. Kaur, J. A comprehensive review on metabolic syndrome. *Cardiol. Res. Pract.* **2014**, *2014*. [CrossRef] [PubMed]

Article

Genome-Wide Identification and Functional Characterization of the Heat Shock Factor Family in Eggplant (*Solanum melongena* L.) under Abiotic Stress Conditions

Jinglei Wang, Haijiao Hu, Wuhong Wang, Qingzhen Wei, Tianhua Hu and Chonglai Bao *

Institute of Vegetables Research, Zhejiang Academy of Agricultural Sciences, Hangzhou 310021, China; syauwjl@163.com (J.W.); huhj0571@126.com (H.H.); hongge5@163.com (W.W.); weiqz@mail.zaas.ac.cn (Q.W.); hutianh@126.com (T.H.)

* Correspondence: baocl@mail.zaas.ac.cn; Tel.: +86-571-8640-9722

Received: 28 June 2020; Accepted: 15 July 2020; Published: 20 July 2020

Abstract: Plant heat shock factors (Hsfs) play crucial roles in various environmental stress responses. Eggplant (*Solanum melongena* L.) is an agronomically important and thermophilic vegetable grown worldwide. Although the functions of Hsfs under environmental stress conditions have been characterized in the model plant *Arabidopsis thaliana* and tomato, their roles in responding to various stresses remain unclear in eggplant. Therefore, we characterized the eggplant SmeHsf family and surveyed expression profiles mediated by the SmeHsfs under various stress conditions. Here, using reported Hsfs from other species as queries to search SmeHsfs in the eggplant genome and confirming the typical conserved domains, we identified 20 SmeHsf genes. The SmeHsfs were further classified into 14 subgroups on the basis of their structure. Additionally, quantitative real-time PCR revealed that SmeHsfs responded to four stresses—cold, heat, salinity and drought—which indicated that SmeHsfs play crucial roles in improving tolerance to various abiotic stresses. The expression pattern of *SmeHsfA6b* exhibited the most immediate response to the various environmental stresses, except drought. The genome-wide identification and abiotic stress-responsive expression pattern analysis provide clues for further analysis of the roles and regulatory mechanism of SmeHsfs under environmental stresses.

Keywords: eggplant; heat shock factor; gene family; expression profile; abiotic stress

1. Introduction

Plants have developed various defense mechanisms that are responsive to different environmental stresses, such as drought, cold, salinity, and heat [1]. Transcription factors, like AP2/ERF, HSP90, WRKY, MYB, NAC, LOX, bZip, and heat shock (Hsfs) [2–8], are activated and regulate multiple genes and signaling pathways that enable plant adaptation to unfavorable conditions. Among them, Hsfs are involved in many aspects of protein homeostasis under stress conditions [9] and are especially involved in responding to high-temperature stress [10]. In addition to stress responses, Hsfs also play important roles in developmental processes in animals and plants [11].

Although the sequences and sizes of *Hsf* genes vary, the basic structures and promoter recognition modes are considerably conserved in higher eukaryotes [12]. Almost all the Hsfs have a highly conserved DNA-binding domain (DBD), located close to the N-terminus and containing an antiparallel four-stranded β-sheet and a three-helical bundle, which are required for specific binding with heat stress promoter elements [13–15]. The oligomerization domain (HR-A/B region), separated from the DBD domain by a flexible linker of a variable length, contributes to the trimerization of Hsfs by forming a coiled-coil structure [16]. Additionally, three other conserved structures—a nuclear localization signal (NLS), nuclear export signal (NES) and activator motif (AHA)—are present. Some Hsfs also contain a repression domain at the C-terminus [17]. On the basis of structural characteristics and phylogenetic comparisons, plant *Hsf* genes can be further divided into A, B, and C classes [12,18], which contain insertions of 21, 0, and 7 amino acid residues, respectively, between the HR-A and HR-B regions [12,18]. In addition, the amino acid length from the DBD to HR-A/B differs among the three classes [12]. The AHA is present in class A, but absent in classes B and C [17]. Class A Hsfs are involved in transcriptional activation and responses to environmental stresses [19], while class B Hsfs function as transcriptional coactivators with class A Hsfs or as gene expression repressors [9,20]. At present, there are few studies on class C; only several studies show that class C Hsfs can be induced by a variety of stresses [21,22].

Hsfs are engaged in responses to abiotic stresses conditions. For example, *Arabidopsis thaliana HSFA1*s and *HSFA2* participate in responses to various abiotic stresses, such as salinity, osmotic pressure, oxidation, and anoxia [23–25], while *HSFA1b* and *HSFA3* are involved in drought-stress responses [17,26]. Tomato *HsfA1a* plays a critical role in the development of thermotolerance and cannot be replaced by other tomato *Hsf*s [27]; *HsfA1b* and *HsfA1e* are likely responding to stress in specific tissues, while *HsfA1c* functions as a co-regulator in mild heat stress response. Tomato *HsfA2* can increase plant heat tolerance by accumulating to high levels [28] and is also involved in protecting maturing and germinating pollen under heat-stress conditions [29]. In addition, wheat *HsfA4a* is involved in cadmium tolerance [19]. Chrysanthemum *HSFA4* confers salinity tolerance as a consequence of Na$^+$/K$^+$ ion and reactive oxygen species homeostasis [30]. *HSFA2* and *A6* from wheat, *HSF3, -18, -24, -32, -37,* and *-40* from cotton and *HSF-06, -10, -14, -20,* and *-21* from maize may be involved in responding to heat stress [22,31,32]. Owing to their essential modulatory functions in plants, *Hsf* gene family members have been studied in several agronomically important plants, such as rice (*Oryza sativa*), maize (*Zea mays*), apple (*Malus domestica*), poplar (*Populus trichocarpa*), and cabbage (*Brassica oleracea*) [31,33–36]. However, the *Hsf* gene family in eggplant (*Solanum melongena* L.) has not been systematically studied.

Eggplant is an economically important vegetable cultivated worldwide. The optimal season for eggplant growth is autumn, when the temperature ranges from 22 to 30 °C [37]. During year-round production in protected cultivation, eggplant encounters various environmental stresses, including heat, cold, drought, and salinity. Here, we performed a genome-wide study to comprehensively analyze the eggplant *Hsf* gene family. We identified 20 *SmeHsf* genes and determined protein properties, phylogenetic relationships, gene structures, and conserved protein domains. We also investigated the expression changes of *Hsf* genes in plants subjected to different abiotic stresses. Our study provides a foundation for further *SmeHsf*s functional investigations and could help better understand the environmental stress-response-related molecular mechanisms of *Hsf* genes in eggplant.

2. Results

2.1. Identification, Classification, and Characterization of the Hsf Gene Family in Eggplant

A total of 20 *Hsf* genes were identified in the eggplant genome (Table S1). This is less than in pepper (25), tomato (26), potato (25), and cultivated tobacco (65). Subsequently, the 20 *SmeHsf* genes were classified into three subgroups—A, B, and C—according to the HEATSTER websites [38]. Most of the SmeHsfs, 14 out of 20, were classified into subgroup A, and these SmeHsfs were further classified into seven subgroups (A1, A3, A4, A5, A6, A8, and A9). Class B had five members from four subgroups (B1, B2, B3, and B4). Subgroups A1, A4, A6, A9, and B2 contained more than one member (Table 1).

Table 1. Physicochemical characteristics and classification of Hsf genes in eggplant.

Number	Gene Name	Gene Code	Subgroup	Protein Length (aa)	Molecular Weight (kDa)	Aromaticity	Instability Index	Isoelectric Point	GRAVY [1]
1	SmeHsfA1b	Sme2.5_02334.1_g00004.1	A1	496	54.84	0.05	52.58	5.14	−0.48
2	SmeHsfA1e	Sme2.5_00204.1_g00007.1	A1	478	53.81	0.06	51.96	5.88	−0.61
3	SmeHsfA3	Sme2.5_00292.1_g00007.1	A3	494	55.08	0.10	53.86	4.60	−0.58
4	SmeHsfA4a	Sme2.5_01013.1_g00005.1	A4	403	46.04	0.07	42.59	5.15	−0.78
5	SmeHsfA4b	Sme2.5_01314.1_g00005.1	A4	421	48.36	0.09	54.15	5.35	−0.74
6	SmeHsfA4c	Sme2.5_04312.1_g00009.1	A4	377	43.01	0.07	47.45	5.14	−0.76
7	SmeHsfA5	Sme2.5_09846.1_g00002.1	A5	475	53.31	0.07	57.57	5.51	−0.78
8	SmeHsfA6a	Sme2.5_00065.1_g00020.1	A6	357	41.70	0.10	45.43	5.15	−0.93
9	SmeHsfA6b	Sme2.5_08000.1_g00008.1	A6	324	37.93	0.08	47.19	5.50	−0.83
10	SmeHsfA6c	Sme2.5_04149.1_g00004.1	A6	343	39.84	0.09	59.73	5.99	−0.79
11	SmeHsfA8	Sme2.5_08951.1_g00003.1	A8	374	43.27	0.10	58.33	4.72	−0.62
12	SmeHsfA9a	Sme2.5_00023.1_g00025.1	A9	410	47.37	0.09	50.08	5.29	−0.55
13	SmeHsfA9b	Sme2.5_03412.1_g00012.1	A9	317	36.77	0.12	41.31	9.00	−0.74
14	SmeHsfA9c	Sme2.5_04312.1_g00005.1	A9	340	38.48	0.07	48.87	5.93	−0.86
15	SmeHsfB1	Sme2.5_00010.1_g00004.1	B1	481	53.57	0.07	30.35	5.24	−0.67
16	SmeHsfB2a	Sme2.5_13301.1_g00001.1	B2	341	38.30	0.07	60.33	6.35	−0.62
17	SmeHsfB2b	Sme2.5_02712.1_g00007.1	B2	331	36.21	0.06	55.45	5.18	−0.44
18	SmeHsfB3a	Sme2.5_00159.1_g00006.1	B3	213	24.62	0.08	52.04	9.44	−0.79
19	SmeHsfB4a	Sme2.5_01029.1_g00008.1	B4	357	40.64	0.09	60.84	7.73	−0.60
20	SmeHsfC1	Sme2.5_04829.1_g00004.1	C1	352	39.42	0.09	68.53	6.14	−0.65

[1] GRAVY—The abbreviation for grand average of hydropathy values.

The physical and chemical properties of the SmeHsf proteins were analyzed and some differences were observed. The lengths of SmeHsf proteins varied from 213 to 496 amino acids, and the molecular weights ranged from 24.62 to 55.07 kDa. Among the 20 SmeHsf proteins, SmeHsfB3a had the shortest length and lowest molecular weight. Additionally, the predicted aromaticity ranged from 0.05 to 0.11, and the isoelectric point ranged from 4.60 to 9.44 (Table 1). The instability index, which provides an estimate of the stability of a protein in a test tube, indicated that all the SmeHsf proteins are unstable (scores greater than 40), except SmeHsfB1, which had an instability index of 30.35. The grand average of hydropathy values of the SmeHsfs was less than 0, suggesting that they are hydrophilic. These differences mostly resulted from variations in the non-conserved regions' amino acid sequences.

2.2. Conserved Domains and Structural Analysis of SmeHsfs

The functional domains of the Hsfs have been studied in some model plants [11]. Detailed information regarding the conserved domains, such as DBD, HR-A/B, NLS, NES, and AHA, are presented in Table 2. As the core functional domain of the Hsfs, the DBD was composed of approximately 90 amino acids and existed in all the predicted SmeHsf proteins. In addition, another conserved domain, HR-A/B, was also present in all the SmeHsfs. Based on the number of amino acid residues inserted into the HR-A/B regions, the 20 SmeHsfs were divided into three major classes. Class A and class C Hsfs contained 21 and 7 amino acid residues between the A and B regions, respectively (Figure 1). This classification confirmed the results of HEATSTER website. Most of the SmeHsf proteins (13 out of 20) included an NLS domain, and the NES domain was detected in seven SmeHsfs. The AHA domain was detected in the A4, A5, A6, and A9 subgroups; however, it was not found in class B or C.

```
                                  Insert
            [  HR-A  ]———————————————————————————[  HR-B  ]
SmeHsfA1b   RLKRDKDILMQELVKLRQQQQETDHQLVTVGQRFQLMENRQQQMMSFLAKAM   52
SmeHsfA1e   RLKIDKNIHREELVKLRQQQQATDHRLENVGQRLQLMEQTEQQAMSFLAKVL   52
SmeHsfA3    KLRKEKSLMMQEVVELQQQQRGTVQQMESVNEKLQAAEQRQKQMVSFLAKVF   52
SmeHsfA4a   KLKHENELLHLELHRHKQDHQELETQLQVLTKRVQQVKDRQMNVLSTLARTI   52
SmeHsfA4b   RLKKENSLLQTSVENQEKFNYEYEFGIKSMEQRLQNVAHRQGKLISLLAQLL   52
SmeHsfA4c   KLKHENGSLHMDLQRHKQDHQALELQMQDLTERVQHAEHRQKTMLSALARTL   52
SmeHsfA5    KLTREKSGLEANVSKFRQQQSAAELQLEELTARLGSIEQRQESLLTFVEKAL   52
SmeHsfA6a   RLKREKQDLMMELVELKRHQQTTKSHIKAMEEKLKRTESKQQQMMNFLAKAM   52
SmeHsfA6b   RLRRDKQVLMIELVKLKQCQKTTKSHLQAMEQKLKGTEIRQQQMMSFLAKAL   52
SmeHsfA6c   RLRRDKQVLMMELVKLRQNQQNTRAYLRSLELRLQGTERKQQQMMNFLARAM   52
SmeHsfA8    NLKYERNVLMQELLKLKEHQQISESQLILLREQLEVREKNQQQMLSFIVMAM   52
SmeHsfA9a   KLKKDHNALRMEILRLKQQQESTDSCLATMKERLQNSETRQKYMVIFMAKTF   52
SmeHsfA9b   KQRDLNNTVKSEFDKLKERQDDMVNGIASLKEYLEKSEAESRKFLCFLAKAV   52
SmeHsfA9c   KLRNDHISLKVELQKLKDGQENLTSFFPNLVGCGKETDIRNIFKLLLEKSQV   52
SmeHsfC1    RLKQEQKALEHEVENMTR..............RLEATEKRPQQMMAFLCKVV   38
SmeHsfB2a   RLRLENLQLNKELNQMKQ.....................LCGNIYGMMSKYA   31
SmeHsfB2b   RLRKENAQLNQELNRLRI.....................LCNNVYNLMSNFT   31
SmeHsfB3a   RLKMENGILSSELSLMKN.....................KCKELIGIVTIFA   31
SmeHsfB4a   RLRRSNNMLMSELAHMRK.....................LYNDIIYFVQNHV   31
SmeHsfB1    KLKKDNQMLSSELVRAKK.....................QCDELVDFLSQYV   31
```

Figure 1. Multiple sequence alignment of the HR-A/B regions (OD) of SmeHsfs.

Table 2. The function domains and their position in SmeHsfs.

Gene	DBD	HR-A/B	NLS	NES	AHA
SmeHsfA1b	13–106	143–196	(205) NNSKKRRLLVSNY	(150) ILM	Na
SmeHsfA1e	10–103	140–193	(211) ITGMNKKRRFP	Na	Na
SmeHsfA3	78–171	196–249	Na	Na	Na
SmeHsfA4a	10–103	136–189	Na	Na	(341) DVFWEQFLTE
SmeHsfA4b	16–109	139–192	Na	(236) LEM	(245) INFWERFLYG; (354) DVFWQQFLTE
SmeHsfA4c	10–104	136–189	(203) NDRKRRFPG	Na	(343) DVFWEQFLTE
SmeHsfA5	13–106	133–186	(202) ISAFSKKRRLP	(193) LAQKLESMDI	(422) DVFWEQFLTE
SmeHsfA6a	34–127	159–212	(220) EIRNKRKRQID	Na	(314) EGFWEDLLNE
SmeHsfA6b	29–122	153–206	(115) LLRTIKRRKTTNF; (226) EINKKRRRPID	(265) VALNM	Na
SmeHsfA6c	33–126	160–213	(119) LLRNIKRRKTP; (222) QQKGKRKEIEEDITKKRRQPI	(191) LRL	(305) MGFWEELFND
SmeHsfA8	10–103	138–191	Na	Na	Na
SmeHsfA9a	46–139	164–217	(134) INIKRRKQYP; (229) KQGKKRKLCDAQF	Na	Na
SmeHsfA9b	29–122	147–200	Na	Na	Na
SmeHsfA9c	36–129	160–213	(219) DTRKRPCLV	Na	(269) REFWEKLFED
SmeHsfB1	6–99	161–193	(254) KEKKKKRGPD	Na	Na
SmeHsfB2a	22–115	177–209	Na	Na	Na
SmeHsfB2b	21–114	196–228	Na	Na	Na
SmeHsfB3a	2–84	137–169	(190) EMERKRKRVEL	Na	Na
SmeHsfB4a	21–114	201–233	(203) NERKRRLPG	(342) LEKNDLGL	Na
SmeHsfC1	1–83	108–147	(169) REKKRRLMIS	(155) LMEKERSKRLSL	Na

DND-binding domain (DBD), oligomerization domain (HR-A/B), nuclear export signal (NES), nuclear localization signal (NLS), activator motifs (AHA). Numbers in brackets reveals the position of the first amino acid of NLS, NES, and AHA domains in the sequence; Na—no domains detected.

To further analyze the motifs and structural variations of SmeHsfs, we constructed a separate phylogenetic tree containing only SmeHsf proteins, and then, compared motif compositions and exon/intron organizations (Figure 2A). Generally, most of the closely related members had similar motif compositions and exon/intron organizations and lengths. The Multiple Em for Motif Elicitation (MEME) web server was used to search for motifs in the SmeHsf proteins. There were 15 potential motifs distributed throughout the Hsf protein sequences (Figure 2B). Motifs 1 and 2, or Motifs 2 and 8, which corresponded to the DBD domain, were found in all the SmeHsfs. Motif 4 was also identified in all the SmeHsfs and corresponded to the HR-A/B region. Different subgroups had similar motifs and contained their own unique motifs. Motif 3 was found in class A and C members, while Motif 12 was only found in class B members.

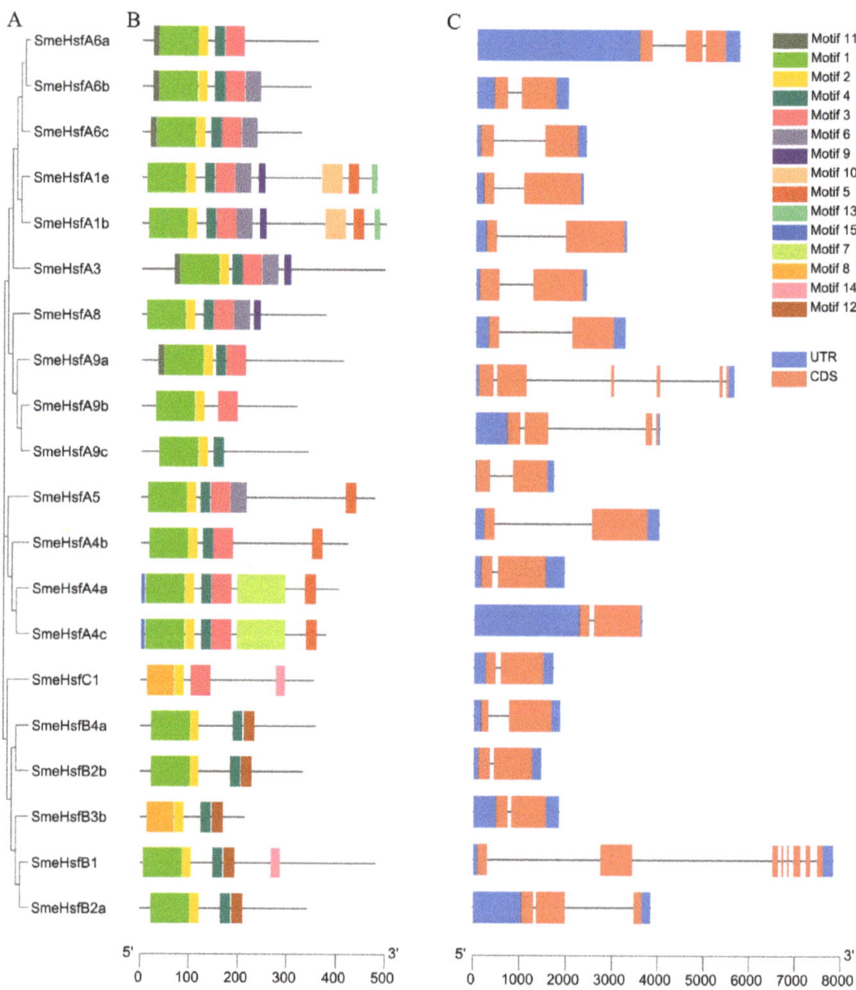

Figure 2. Phylogenetic, motif and structural analyses of SmeHsfs. (**A**) Phylogenetic tree of SmeHsf proteins. (**B**) Schematic representation of the motif compositions of SmeHsfs. (**C**) Exon/intron structures of *SmeHsf* genes.

The exon/intron structures exhibited a highly conserved organization in 14 out of 20 *SmeHsfs* possessing strictly two exons, which was similar to the structures of *Hsf* genes in other plants [39,40].

In addition, we identified two genes possessing three exons (*SmeHsfA6a* and *SmeHsfB2a*), one gene containing four exons (*SmeHsfA9b*), and two genes (*SmeHsfA9a* and *SmeHsfB1*) having more than six exons (Figure 2C).

2.3. Phylogenetic Analysis of SmeHsfs

To study the evolutionary characteristics of SmeHsf proteins, we selected three other well-studied and representative plant species, including one related species (*Solanum lycopersicum*), a monocot (*O. sativa*), and a eudicot (*A. thaliana*). The full-length amino acid sequences of Hsf proteins in eggplant and these three species were used to construct a phylogenetic tree (Figure 3 and Table S2). The *SmeHsf*s in the same subgroup were classified together, which indicated that the *SmeHsf*s in the same subgroup not only have similar domain structures, and also, have similar sequences. The phylogenetic analysis also showed that the number of *Hsf* genes in different subclasses varied among land plants. For example, eggplant has no subclass A2 members, while rice has no subclass A9 and B3 members. Besides, the *Hsf* genes only varied slightly in different subclasses between eggplant and tomato, indicating an even distribution within the family Solanaceae.

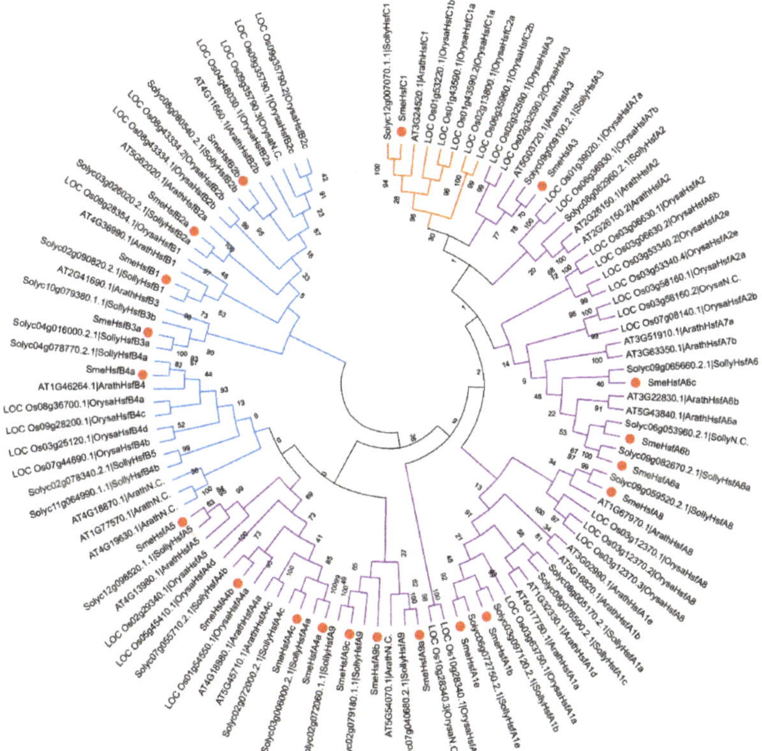

Figure 3. The phylogenetic tree of the *Hsf* genes from four plant species. Individual species are distinguished by different gene code prefixes. The prefixes Arath, Orysa, Solyc, and Sme indicate that these genes are from *A. thaliana*, rice, tomato, and eggplant, respectively. Red circles indicate eggplant genes. Additionally, purple, blue, and yellow branches indicate classes A, B, and C, respectively.

2.4. Putative Regulatory cis-Elements of the SmeHsf Promoters

To further explore the potential regulatory mechanisms of *SmeHsf*s during stress responses, we used the PlantCARE database [41] to detect the *cis*-elements in the promoters (Table S3). In total, 88 *cis*-elements were identified, with 55 having known functions. The most commonly known function was responsiveness to light (26 out of 55), followed by other regulatory functions (14 out of 55), and responsiveness to hormones (9 out of 55). In addition, four abiotic stress-response elements—LTRs, MBSs, TC-rich repeats, and WUN motifs—were identified. The SmeHsfs, except for *SmeHsfA4c*, *SmeHsfB2b*, *SmeHsfB3a*, and *SmeHsfC1*, possessed at least one stress-response-related *cis*-element (Figure 4). In total, six *SmeHsf*s had one or more LTR, suggesting a potential cold-stress response under low temperature conditions. Additionally, MBSs, TC-rich repeats, and WUN motifs were found in 9, 2, and 10 *SmeHsf*s, respectively. The *cis*-element analysis indicated that *SmeHsf* genes could respond to different abiotic stresses.

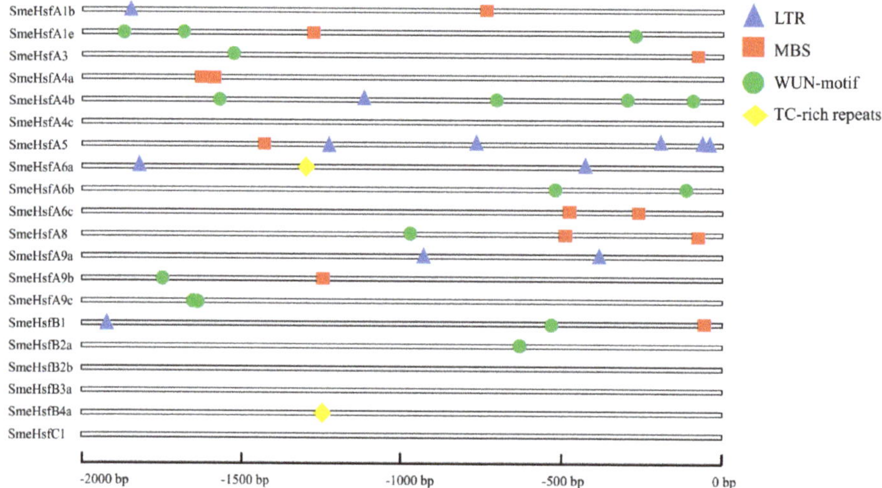

Figure 4. The position of abiotic stress-response *cis*-elements on *SmeHsf* promoters.

2.5. qRT-PCR Analysis of SmeHsf Responses to Different Abiotic Stresses

The expression levels of *Hsf* genes are affected by heat and other abiotic stresses in plants [42]. In this study, we analyzed the expression levels of *SmeHsf*s under different stress conditions, including cold, heat, salinity, and drought, to determine the stress-responsive candidates (Figure 5). The expression level of *SmeHsfA6b* dramatically increased 43-, 54-, and 8-fold under cold, salinity, and heat treatments, respectively, indicating its function in increasing plant adaptability to these abiotic stresses. In total, 14, 10, 9, and 8 *SmeHsf*s showed significant differential expression levels under cold, heat, salinity, and drought treatments, respectively. Thus, the functions of these stress-induced *SmeHsf*s should be analyzed in further studies. Overall, the average ranges of expression level changes of these *SmeHsf*s under cold conditions were greater than those identified under other stress conditions. Under cold-stress conditions, the expression levels of *SmeHsfC1* and *SmeHsfA1b* increased more than 10-fold, while those of *SmeHsfA3*, *SmeHsfA4c*, and *SmeHsfB3a* increased 3–5-fold. In addition, the expression levels of *SmeHsfA5* and *SmeHsfA6a* were upregulated approximately 3–4-fold in response to a heat treatment.

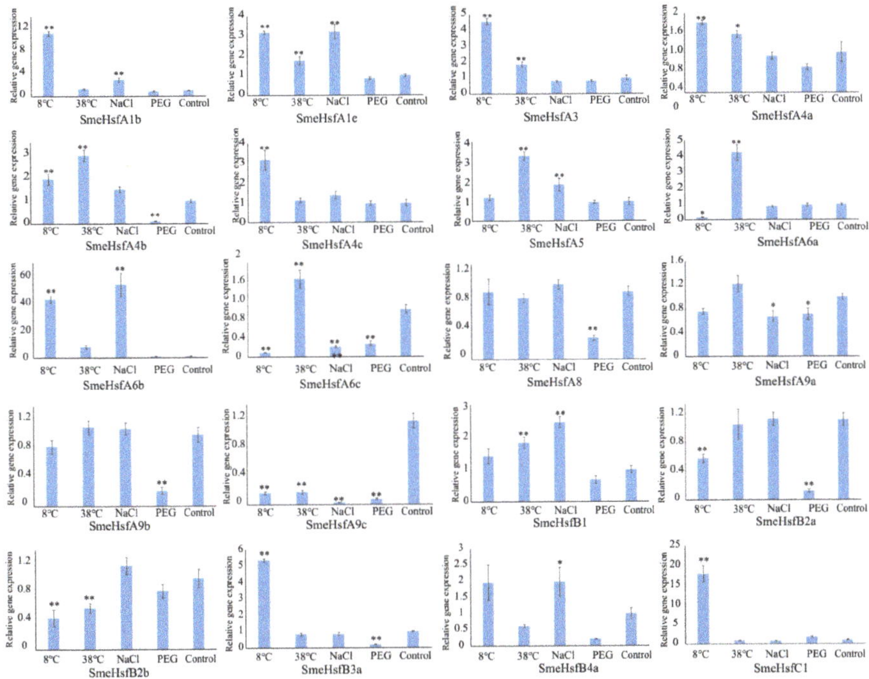

Figure 5. qRT-PCR analysis of *SmeHsf* genes under various abiotic stress conditions. The expression level of the control CK treatment was normalized as 1.0. The results are shown as means ± SDs of three independent experiments. The significant differences at $p \leq 0.05$ and $p \leq 0.01$ are represented by one and two asterisks, respectively.

3. Discussion

Eggplant is an important vegetable belonging to the Solanaceae family, which encompasses crops like tobacco, tomato, potato, and pepper. The Hsfs act as terminal components of signal networks that participate in various abiotic stress responses [43]. Hsfs can regulate the expression of molecular chaperones, such as heat shock proteins, which are involved in heat-stress responses [44] and regulate the signaling networks of stress-related phytohormones, such as salicylic and abscisic acids [32,45]. However, a comprehensive characterization of the *Hsf* gene family in eggplant is lacking.

In this study, we identified 20 *SmeHsf* genes. In Solanaceae species, the genome sizes and gene numbers of eggplant (833.1 Mb and 42,035 coding genes, respectively), potato (844 Mb and 35,119 coding genes), and tomato (950 Mb and 34,727 coding genes) are similar [46–49]; however, eggplant has the lowest number of *Hsf* genes. Notably, the genome size of pepper (3.48 Gb) was approximately fourfold larger than those of these three species, but the coding genes (34,899) and *Hsf* gene numbers did not vary significantly [49]. However, cultivated tobacco, which has an almost fivefold larger genome size (4.41–4.57 Gb) than that of eggplant and has a high number of coding genes (85,439 coding genes) [50], has more than twice the number of *Hsf* genes than eggplant. Thus, the number of *Hsf* genes is not correlated with the genome size, but is proportionally related to the total number of coding genes. Consequently, because pepper is a diploid species and contains a large number of repetitive sequences [49], it has less *Hsf* genes compared with the tetraploid cultivated tobacco, which has undergone an allopolyploidization event.

An unrooted phylogenetic tree was constructed using previously reported Hsfs and SmeHsfs. The SmeHsfs in the same subgroups clustered together, corresponding to other *Hsf* genes, which confirmed the SmeHsf classification. Class A was the predominant class in both monocots and dicots.

Like the Hsfs in other plants, all 20 SmeHsfs contained conserved DBD and HR-A/B domains, which are essential for their transcriptional functions. Although the overall gene structure of SmeHsfs in the A, B, and C classes were similar, the different groups contained characteristic domains.

Expression profile changes of Hsf genes that occur under various abiotic stresses have been extensively analyzed in different plants [40,51–53]. Investigating the expression changes of *SmeHsf*s under different stresses provides clues to their functions. *SmeHsf*s responded to the four stresses including heat, cold, salinity, and drought, which indicated that *SmeHsf*s increase tolerance levels to various abiotic stresses. Up to the present, many researchers found that class A and B Hsfs are involved in responding to environmental stresses, and few focused on the class C. In our study, we found that all class A and B SmeHsfs up- or downregulated under at least one stress, which indicated their functions in responding to environmental stresses. Interestingly, the expression level change of *SmeHsfC1* was only less than that of *SmeHsfA6b* under low temperature treatment, which indicated the class C member also plays a role in responding to low temperature stress in eggplant.

Hsf genes are expected to always respond to heat stress [54]. However, more *SmeHsf*s showed significant differential expression levels and greater ranges in expression changes under cold conditions than under other stress conditions, which might be because eggplant is a warm-weather plant that is more sensitive to low temperature [55]. More *SmeHsf*s showed significant differential expression levels under cold- and heat-stress conditions than under saline and drought conditions, which might be because leaf tissue was detected in this study and the leaves being the first organs to perceive heat and cold stresses, while the roots are the first organs to sense drought and salinity stresses [40]. *AtHsfA6a* and *AtHsfA6b* participate in abscisic acid-mediated thermotolerance and drought tolerance [32]; however, the wheat *HsfA6*, which is the most inducible wheat *Hsf* gene, is only responsive to the oxidative stress-signaling pathway [40]. In our study, *SmeHsfA6b* was also the most inducible *SmeHsf* gene, being upregulated by cold, heat and salinity treatments in the leaves, but not in response to the drought stress, which indicated that homologous *Hsf* genes have different functions in different plants. Moreover, in tomato, *HsfA1* plays a leading role in the heat-shock reaction and combines with *HsfA2* to form a complex that increases plant heat tolerance [28]. However, *HsfA2* was not identified in eggplant. Thus, the expression analysis indicated that *SmeHsf*s respond in unique manners to various environmental stresses, and the responses of these *SmeHsf*s are different in both magnitude and sensitivity to the above stresses.

4. Materials and Methods

4.1. Identification and Characterization of Hsfs in Eggplant

The eggplant genome (version SME_r2.5.1) and annotation data were downloaded from the Sol Genomics Network database [48]. Hmmsearch methods and BLAST searches were combined to identify *Hsf* genes in eggplant. Briefly, 325 *Hsf* gene sequences from *A. thaliana* (25), *Capsicum annuum* (25), *Carica papaya* (18), *Glycine max* (81), *M. domestica* (47), *Nicotiana tabacum* (65), *O. sativa* (36), and *Solanum lycopersicum* (26) were downloaded from PlantTFBD [56]. The downloaded Hsf proteins from different species were used as queries to search for all the possible Hsf protein sequences in the eggplant proteome file with an E-value of le-10 and identity of 60% as the thresholds. Then, Hmmsearch software was used to search for the Hsf domain (PF00447), which was downloaded from the Pfam database 32.0 [57], in the set of BLAST-identified proteins. The Hsf proteins were filtered with an E-value cutoff of 1×10^{-5} and at least a 60% coverage of the Pfam Hsf domain from the raw screening proteins. Furthermore, all the candidate Hsf protein sequences were analyzed to detect the DBD and coiled-coil structures using the SMART [58] and MARCOIL programs [59]. Those protein sequences, containing both a DBD and coiled-coil structure, were regarded as credible Hsf proteins. Moreover, the HEATSTER website [38] was used to confirm the 20 *SmeHsf* genes and classified them into subgroups. Finally, the Biopython module [60] was used to predict the molecular weight, isoelectric point, and other physical and chemical properties of the SmeHsf proteins. All the *SmeHsf* genes were renamed on the basis of their classifications and their phylogenetic relationships to *S. lycopersicum* and other species.

4.2. Gene Structure, Domain and Motif Analyses

Gene structural information was obtained from GFF3 files and visualized using TBtools software [61]. All the full-length amino acid sequences of the SmeHsfs were used to search for conserved motifs using the MEME tool [62]. The MEME parameters were set as follows: the maximum number to be found was set to 15 and the motif window length was set 8 to 100 bp. Additionally, the conserved NLS and NES domains were predicted using cNLS Mapper software [63] and NetNES 1.1 server software, respectively. The AHA domain was identified using the conserved motif FWxxF/L, F/I/L [64].

4.3. Phylogenetic Analysis and Classification of SmeHsf Genes

The amino acid sequences of SmeHsf proteins identified in this study and other Hsfs from *A. thaliana*, *O. sativa*, and *S. lycopersicum* downloaded from the HEATSTER website [38] were used in the phylogenetic analysis. The complete amino acid sequences of Hsf proteins and HR-A/B domain were aligned using the MUSCLE program [65]. Subsequently, the MEGA-X program was used to construct an unrooted maximum likelihood phylogenetic tree with the Jones–Taylor–Thornton model. Additionally, a bootstrap test was replicated 500 times and a partial deletion with a site coverage cutoff of 70% was used for gap treatment.

4.4. cis-Element Analysis of SmeHsf Promoters

The upstream 2000 bps of *SmeHsf* genes were abstracted as the promoter sequences from the eggplant genome file. Then, the PlantCARE database was used to determine the *cis*-regulatory elements present in each gene's promoter [41]. Besides, the upstream 2000 bps of random selected 500 eggplant genes were also used to determine the *cis*-regulatory elements using PlantCARE and compared with *SmeHsf*.

4.5. Plant Materials and Stress Treatments

The seeds of eggplant inbred line 'E22' were grown in plastic pots on the horticultural farm of the Zhejiang Academy of Agriculture Science (Hangzhou, China). At the four true-leaf stage, seedlings were moved to a growth chamber set at 16 h day (28 °C)/8 h night (24 °C) and used for experiments. For drought- and salt-stress treatments, seedlings were subjected to 100 mL of 30% PEG6000 and 300 mM NaCl, respectively, for 48 h, and for heat- and cold-stress treatments, seedlings were subjected to 38 °C and 8 °C, respectively, for 24 h. Plants were cultured under normal conditions for the control. The new leaves of five seedlings were collected as biological replicates, and each treatment had three replicates. The freshly collected samples were immediately frozen in liquid nitrogen stored at −80 °C for RNA isolation.

4.6. RNA Extraction and qRT-PCR Analysis

TRIzol reagent (Invitrogen, Carlsbad, CA, USA) was used to independently extract total RNAs of all the samples, and genomic DNA contamination was removed using DNase I. Then, RNA concentrations were measured using a NanoDrop2000 microvolume spectrophotometer (Thermo Fisher Scientific, Wilmington, DE, USA), and the RNA integrity was checked by 1.5% agarose gel electrophoresis. PrimeScript RTase (TaKaRa Biotechnology, Dalian, China) was used for first-strand cDNA synthesis following the manufacturer's instructions. The primers for qRT-PCR reactions were designed using Primer Premier 5.0, and the *SmEF1a* gene was used as a stable reference gene [66]. The qRT-PCR reactions were performed on a TIB8600 machine using AceQ® qPCR SYBR® Green Master Mix kits (Vazyme Biotechnology, Nanjing, China) with the following settings: 95 °C for 5 min; followed by 40 cycles of 95 °C for 15 s and 60 °C for 30 s. The relative expression levels of *SmeHsf* genes were calculated using the $2^{-\Delta\Delta Ct}$ method [67]. The analysis included three biological replicates for each sample. All the primer sequences are listed in Table S4.

4.7. Statistical Analyses

The statistical analysis was carried out by calculating the average values and standard errors for the three replicates. SPSS software version 16.0 was used to determine the significant differences between controls and stress treatments using a one-way ANOVA procedure and post hoc analysis. A p value ≤ 0.05 indicates a significant difference and is represented by an asterisk (*); a p value ≤ 0.01 indicates a very significant difference and is represented by two asterisks (**).

5. Conclusions

In the present study, 20 full-length *SmeHsf* genes were identified in the eggplant genome. These *SmeHsf*s were comprehensively characterized using a systematic approach comprising analyses of sequence characteristics, phylogeny, classifications, gene structures, and motif compositions. Moreover, a qRT-PCR analysis of *SmeHsf* expression levels in response to various abiotic stresses indicated that *SmeHsf*s not only play crucial roles in heat tolerance, but also increase the tolerance levels to various abiotic stresses. This comprehensive analysis provides candidate genes for future functional analyses under stress conditions and also lays the foundation for investigating molecular mechanisms of abiotic stress tolerance in plants.

Supplementary Materials: Supplementary Materials can be found at http://www.mdpi.com/2223-7747/9/7/915/s1. Table S1: The alignment information and scaffold position of the SmeHsfs, Table S2: The protein sequences of the SmeHsfs, Table S3: cis-Element analysis of SmeHsf promoters, Table S4: qRT-PCR primers for the SmeHsf genes.

Author Contributions: Conceptualization, C.B.; methodology, J.W.; software, J.W.; validation, J.W., C.B.; formal analysis, J.W.; investigation, T.H.; resources, T.H.; data curation, W.W.; writing—original draft preparation, J.W.; writing—review and editing, J.W.; visualization, Q.W.; supervision, C.B., T.H.; project administration, H.H.; funding acquisition, C.B.; H.H. All authors have read and agreed to the published version of the manuscript.

Funding: This work was supported by grant from the New Variety Breeding Project of the Major Science Technology Projects of Zhejiang (2016C02051-2), Zhejiang Science and Technology Program (2018C02057) and the Natural Science Foundation of Zhejiang (LQ18C150004).

Conflicts of Interest: The authors declare no conflict of interest.

References

1. Mittler, R. Abiotic stress, the field environment and stress combination. *Trends Plant Sci.* **2006**, *11*, 15–19. [CrossRef] [PubMed]
2. Wu, C. Heat shock transcription factors: Structure and regulation. *Annu. Rev. Cell. Dev. Biol.* **1995**, *11*, 441–469. [CrossRef] [PubMed]
3. Brash, A.R. Lipoxygenases: Occurrence, functions, catalysis, and acquisition of substrate. *J. Biol. Chem.* **1999**, *274*, 23679–23682. [CrossRef] [PubMed]
4. Dubos, C.; Stracke, R.; Grotewold, E.; Weisshaar, B.; Martin, C.; Lepiniec, L. MYB transcription factors in *Arabidopsis. Trends Plant Sci.* **2002**, *15*, 573–581. [CrossRef] [PubMed]
5. Rushton, P.J.; Somssich, I.E.; Ringler, P.; Shen, Q.J. WRKY transcription factors. *Trends Plant Sci.* **2010**, *15*, 247–258. [CrossRef]
6. Puranik, S.; Sahu, P.P.; Srivastava, P.S.; Prasad, M.N.V. NAC proteins: Regulation and role in stress tolerance. *Trends Plant Sci.* **2012**, *17*, 369–381. [CrossRef] [PubMed]
7. Xu, F.; Park, M.; Kitazumi, A.; Herath, V.; Mohanty, B.; Yun, S.J.; Reyes, B.G.D.L. Cis-regulatory signatures of orthologous stress-associated bZIP transcription factors from rice, sorghum and *Arabidopsis* based on phylogenetic footprints. *BMC Genom.* **2012**, *13*, 497. [CrossRef] [PubMed]
8. Agarwal, G.; Garg, V.; Kudapa, H.; Doddamani, D.; Pazhamala, L.T.; Khan, A.W.; Thudi, M.; Lee, S.; Varshney, R.K. Genome-wide dissection of AP2/ERF and HSP90 gene families in five legumes and expression profiles in chickpea and pigeonpea. *Plant Biotechnol. J.* **2016**, *14*, 1563–1577. [CrossRef]

9. Wang, J.; Sun, N.; Deng, T.; Zhang, L.; Zuo, K. Genome-wide cloning, identification, classification and functional analysis of cotton heat shock transcription factors in cotton (*Gossypium hirsutum*). *BMC Genom.* **2014**, *15*, 961. [CrossRef]

10. Guo, M.; Liu, J.; Ma, X.; Luo, D.; Gong, Z.; Lu, M. The plant heat stress transcription factors (HSFs): Structure, regulation, and function in response to abiotic stresses. *Front. Plant Sci.* **2016**, *7*, 114. [CrossRef]

11. Akerfelt, M.; Morimoto, R.I.; Sistonen, L. Heat shock factors: Integrators of cell stress, development and lifespan. *Nat. Rev. Mol. Cell Biol.* **2010**, *11*, 545–555. [CrossRef] [PubMed]

12. Nover, L.; Bharti, K.; Doring, P.; Mishra, S.K.; Ganguli, A.; Scharf, K. Arabidopsis and the heat stress transcription factor world: How many heat stress transcription factors do we need? *Cell Stress Chaperon.* **2001**, *6*, 177–189. [CrossRef]

13. Bienz, M.; Pelham, H.R.B. Mechanisms of heat-shock gene activation in higher eukaryotes. *Adv. Genet.* **1987**, *24*, 31–72.

14. Damberger, F.F.; Pelton, J.G.; Harrison, C.J.; Nelson, H.C.M.; Wemmer, D.E. Solution structure of the DNA-binding domain of the heat shock transcription factor determined by multidimensional heteronuclear magnetic resonance spectroscopy. *Protein Sci.* **1994**, *3*, 1806–1821. [CrossRef] [PubMed]

15. Vuister, G.W.; Kim, S.; Orosz, A.; Marquardt, J.L.; Wu, C.; Bax, A. Solution structure of the DNA-binding domain of Drosophila heat shock transcription factor. *Nat. Struct. Mol. Biol.* **1994**, *1*, 605–614. [CrossRef]

16. Peteranderl, R.; Rabenstein, M.; Shin, Y.; Liu, C.W.; Wemmer, D.E.; King, D.S.; Nelson, H.C.M. Biochemical and biophysical characterization of the trimerization domain from the heat shock transcription factor. *Biochemistry* **1999**, *38*, 3559–3569. [CrossRef]

17. Scharf, K.; Berberich, T.; Ebersberger, I.; Nover, L. The plant heat stress transcription factor (Hsf) family: Structure, function and evolution. *BBA-Gene Regul. Mech.* **2012**, *1819*, 104–119. [CrossRef]

18. Baniwal, S.K.; Bharti, K.; Chan, K.Y.; Fauth, M.; Ganguli, A.; Kotak, S.; Mishra, S.K.; Nover, L.; Port, M.; Scharf, K. Heat stress response in plants: A complex game with chaperones and more than twenty heat stress transcription factors. *J. Biosci.* **2004**, *29*, 471–487. [CrossRef]

19. Shim, D.; Hwang, J.; Lee, J.; Lee, S.; Choi, Y.; An, G.; Martinoia, E.; Lee, Y. Orthologs of the class A4 heat shock transcription factor HsfA4a confer cadmium tolerance in wheat and rice. *Plant Cell* **2009**, *21*, 4031–4043. [CrossRef]

20. Ikeda, M.; Mitsuda, N.; Ohmetakagi, M. *Arabidopsis* HsfB1 and HsfB2b act as repressors of the expression of heat-inducible hsfs but positively regulate the acquired thermotolerance. *Plant Physiol.* **2011**, *157*, 1243–1254. [CrossRef]

21. Ma, J.; Xu, Z.; Wang, F.; Tan, G.; Li, M.; Xiong, A. Genome-wide analysis of HSF family transcription factors and their responses to abiotic stresses in two Chinese cabbage varieties. *Acta Physiol. Plant.* **2014**, *36*, 513–523. [CrossRef]

22. Xue, G.P.; Sadat, S.; Drenth, J.; Mcintyre, C.L. The heat shock factor family from *Triticum aestivum* in response to heat and other major abiotic stresses and their role in regulation of heat shock protein genes. *J. Exp. Bot.* **2014**, *65*, 539–557. [CrossRef]

23. Ogawa, D.; Yamaguchi, K.; Nishiuchi, T. High-level overexpression of the *Arabidopsis* HsfA2 gene confers not only increased themotolerance but also salt/osmotic stress tolerance and enhanced callus growth. *J. Exp. Bot.* **2007**, *58*, 3373–3383. [CrossRef] [PubMed]

24. Banti, V.; Mafessoni, F.; Loreti, E.; Alpi, A.; Perata, P. The heat-inducible transcription factor HSFA2 enhances anoxia tolerance in *Arabidopsis*. *Plant Physiol.* **2010**, *152*, 1471–1483. [CrossRef] [PubMed]

25. Liu, H.; Liao, H.; Charng, Y. The role of class A1 heat shock factors (HSFA1s) in response to heat and other stresses in *Arabidopsis*. *Plant Cell Environ.* **2011**, *34*, 738–751. [CrossRef]

26. Bechtold, U.; Albihlal, W.S.; Lawson, T.; Fryer, M.J.; Sparrow, P.A.C.; Richard, F.; Persad, R.; Bowden, L.; Hickman, R.; Martin, C. *Arabidopsis HEAT SHOCK TRANSCRIPTION FACTORA1b* overexpression enhances water productivity, resistance to drought, and infection. *J. Exp. Bot.* **2013**, *64*, 3467–3481. [CrossRef] [PubMed]

27. Mishra, S.K.; Tripp, J.; Winkelhaus, S.; Tschiersch, B.; Theres, K.; Nover, L.; Scharf, K.-D. In the complex family of heat stress transcription factors, HsfA1 has a unique role as master regulator of thermotolerance in tomato. *Genes Dev.* **2002**, *16*, 1555–1567. [CrossRef]

28. Scharf, K.-D.; Heider, H.; Höhfeld, I.; Lyck, R.; Schmidt, E.; Nover, L. The tomato Hsf system: HsfA2 needs interaction with HsfA1 for efficient nuclear import and may be localized in cytoplasmic heat stress granules. *Mol. Cell Biol.* **1998**, *18*, 2240–2251. [CrossRef]

29. Giorno, F.; Woltersarts, M.; Grillo, S.; Scharf, K.; Vriezen, W.H.; Mariani, C. Developmental and heat stress-regulated expression of HsfA2 and small heat shock proteins in tomato anthers. *J. Exp. Bot.* **2010**, *61*, 453–462. [CrossRef] [PubMed]

30. Li, F.; Zhang, H.; Zhao, H.; Gao, T.; Song, A.; Jiang, J.; Chen, F.; Chen, S. Chrysanthemum *CmHSFA4* gene positively regulates salt stress tolerance in transgenic chrysanthemum. *Plant Biotechnol. J.* **2018**, *16*, 1311–1321. [CrossRef] [PubMed]

31. Giorno, F.; Guerriero, G.; Baric, S.; Mariani, C. Heat shock transcriptional factors in *Malus domestica*: Identification, classification and expression analysis. *BMC Genom.* **2012**, *13*, 639. [CrossRef] [PubMed]

32. Huang, Y.; Niu, C.; Yang, C.; Jinn, T. The heat stress factor *HSFA6b* Connects ABA signaling and ABA-mediated heat responses. *Plant Physiol.* **2016**, *172*, 1182–1199. [CrossRef] [PubMed]

33. Guo, J.; Wu, J.; Ji, Q.; Wang, C.; Luo, L.; Yuan, Y.; Wang, Y.; Wang, J. Genome-wide analysis of heat shock transcription factor families in rice and *Arabidopsis*. *J. Genet. Genom.* **2008**, *35*, 105–118. [CrossRef]

34. Chauhan, H.; Khurana, N.; Agarwal, P.; Khurana, P. Heat shock factors in rice (*Oryza sativa* L.): Genome-wide expression analysis during reproductive development and abiotic stress. *Mol. Genet. Genom.* **2011**, *286*, 171–187. [CrossRef]

35. Wang, F.; Dong, Q.; Jiang, H.; Zhu, S.; Chen, B.; Xiang, Y. Genome-wide analysis of the heat shock transcription factors in *Populus trichocarpa* and *Medicago truncatula*. *Mol. Biol. Rep.* **2012**, *39*, 1877–1886. [CrossRef]

36. Lohani, N.; Golicz, A.A.; Singh, M.; Bhalla, P.L. Genome-wide analysis of the Hsf gene family in *Brassica oleracea* and a comparative analysis of the Hsf gene family in *B. oleracea*, *B. rapa* and *B. napus*. *Funct. Integr. Genom.* **2019**, *19*, 515–531. [CrossRef]

37. Li, Y.; Li, Z.; Li, Z.; Luo, S.; Sun, B. Effects of heat stress on gene expression in eggplant (*Solanum melongema* L.) seedlings. *Afr. J. Biotechnol.* **2011**, *10*, 18078–18084.

38. Berz, J.; Simm, S.; Schuster, S.; Scharf, K.; Schleiff, E.; Ebersberger, I. HEATSTER: A database and web server for identification and classification of heat stress transcription factors in plants. *Bioinform. Biol. Insights* **2019**, *13*, 117793221882136. [CrossRef]

39. Agarwal, P.; Khurana, P. Functional characterization of HSFs from wheat in response to heat and other abiotic stress conditions. *Funct. Integr. Genom.* **2019**, *19*, 497–513. [CrossRef]

40. Duan, S.; Liu, B.; Zhang, Y.; Li, G.; Guo, X. Genome-wide identification and abiotic stress-responsive pattern of heat shock transcription factor family in *Triticum aestivum* L. *BMC Genom.* **2019**, *20*, 257. [CrossRef] [PubMed]

41. Lescot, M.; Dehais, P.; Thijs, G.; Marchal, K.; Moreau, Y.; De Peer, Y.V.; Rouze, P.; Rombauts, S. PlantCARE, a database of plant cis-acting regulatory elements and a portal to tools for in silico analysis of promoter sequences. *Nucl. Acids Res.* **2002**, *30*, 325–327. [CrossRef] [PubMed]

42. Zhang, Y.; Chou, S.; Murshid, A.; Prince, T.L.; Schreiner, S.; Stevenson, M.A.; Calderwood, S.K. The role of heat shock factors in stress-induced transcription. *Methods Mol. Biol.* **2011**, *787*, 21–32. [PubMed]

43. Ohama, N.; Sato, H.; Shinozaki, K.; Yamaguchi-Shinozaki, K. Transcriptional regulatory network of plant heat stress response. *Trends Plant Sci.* **2017**, *22*, 53–65. [CrossRef] [PubMed]

44. Jung, K.H.; Gho, H.J.; Nguyen, M.X.; Kim, S.-R.; An, G. Genome-wide expression analysis of HSP70 family genes in rice and identification of a cytosolic HSP70 gene highly induced under heat stress. *Funct. Integr. Genom.* **2013**, *13*, 391–402. [CrossRef] [PubMed]

45. Snyman, M.; Cronjé, M. Modulation of heat shock factors accompanies salicylic acid-mediated potentiation of Hsp70 in tomato seedlings. *J. Exp. Bot.* **2008**, *59*, 2125–2132. [CrossRef] [PubMed]

46. Xu, X.; Pan, S.K.; Cheng, S.F.; Zhang, B.; Bachem, C.W.B.; De Boer, J.M.; Borm, T.J.A.; Kloosterman, B.; Van Eck, H.J.; Datema, E. Genome sequence and analysis of the tuber crop potato. *Nature* **2011**, *475*, 189–195.

47. Sato, S.; Tabata, S.; Hirakawa, H.; Asamizu, E.; Shirasawa, K.; Isobe, S.; Kaneko, T.; Nakamura, Y.; Shibata, D.; Aoki, K. The tomato genome sequence provides insights into fleshy fruit evolution. *Nature* **2012**, *485*, 635–641.

48. Hirakawa, H.; Shirasawa, K.; Miyatake, K.; Nunome, T.; Negoro, S.; Ohyama, A.; Yamaguchi, H.; Sato, S.; Isobe, S.; Tabata, S. Draft genome sequence of eggplant (*Solanum melongena* L.): The representative solanum species indigenous to the old world. *DNA Res.* **2014**, *21*, 649–660. [CrossRef]

49. Kim, S.; Park, M.; Yeom, S.; Kim, Y.; Lee, J.M.; Lee, H.; Seo, E.; Choi, J.Y.; Cheong, K.; Kim, K. Genome sequence of the hot pepper provides insights into the evolution of pungency in Capsicum species. *Nat. Genet.* **2014**, *46*, 270–278. [CrossRef]

50. Sierro, N.; Battey, J.N.D.; Ouadi, S.; Bakaher, N.; Bovet, L.; Willig, A.; Goepfert, S.; Peitsch, M.C.; Ivanov, N.V. The tobacco genome sequence and its comparison with those of tomato and potato. *Nat. Commun.* **2014**, *5*, 3833. [CrossRef]

51. Mittal, D.; Chakrabarti, S.; Sarkar, A.; Singh, A.; Grover, A. Heat shock factor gene family in rice: Genomic organization and transcript expression profiling in response to high temperature, low temperature and oxidative stresses. *Plant Physiol. Bioch.* **2009**, *47*, 785–795. [CrossRef] [PubMed]

52. Dossa, K.; Diouf, D.; Cisse, N. Genome-wide investigation of hsf genes in sesame reveals their segmental duplication expansion and their active role in drought stress response. *Front. Plant Sci.* **2016**, *7*, 1522. [CrossRef]

53. Zhou, M.; Zheng, S.; Liu, R.; Lu, J.; Lu, L.; Zhang, C.; Liu, Z.; Luo, C.; Zhang, L.; Yant, L. Genome-wide identification, phylogenetic and expression analysis of the heat shock transcription factor family in bread wheat (*Triticum aestivum* L.). *BMC Genom.* **2019**, *20*, 505. [CrossRef] [PubMed]

54. Chen, S.; Jiang, J.; Han, X.; Zhang, Y.; Zhuo, R. Identification, expression analysis of the Hsf family, and characterization of class A4 in *Sedum Alfredii* hance under cadmium stress. *Int. J. Mol. Sci.* **2018**, 1216. [CrossRef]

55. Lv, L.L.; Feng, X.F.; Li, W.; Li, K. High temperature reduces peel color in eggplant (*Solanum melongena*) as revealed by RNA-seq analysis. *Genome* **2019**, *62*, 503–512. [CrossRef] [PubMed]

56. Jin, J.; Tian, F.; Yang, D.; Meng, Y.; Kong, L.; Luo, J.; Gao, G. PlantTFDB 4.0: Toward a central hub for transcription factors and regulatory interactions in plants. *Nucl. Acids Res.* **2017**, *45*. [CrossRef]

57. Elgebali, S.; Mistry, J.; Bateman, A.; Eddy, S.R.; Luciani, A.; Potter, S.C.; Qureshi, M.; Richardson, L.J.; Salazar, G.A.; Smart, A. The Pfam protein families database in 2019. *Nucl. Acids Res.* **2019**, *47*.

58. Letunic, I.; Bork, P. 20 years of the SMART protein domain annotation resource. *Nucleic Acids Res.* **2018**, *46*, 193–196. [CrossRef]

59. Zimmermann, L.; Stephens, A.; Nam, S.; Rau, D.; Kübler, J.; Lozajic, M.; Gabler, F.; Söding, J.; Lupas, A.; Alva, V. A completely reimplemented MPI bioinformatics toolkit with a new HHpred server at its core. *J. Mol. Biol.* **2018**, *430*, 2237–2243. [CrossRef]

60. Peter, J.A.C.; Tiago, A.; Jeffrey, T.C.; Brad, A.C.; Cymon, J.C.; Andrew, D.; Iddo, F.; Thomas, H.; Frank, K.; Bartek, W.; et al. Biopython: Freely available Python tools for computational molecular biology and bioinformatics. *Bioinformatics* **2009**, *25*, 1422–1423.

61. Chen, C.; Xia, R.; Chen, H.; He, Y. TBtools, a Toolkit for Biologists integrating various HTS-data handling tools with a user-friendly interface. *bioRxiv* **2018**, 289660.

62. Bailey, T.L.; Boden, M.; Buske, F.A.; Frith, M.C.; Grant, C.E.; Clementi, L.; Ren, J.; Li, W.W.; Noble, W.S. MEME Suite: Tools for motif discovery and searching. *Nucl. Acids Res.* **2009**, *37*, 202–208. [CrossRef] [PubMed]

63. Kosugi, S.; Hasebe, M.; Tomita, M.; Yanagawa, H. Systematic identification of cell cycle-dependent yeast nucleocytoplasmic shuttling proteins by prediction of composite motifs. *Proc. Natl. Acad. Sci. USA* **2009**, *106*, 10171–10176. [CrossRef]

64. Kotak, S.; Port, M.; Ganguli, A.; Bicker, F.; Von Koskulldoring, P. Characterization of C-terminal domains of Arabidopsis heat stress transcription factors (Hsfs) and identification of a new signature combination of plant class A Hsfs with AHA and NES motifs essential for activator function and intracellular localization. *Plant J.* **2004**, *39*, 98–112. [CrossRef]

65. Edgar, R.C. MUSCLE: Multiple sequence alignment with high accuracy and high throughput. *Nucl. Acids Res.* **2004**, *32*, 1792–1797. [CrossRef]

66. Pang, Q.; Li, Z.; Luo, S.; Chen, R.; Jin, Q.; Li, Z.; Li, D.; Sun, B.; Sun, G. Selection and stability analysis of reference gene for qRT-PCR in eggplant under high temperature stress. *Acta Hortic. Sin.* **2017**, *44*, 475–486.

67. Livak, K.J.; Schmittgen, T.D. Analysis of relative gene expression data using real-time quantitative PCR and the 2(-Delta Delta C(T)) Method. *Methods* **2001**, *25*, 402–408. [CrossRef]

Article

Effects of Rhizome Integration on the Water Physiology of *Phyllostachys edulis* Clones under Heterogeneous Water Stress

Xiong Jing, Chunju Cai *, Shaohui Fan *, Guanglu Liu, Changming Wu and Benxue Chen

International Center for Bamboo and Rattan, State Forestry and Grassland Administration Key Laboratory of Bamboo and Rattan, Beijing 100102, China; jingx@icbr.ac.cn (X.J.); liuguanglu@icbr.ac.cn (G.L.); 15761635165@163.com (C.W.); benxuechen@126.com (B.C.)

* Correspondence: caicj@icbr.ac.cn (C.C.); fansh@icbr.ac.cn (S.F.); Tel.: +86-010-8478-9806 (C.C.); +86-010-8478-9720 (S.F.)

Received: 24 February 2020; Accepted: 15 March 2020; Published: 18 March 2020

Abstract: Water is crucial to plant growth and development. Under heterogeneous environmental water deficiency, physiological integration of the rhizomatous clonal plant triggers a series of physiological cascades, which induces both signaling and physiological responses. It is known that the rhizome of *Phyllostachys edulis*, which connects associated clonal ramets, has important significance in this physiological integration. This significance is attributed to the sharing of water and nutrients in the vascular bundle of clonal ramets under heterogeneous water conditions. However, the physiological characteristics of physiological integration under heterogeneous water stress remain unclear. To investigate these physiological characteristics, particularly second messenger Ca^{2+} signaling characteristics, long-distance hormone signaling molecules, antioxidant enzyme activity, osmotic adjustment substance, and nitrogen metabolism, ramets with a connected (where integration was allowed to take place) and severed rhizome (with no integration) were compared in this study. The vascular bundle structure of the rhizome was also observed using laser confocal microscopy. Overall, the results suggest that interconnected rhizome of *P. edulis* can enhance its physiological function in response to drought-induced stress under heterogeneous water deficiency. These measured changes in physiological indices serve to improve the clonal ramets' drought adaptivity through the interconnected rhizome.

Keywords: Heterogeneous water stress; *Phyllostachys edulis*; Rhizome; Vascular bundle; Stress Signal; Physiological characteristics

---

1. Introduction

Recently, drought problems caused by global climate change have been on the rise. Drought is a common and severe problem in agriculture—in fact, damage caused by drought causes a much larger loss of livelihood and yield than other natural disasters [1,2]. Drought is a multidimensional stress and can significantly constrain plant productivity and trigger a wide variety of responses at a level concerning physiological, biochemical, and molecular aspects [3,4].

In China, the bamboo resources and the total area of bamboo forest rank among the largest in the world. There are more than 500 species of bamboo comprising 39 genera [5,6]. Bamboo is also one of the most important clonal plant resources [7]. Clonal plants are those in which vegetative growth is accompanied by cloned ramets, which can self-replicate and spread quickly under natural conditions by asexual reproduction [8,9].

In nature, water and nutrients needed for plant growth, are usually heterogeneously distributed, even at a very small scale [10]. In heterogeneous small-scale environments, physiological integration and foraging behavior are important characteristics of clonal plants, among which physiological integration is the internal driver of clonal growth in heterogeneous habitats. In these plants substances can be transferred via the vascular system of the clonal ramets [11,12]. Clonal plants depend on an extensive rhizome-root system, which has the ability to extend through a large area and create many clones. The rhizome acts as a spacer and is an important vascular system for physiological integration of clonal plants [13,14]. The rhizome of moso bamboo (*Phyllostachys edulis*) acts as a spacer for resource sharing between cloned ramets of *P. edulis* [15]. In a heterogeneously distributed resource environment, the *P. edulis* rhizome system plays an important role in transporting water, photo-assimilates, and carbohydrates under a source-sink gradient [16–18].

Clonal growth and physiological integrations in clonal plants have been an emerging field of research in recent years [11]. Existing research on clonal plants mainly focuses on dwarf herbaceous clonal plants such as *Zoysia japonica* and *Buchloe dactyloides* [19,20] It is worth noting that studies have found that amphibious clonal plants with a high capacity for clonal integration are useful for re-vegetation of degraded aquatic habitats caused by Cd contamination [8]. This provides precedent for the use of clonal plants to reduce heavy metal pollution in other habitats. However, research into the physiological integration of bamboo is still in the preliminary stage, and related research exists mainly focuses on describing the phenomenon and on the intensity of integration.

Under heterogeneous water stress, the relationship between rhizome length and physiological integration of *Indocalamus decoru* has been studied. It was found that as the rhizome length increased, the water potential gradient decreased, that is to say, the integration of water between the donor ramet and the recipient ramet decreased [21]. Clonal integration (resource translocation) between connected ramets of clonal plants can increase their tolerance to stress. However, there are few studies on the physiological characteristics of water translocation in clonal integration. We postulate that both the signaling and physiological responses play a significant role in modulating the physiological integration of *P. edulis*.

To fill this research gap, an experiment was conducted under heterogeneous water content of the soil (RWC = 25 ± 5%, 80 ± 5%) on *P. edulis* with both connected and disconnected rhizomes. Concentrations of the long-distance signaling molecules abscisic acid (ABA) and jasmonic acid (JA) were monitored. Instant flux of the second messenger Ca^{2+} in pilorhiza, rhizomes, and leaves was considered the major signaling procedure. In addition, physiological responses such as endogenous hormone concentration, antioxidant enzyme activity, osmotic adjustment substances, NO^{3-}, and NH^{4+} flux in different water patches in the clonal ramets were investigated, and rhizome and stem structure were observed under laser confocal microscopy (CLSM) to gain a better understanding of the vascular bundle function.

2. Results and Discussion

2.1. Morphological Basis of P. edulis Physiology Integration and Analysis of Its Influencing Factor

P. edulis is a monocotyledonous plant, and the vascular bundle does not contain cambium. The vascular bundle is instead composed of primary phloem and primary xylem, wherein the primary xylem includes vessel or tracheid cells, parenchyma, and fiber. The primary phloem includes sieve elements, sieve cells, and companion cells. Lignin deposition is an important step in the formation of the vascular bundle, so we took advantage of the high lignin content in the vascular bundle and recorded lignin autofluorescence using laser confocal microscopy.

As shown in Figure 1, the structure of both the rhizome and the stem vascular bundles was observed by lignin autofluorescence using a laser confocal microscope. Both the structure in the rhizome vascular bundle (Figure 1c VB) and that of the stem (Figure 1b VB) were composed of two vessel elements (VE) and a sieve tube (ST) containing sieve holes. The vascular bundles are arranged in a V-shaped scattered arrangement on the transverse section. The bottom of the V-shape is a sieve tube (ST) and the arms of the V-shape each have two vessels (VE). The bottom of the V-shape vascular bundles on the rhizome and stem respectively are oriented towards their own cortex, meanwhile the V-shaped arms are each oriented towards their respective medullary canal.

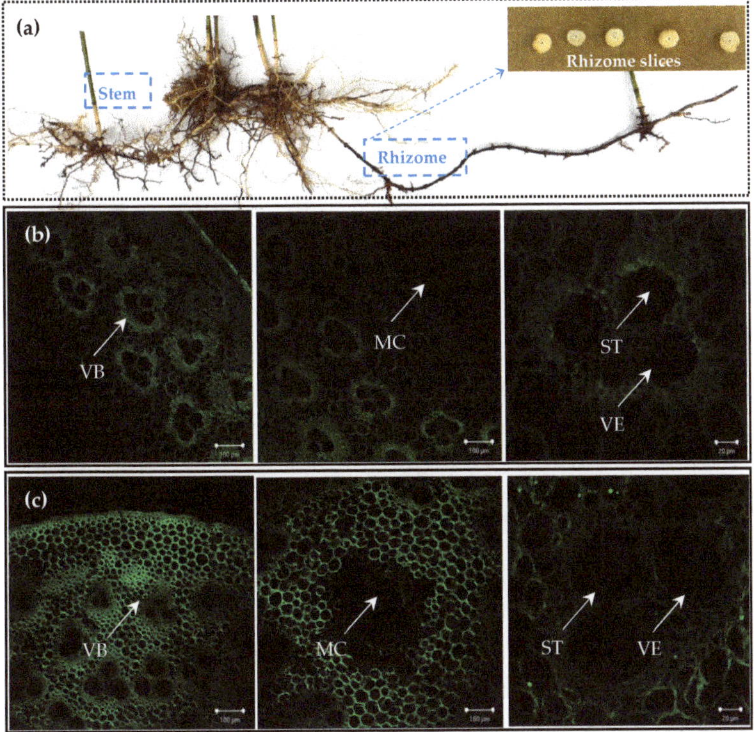

Figure 1. The Laser Confocal Scanning Microscop (LCSM) images of the *P. edulis* stem and rhizome structure. (**a**) The display of *P. edulis* sprouting seedling test material; (**b**) represents the structure of the stem slices; (**c**) represents the structure of the rhizome slices. The magnification is set a ×10, ×10, and ×40 times (left to right) the original size. Vascular bundles were denoted as (VB), medullary cavities were denoted as (MC), sieve tubes were denoted as (ST), and vessel elements were denoted as (VE).

However, in the stem, a medullary ring and a large medullary cavity was found whilst in the rhizome, there was no medullary ring and only a small medullary cavity. From the perspective of the xylem developmental structure, the primary xylem of both the rhizome and stem belonged to the endarch type (referring to proto xylem being directed towards the inner center). In botany, the underground spacing mechanism of the *P. edulis* plant is the rhizome, which is a metamorphosis of the stem, so in effect, the rhizome and stem share the same vascular bundle structure. The vascular system provides objective conditions for the physiological integration of cloned ramets. The interconnected vascular system of *P. edulis* facilitates the transfer and sharing of substances under heterogeneous resource distribution, such as water, nutrients, and photosynthetic products, which are carried along the gradient of material in source-sink relationships. Therefore, physiological integration of the cloned ramets of the *P. edulis plant* is achieved by the rhizomes connected to each other. The rhizome is used as the organ for material transfer between the cloned ramets, and the shared distribution of resources in the heterogeneous plaque environment plays an important role.

The Munch pressure flow theory stipulates that the transport flux of assimilates in the vascular bundle is driven by the swell gradient, which can be understood by means of the flux calculation formula ($J = \delta P \cdot C \cdot k$, where δP is the gradient of turgor between source and sink, C is the concentration of assimilates, and k is equal to the conductivity coefficient) of the assimilates in the vascular bundle [22]. If NC (clonal ramet in normal water treatment) is considered to be a metabolic source, then water and organic matter can be transferred to the metabolic sink (SC, water deficiency of clonal ramet). The transport capacity of the phloem is mainly affected by the cross-sectional area of the sieve tubes and the distance of the source-sink. How regularly the rhizome grows is also related to the source- sink relationship [23]. In a heterogeneous water environment, the vascular bundle in the rhizome plays an important role in the resource and material distribution of the *P. edulis* plant, hence the age and length of the rhizome are important factors.

2.2. Effects of Rhizome Integration on the Related Physiological Indexes of P. edulis Clones under Heterogeneous Water Stress

Under the influence of varying soil relative water content (RWC = 25 ± 5%, 80 ± 5%), rhizome status, a variance analysis of the leaf related physiological indexes of three-year-old *P. edulis* sprouting seedlings is shown in Table 1. A model was designed for the analysis of variance of nine physiological indicators such as leaf enzyme activity (SOD, CAT, and POD), hormone content (JA, ABA, and IAA) and osmotic adjustment substance (MDA) content, etc. The model is: $Y=AX_1+BX_2+CX_1X_2+D+e$, among which X_1, X_2 represents water and rhizome status respectively; D indicates the intercept; e is the error; and ABC is the coefficient (Table 1).

Table 1. Analysis of the variance in various physiological indexes of *P. edulis* sprouting seedlings, showing the correlation between these indexes and of rhizome status (R), water stress (W) and a combination of the two (W × R) on *P. edulis*.

	W	R	W × R
SOD activity	2947.21 ***	4.11 ns	979.94 ***
CAT activity	322.15 ***	77.69 ***	155.23 ***
POD activity	137.26 ***	6.54 ns	26.91 ***
JA concentration	999.70 ***	40.41 ***	105.17 ***
ABA concentration	1076.62 ***	303.77 ***	1040.69 ***
IAA concentration	553.84 ***	73.67 ***	243.18 ***
Pro concentration	56.84 ***	0.02 ns	12.87 **
Bet concentration	1860.51 ***	605.20 ***	1898.34 ***
MDA concentration	518.35 ***	8.93 **	430.09 ***

Note: Values are F and symbols show *p* values (*** $p < 0.001$, ** $p < 0.01$, * $p < 0.05$, and ns $p \geq 0.05$).

It can be seen from Table 1 that both water stress and a combination of water and rhizome status factors had a significant effect on each relevant physiological index ($p < 0.001$). However, rhizome status had a significant impact on all indicators except for SOD, POD, and Pro. This means, the effect of water on the physiological function of the cloned ramets of *P. edulis* was affected by the status of the rhizome. The influence of the rhizome status on the physiology of the plant was also closely related to the existence of heterogeneous water treatment. It can be concluded that the rhizome plays an important role in heterogeneous drought stress physiology.

Principal component analysis (PCA) was used to analyze the correlation between the nine physiological factors including leaf enzyme activity (SOD, CAT and POD), hormone content (JA, ABA and IAA) and osmotic adjustment substance (MDA) content (Figure 2). The results show that the contribution rate of the variance of the first principal component is 88.5%, the second principal component variance contribution rate was 5.3%, making a total contribution of 93.8%. The first and second principal components could reflect the differences of the nine physiological indexes among different groups. There is a significant correlation among various physiological indicators, IAA was significantly negatively correlated with other physiological indicators, and all physiological indicators except IAA were significantly positively correlated. In addition, each physiological index is clearly divided into four groups (water and rhizome status) on the principal component axis, and the physiological indexes between different groups are significantly different, that is, water and rhizome status have a significant impact on physiological indexes of *P. edulis*.

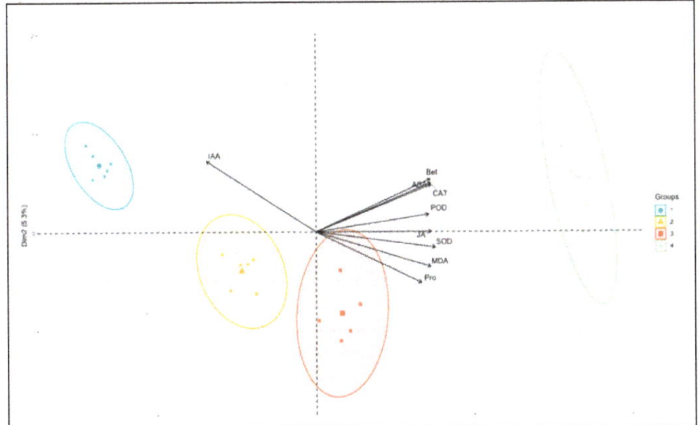

Figure 2. PCA analysis between different physiological indicators of *P. edulis* sprouting seedlings. ND treatments were denoted as 1 Groups, NC treatments were denoted as 2 Groups, SC treatments were denoted as 3 Groups, and SD treatments were denoted as 4 Groups.

2.3. Effect of the Rhizome on the Ca^{2+} Flux as the Second Messenger of P. edulis under Heterogeneous Water Stress

NMT was used to monitor Ca^{2+} flux in the pileorhiza, mesophyll transverse section, and rhizome transverse section in vivo, the flux rule is shown in Figure 3. Analysis of significant difference showed that Ca^{2+} fluxes in cloned ramets ND (treatment of disconnected clonal ramet in normal water) and SD (treatment of disconnected clonal ramet in water stress) were significantly different in pileorhiza, in the rhizome transverse section and in the mesophyll transverse section ($p < 0.05$), which showed that the pileorhiza of the SD had stronger Ca^{2+} absorption capacity than the ND. However, there was no significant difference in Ca^{2+} flux between the pileorhiza and mesophyll transverse section of the NC (treatment of connected clonal ramet in normal water) and SC (treatment of connected clonal ramet in water stress) cloned ramets in the connected rhizome treatment under heterogeneous water.

The rhizome of connected clonal ramets shared water, which is assumed to be due to the source-sink gradient between cloned ramets [21]. There was a smaller difference in Ca^{2+} signal intensity in difference organs of the cloned ramets in the connected rhizome treatment, which might be related to the water "equalization" effect of inter-clonal ramets on the heterogeneous drought stress. This is that the water or substance of the sink-clonal ramet SC was transported by the ramet NC. The difference in the Ca^{2+} flux between the leaves of the NC and SC treated plants and the associated transverse section of the rhizome was not obvious in this shared transport mechanism.

It is understood that ABA can induce the intracellular production of reactive oxygen species (ROS) such as active hydrogen peroxide (H_2O_2) under drought stress. ROS in turn act as post-stress messengers, activating Ca^{2+} channels in the plasma membrane to release Ca^{2+} from the vacuole to the cytosol. The influx of extracellular Ca^{2+} can also initiate intracellular Ca^{2+} oscillations and promote the release of Ca^{2+} from the vacuole to the cytosol [24].

However, due to the complexity of Ca^{2+} flux signaling in the rhizome of *P. edulis*, related research is limited at present. It is possible that there the lack of difference in the Ca^{2+} signal flux in the cloned ramets NC and SC under heterogeneous water stress was due to other factors. Studies have shown that Ca^{2+} signaling can regulate plant physiology to improve the physiological response to drought stress; Ca^{2+} is also related to long-distance signaling substances such as ABA [25,26]. A study conducted showed that the differentiation of ductal molecules required the initiative of a Ca^{2+} influx pathway [27]. Meanwhile, the concentration of free Ca^{2+} in the sieve-tube was nearly 20–100 times higher than that in general cells. It was speculated that the concentration of Ca^{2+} is related to the transport direction of organic matter [28]. In research conducted on bamboo, Yu found that the phloem ganglion showed extracellular Ca^{2+} transfer into the cell as the phloem ganglion developed, and Ca^{2+} regulated the physiological function of the phloem ganglion [29].

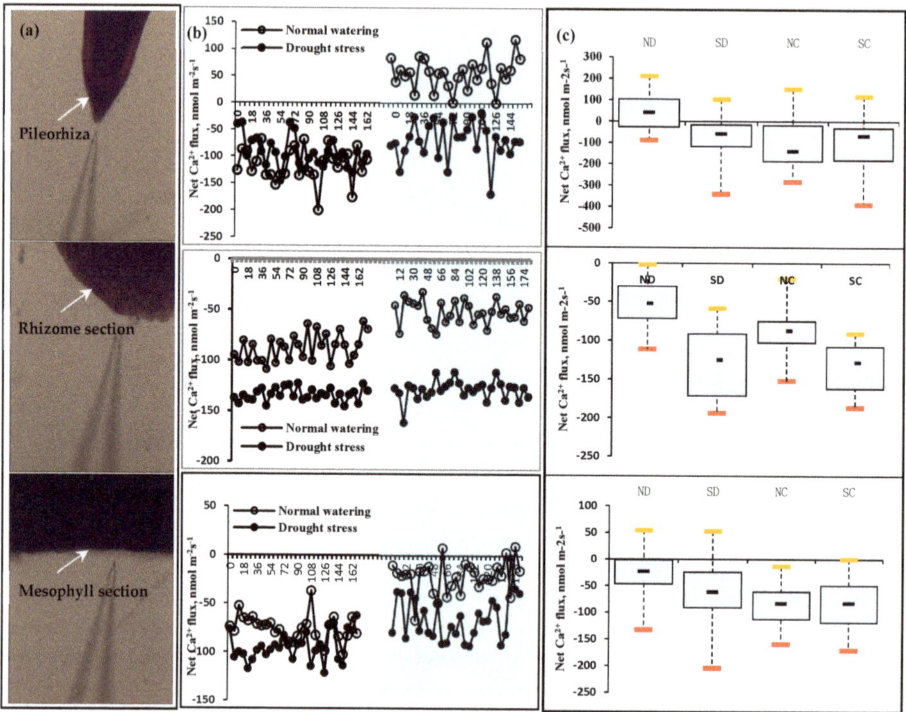

Figure 3. Effect of the rhizome on Ca^{2+} flux in different organs of clonal ramets under heterogeneous water stress. (**a**) Schematic of Ca^{2+} fluxes test in the pileorhiza, rhizome transection, and leaf transection of *P. edulis* seedling, respectively; (**b**) Ca^{2+} flux dynamic in clonal remats of *P. edulis* sprouting seedlings under heterogeneous water stress. The white area represents the Ca^{2+} oscillation in the connected rhizome treatment of the *P. edulis* seedlings, and the blue region indicates Ca^{2+} oscillation under disconnected rhizome treatment; (**c**) Box plot shows Ca^{2+} flux range in the different organs tested.

The box plot shows that Ca^{2+} oscillation in the rhizome of the cloned ramets NC and SC was smaller than that of ND and SD. By comparing the four treatments, it is apparent that the oscillation amplitude of Ca^{2+} flux in the NC cloned ramet was the smallest of the four treatments and that the overall order of Ca^{2+} oscillation amplitude was NC < ND < SC < SD. However, this rule does not apply to the other plant organs tested, such as the pileorhiza and mesophyll layers (Figure 3c).

2.4. Effect of Rhizome Integration on Endogenous Hormone Concentration in the Leaves of P. edulis under Heterogeneous Water Stress

Plant hormones, such as JA and ABA, play an important regulatory role under drought stress, and also act as important signal regulators. Figure 4 shows the concentration of JA, ABA, and IAA in the leaves of *P. edulis* with rhizomes treatment by comparing severed (preventing integration) and connected rhizomes (allowing integration) under heterogeneous water stress. The concentration of

JA and ABA in the leaves of the cloned ramets were significantly correlated ($p < 0.01$, r = 0.871), indicating that the endogenous hormones JA and ABA act as adversity sensors and initiate the stress metabolism under heterogeneous drought stress [30]. Moreover, ABA and JA may have synergistic effects in response to drought stress. ABA and JA may play a key role in perceiving the heterogeneous water environment between two clonal ramets. Interestingly, changes in ABA and JA concentration in the leaves of the connected cloned ramets were different; there was no significant difference in ABA concentration, but there was a significant difference in JA concentration.

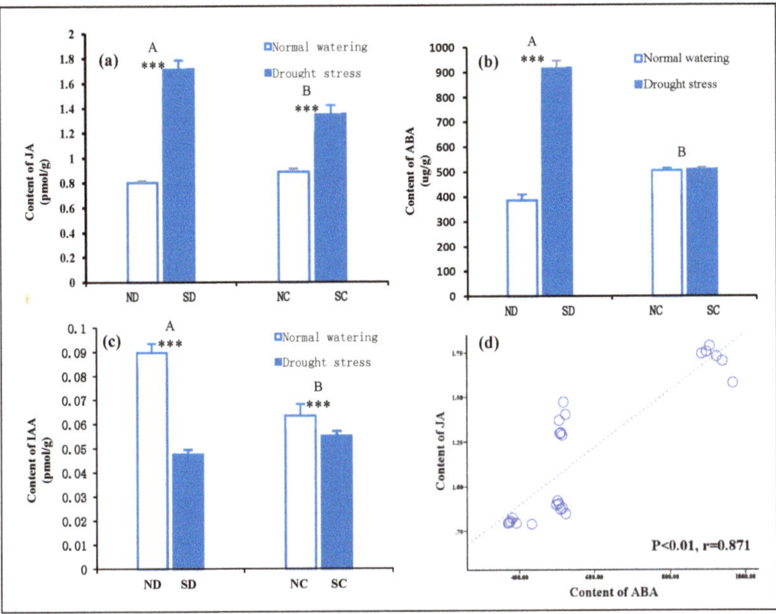

Figure 4. Effect of the rhizome on the changes of endogenous hormones in leaves of clonal ramets under heterogeneous water stress conditions. Letters (**a**–**c**) indicate JA, ABA, and IAA concentrations in the leaves of the four treatments respectively. Bars are mean values (±SE, n = 4). Different capital letters in the figure indicate significant differences between connected and disconnected ($p < 0.01$). Symbols indicate levels of statistical significance between normal watering (RWC = 80 ± 5%) and drought stress (RWC = 25 ± 5%) treatments for rhizome treatment: no symbol $p > 0.1$; *** $p < 0.001$. (**d**) Statistical correlation analysis between JA and ABA concentration in the leaves (y = 0.0016x + 0.2549, R^2 = 0.759; $p < 0.01$; Pearson correlation coefficient, r = 0.871.).

It can be seen from Figure 4 that under heterogeneous water stress, abscisic acid (ABA) and JA concentration in the *P. edulis* plant with a connected rhizome were significantly lower than that of the disconnected rhizome. After the plant was subjected to drought stress, ABA was found to act as a long-distance signal produced in the pileorhiza, which was transmitted along with water to the aerial part through the xylem vessel to regulate the stomatal opening of the plant leaves under drought stress. This mechanism reduced transpiration [31,32].

JA is transported from the top of the plant to the bottom of the plant by the phloem. Studies have shown that JA is produced in the plant leaves, then transported from the leaves by the phloem (screen) to the roots. JA transfer accelerated under water stress and the upward transfer of ABA was in turn promoted [33–36]. This indicates that the response sites of ABA and JA under drought stress are different, which may be one of the reasons that the difference in JA concentration between NC and SC leaves of cloned plants is significant, but the difference in ABA concentration is not significant.

Polar transport of IAA is observed in the plant from the top to the bottom of the stalk, and the plant can produce a variety of physiological effects as the concentration of IAA changes [37]. Under heterogeneous water stress, the IAA concentration of the clonal SC was significantly lower than that of the NC. However, when compared with the difference between ND and SD leaves, the difference in IAA concentration between cloned NC and SC leaves was much lower (3.9 times lower) (Figure 4). The effect of the rhizome treatment (connected rhizome and disconnected rhizome) on the endogenous hormone concentration in a heterogeneous water environment is considered to be the result of the rhizome vascular system participating in the sharing of water and material distribution. From the perspective of the source-sink relationship, the NC-treated clonal ramets supplied water to the drought stress treatment clonal SC, which not only relieved the stress of the clonal strain SC, but also improved the supply equilibrium fitness of the whole plant to the heterogeneous stress environment.

2.5. Effect of Rhizome Severance on the Antioxidant Enzyme System in the Leaves of P. edulis under Heterogeneous Water Stress

Abscisic acid (ABA) was induced by biological and abiotic stress stimulation and the production and accumulation of reactive oxygen species (ROS) [38]. When the plant was subjected to drought stress, a large amount of ROS, such as superoxide anion radicals ($\cdot O^{2-}$), hydrogen peroxide (H_2O_2), and hydroxyl radicals ($\cdot OH$) were produced in vivo. ROS may have toxic effects on the growth and development of plants. The plant removes ROS through the antioxidant enzyme system. In the cellular antioxidant defense system, SOD, CAT, and POD are all regulated by ROS, and their reaction effects can be synergistic [39]. Excess $\cdot O^{2-}$ produced in the plant cells is disproportionated into H_2O_2 and O_2 under the catalysis of SOD, and H_2O_2 is further dismutated into O_2 and H_2O under the catalysis of hydrogen peroxide dismutase CAT and peroxidase POD. Through the interaction of these three enzymes, the plant can effectively control ROS accumulation and ensure the continuation of normal physiological functions under drought stress [40].

As shown in Figure 5, SOD, CAT, and POD activity in the leaves of ND and SD were significantly different ($p < 0.001$). The leaves of ND showed low activity of these substances, and the SD showed high activity. When only taking the connected/disconnected rhizome factor into account, it was found that CAT and POD activity was considerably weaker in the connected rhizome treatment than in the disconnected treatment. SOD activity was not significantly different across the two treatments. This may be due to the fact that the clonal ramets of the *P. edulis* plant are able to perceptively sense heterogeneous environmental stress and affect physiological metabolism through the rhizome itself. The presence of the rhizome significantly affects the antioxidant enzyme metabolism of the source-sink clonal ramets, and improves the anti-ROS ability of the NC cloned ramets.

One of the inevitable consequences of drought stress is an increase in reactive oxygen species (ROS) production in different cellular compartments [4]. When the concentration of intracellular ROS increases, an increase in antioxidant enzyme activity and plant antioxidant capacity is induced. However, when the plant is under environmental stress for a long time, ROS production and scavenging by the antioxidant enzyme system become unbalanced, which causes the accumulation of MDA in the plant [41]. The results of this study showed that under heterogeneous drought stress, the MDA concentrations in the connected rhizome between clone ramets NC and SC were not significantly different. However, there were significant differences in the MDA concentration between the disconnected rhizome cloned ramets ND and SD (Figure 5d).

The results indicated that under the same degree of heterogeneous drought stress treatment, the structural function of the connected rhizome may be the cause of these physiological effects. On the other hand, under heterogeneous water stress conditions, the rhizome does have an effect on ROS accumulation in cloned plants. The rhizome was an important hub for the source-sink relationship of *P. edulis*.

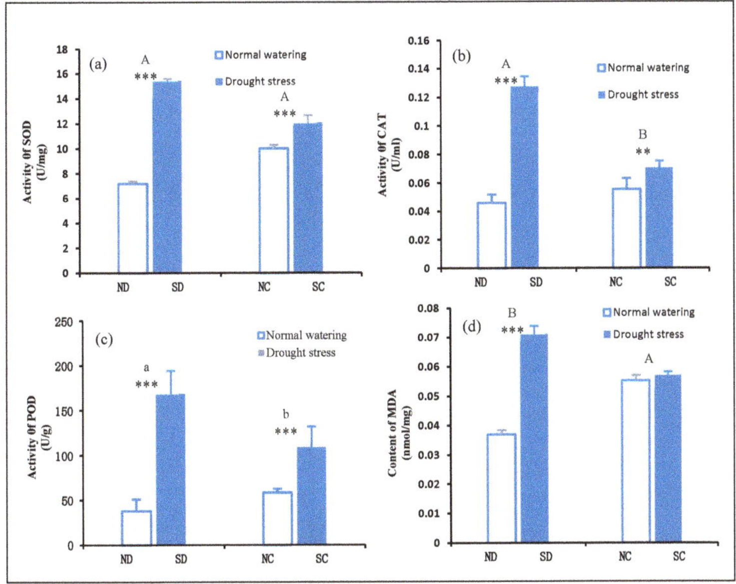

Figure 5. Effects of the rhizome on the changes of antioxidant enzyme activities and malondialdehyde concentration in leaves of clonal ramets under heterogeneous water stress. (**a–c**) represent SOD, CAT, POD activity of *P. edulis* leaves among the four treatment, respectively. (**d**) MDA concentration of *P. edulis* leaves under different treatment. Bars are mean values (±SE, n = 4). Different capital letters in the figure indicate significant differences between connected and disconnected ($p < 0.01$). Different lowercase letters indicate statistical significance differences ($p = 0.01–0.05$). Symbols indicate levels of statistical significance between normal watering (RWC = 80 ± 5%) and drought stress (RWC = 25 ± 5%) treatments for rhizome treatment: no symbol P > 0.1; ** $p = 0.001–0.01$; *** $p < 0.001$.

2.6. Effects of the Rhizome on Osmotic Adjustment Substances in P. edulis Leaves under Heterogeneous Water Stress

Osmotic adjustment is an important physiological regulation mechanism for plants to adapt to drought stress [42]. ROS produced in the leaves under drought stress can lead to increased membrane lipid permeability and electrolyte extravasation. By accumulating osmotic adjustment substances such as Pro and Bet, the cell's osmotic potential can be reduced and cell swell pressure maintained [2].

Under heterogeneous drought stress, the difference of Pro concentration between the ND and SD leaves of the disconnected rhizome by severing was significant ($p < 0.001$), while the clonal ramets NC and SC, with a connected rhizome, have a difference in the leaves of Pro concentration which was found to be significant at $p < 0.05$. At the same time, Bet concentration under the disconnected rhizome treatment was extremely significant, but there was no significant difference in the Bet concentration under the connected rhizome treatment.

The difference of osmotic adjustment substances (Pro and Bet) between the NC and SC leaves of the clonal ramets is smaller than that of ND and SD ramets. This difference in the concentration of osmotic adjustment substances caused by the rhizome may be related to the physiological response of the rhizome in a heterogeneous water environment. Studies have shown that drought stress can induce the increase of Pro and ABA concentration in plants, and that Pro accumulation depends on ABA, while ABA accumulation is independent of Pro [43]. Under heterogeneous drought stress, ABA concentration in SD treated leaves was significantly higher than that in ND, but there was no significant difference in ABA concentration between NC and SC treated leaves (Figure 6b). According to Figure 6, this is similar to the change of Pro and Bet concentration in NC and SC treated leaves under the same treatment. Increase in intracellular Pro- and Bet- concentration, which act as a driver of permeability in vivo, can maintain a high level of osmotic potential in the cell. Their accumulation can in turn stabilize the structure and function of biomacromolecules.

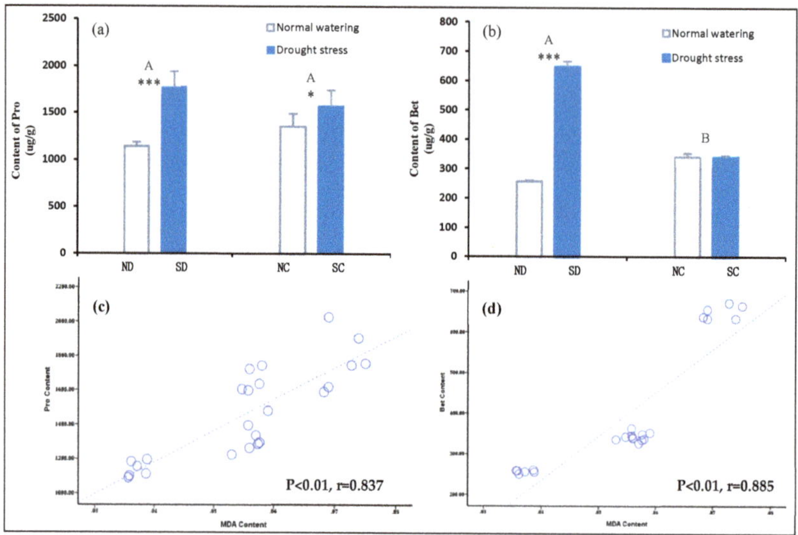

Figure 6. Effects of the rhizome on the changes of the concentration of substance in of clonal ramets under heterogeneous water stress. (**a,b**) represent Pro, Bet concentration in *P. edulis* leaves among the four treatment, respectively. Bars are mean values (±SE, n = 4). Different capital letters in the figure indicate significant differences between connected and disconnected ($p < 0.01$). Symbols indicate levels of statistical significance between normal watering (RWC = 80 ± 5%) and drought stress (RWC = 25 ± 5%) treatments for rhizome treatment: no symbol $p > 0.1$; * $p = 0.01$–0.05; *** $p < 0.001$. In (**c**) dashed (y = 18406.8x + 445.9, R^2 = 0.687) and (**d**) dashed (y = 10770.8x – 194.0, R^2 = 0.783) lines represent the linear regression for Pro concentration and MDA concentration (r = 0.837, $p < 0.01$), Bet concentration and MDA concentration (r = 0.885, $p < 0.01$), respectively. When $p < 0.05$, Pearson correlation coefficient (r) indicates the correlation between Pro or Bet concentration and MDA concentration.

The correlation between proline (Pro) and malondialdehyde concentration and betaine and malondialdehyde (MDA) concentration was analyzed (Figure 6c,d). It was found concentrations of osmotic adjustment substances Pro and Bet were strongly correlated with MDA concentration. Furthermore, it indicated that under heterogeneous water stress, the interaction of water availability and rhizome treatment affected the osmotic adjustment substances of each cloned ramet. The rhizome can thus be said to improve the adaptability of the clonal ramets to heterogeneous drought stress.

2.7. Effects of the Rhizome on the Nitrogen Metabolism in the Leaves of P. edulis under Heterogeneous Water Stress

NH^{4+} and NO^{3-} absorbed by the roots of the plant are transported to the aboveground organs through the vascular tissue, and the absorption of nitrogen can then be used for the synthesis of substances in various plant organs. Nitrogen metabolism in leaves is closely related to chlorophyll, protein, free amino acid concentration, and photosynthetic nitrogen-use efficiency (PNUE) [44].

Figure 7 indicates that under different treatments, the absorption of NH^{4+} and NO^{3-} by leaves showed an antagonistic trend. Compared with other treatments, the NH^{4+} efflux of SD leaves was the smallest, that is, the metabolic utilization of NH^{4+} in SD leaves under drought stress was increased, and the NO^{3-} metabolism absorption trend was opposite to that of NH^{4+} under the same treatment.

From the NH_4^+ and NO_3^- absorption flux, it was found that the NH_4^+ and NO_3^- fluxes in the SC and NC leaves were not significantly different, while the difference in NH_4^+ and NO_3^- between the SD and ND leaves in which the rhizome had been severed was larger.

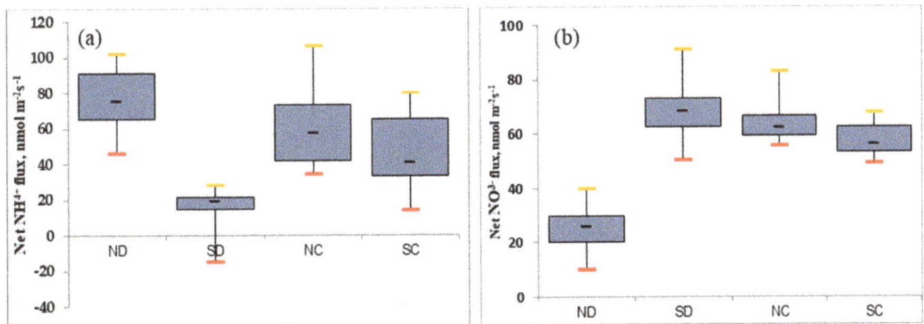

Figure 7. Nitrogen metabolism in leaves with either a connected or disconnected (severed) rhizome under heterogeneous water stress. (**a,b**) box plot shows NH_4^+ and NO_3^- flux respectively.

Nitrogen metabolism in leaves has an important role in photosynthesis and protein synthesis. Under different treatments, the changes of NH_4^+ and NO_3^- in leaves may be related to the absorption and metabolism of nitrogen in leaves. Combined with the above analysis, this shows that rhizome structure and water stress can both affect the nitrogen metabolism of *P. edulis* leaves.

3. Materials and Methods

3.1. Materials

P. edulis is a gramineae, perennial grass. *P. edulis* sprouting seedlings used in this experiment were obtained by seed sowing from seed collected from one mother line, and cultivation in 2015 under the same environmental conditions. They were cultivated in a greenhouse located in the International Center for Bamboo and Rattan's Anhui Taiping Experimental Station, located in Huangshan, Anhui, China (Figure 8).

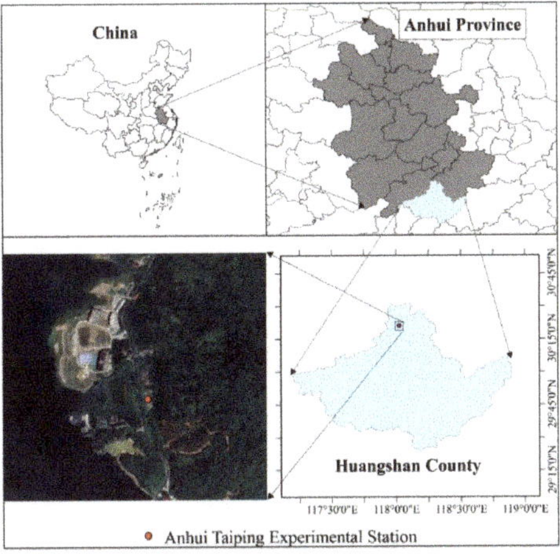

Figure 8. Schematic diagram of the location of *P. edulis* sprouting seedling (test materials) cultivation (30°21′N, 118°01′E).

This experimental area experiences a humid subtropical monsoon climate. The average temperature in July is around 25 °C, and the annual average temperature is 15–16 °C. Three-year-old *P. edulis* sprouting seedlings were selected from the greenhouse nursery. Clonal ramets of *P. edulis* consisting of two ramets connected by a rhizome were selected. Selected clonal ramets were all at the same development stage. To obtain similar clones, we selected plants with similar rhizome length and plant growth, consisting of integrated ramets of the same strain. The roots of each clonal branch thus had the same ability to absorb water and nutrients, therefore ensuring their functionally autonomous ability.

3.2. Experiment Design

Three-year-old sprouted seedlings with similar rhizome characteristics as described in *2.1. Materials* were selected. In the experiment, a two-factor design was used, with two levels of soil relative water content (RWC = 25 ± 5%, 80 ± 5%) crossed with two levels of rhizome treatment (connected or disconnected), resulting in a total of four different treatments (Figure 9). The *P. edulis* sprouting seedlings were henceforth denoted as connected rhizome treatment (NC, SC) and disconnected rhizome treatment (ND, SD) in experiments, where C denoted connected rhizome (with integration), D represented disconnected, severed rhizome (no integration), and N and S were short for normal water treatment or stress (drought) treatment respectively. Each treatment was treated in six replicates. The seedlings to be tested were placed in a plastic box (size 50 cm × 25 cm × 30 cm, long × wide × high). In order to prevent water from seeping between the two ramets, a hole was drilled in the wall of the plastic box for the rhizome to be placed (Figure 9c). After three weeks for recovery (on 10 October 2018), 30 clonal fragments of *P. edulis* were selected and used in the experiment as described below.

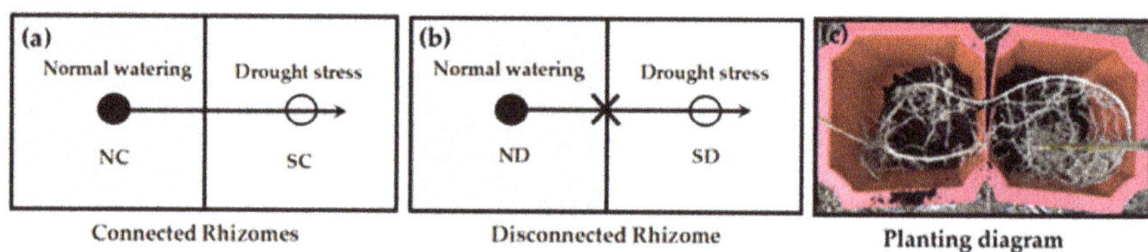

Figure 9. Schematic representation of the experimental design. (**a,b**) The line with the arrow represents the rhizome, the arrow represents the rhizome apex. Cloned ramets grown under normal watering (RWC: 80 ± 5%) and drought stress (RWC: 25 ± 5%) are represented by filled and open circles respectively. (**c**) Clonal strains of *P. edulis* consist of two clonal ramets that were growing in soil with different soil RWC.

Water was added to the *P. edulis* sprouting seedlings test material to saturate the soil. ND and NC treatments were continuously watered to maintain RWC = 80 ± 5%, and the SD and SC parts were subjected to natural transpiration treatment (not adding water), and in line with the drought stress threshold of *P. edulis* sprouting seedlings (RWC was about 30% measured by our research group). All test treatments were sampled and tested until natural transpiration consumption reached a relative water content of 25 ± 5%. RWC was monitored by a soil temperature and humidity meter EM-50 (METER Group, Inc., Pullman, Washington, DC, USA).

3.3. Methods

3.3.1. Laser Confocal Microscopy Luminescence Imaging to Observe the Structural Characteristics of Vascular Bundles of the Stem and Rhizome

Using the characteristic high lignin concentration in the *P. edulis* vascular bundles, the stem and rhizome were observed. Lignin concentration of vascular bundles is directly proportional to the fluorescence intensity of the laser, the structural characteristics of the vascular bundles in the rhizome and stem were observed using Laser confocal microscopy (LSM510, LeicaDM4, Berlin, Germany) at a 488 nm excitation wavelength.

3.3.2. Measurement of Ca^{2+}, NH^{4+}, and NO^{3-} Flux

Net Ca^{2+}, NH^{4+}, and NO^{3-} flux was measured using Non-Invasive Micro-Test Technology (NMT Physiolyzer®, Younger USA LLC, Amherst, MA 01002, USA; Xuyue (Beijing) Sci.&Tech. Co., Ltd., Beijing, China) at the Eco-instrument Analysis Lab, College of Biology and the Environment, Nanjing Forestry University (Nanjing, Jiangsu 210037, China). Non-invasive Micro-Test Technology (NMT) can non-invasively measure Ca^{2+}, NH^{4+}, and NO^{3-} flux with a high level of both temporal and spatial resolution. It measures the concentration gradient of Ca^{2+}, NH^{4+}, and NO^{3-} by means of a flux microsensor "vibrating" repeatedly between two points in the sample surface. Measurement of Ca^{2+}, NH^{4+}, and NO^{3-} flux in the test was reported by voltage output [45].

After different test treatments, living samples were incubated in the measuring solution to equilibrate for 10 min (Pileorhiza), 3 h (Rhizome), and 30 min (Mesophyll). Then, samples were transferred to a measuring chamber containing 10–15 mL of a fresh measuring solution. Measuring solution makeup and position site are shown Table 2.

Table 2. Non-invasive Micro-Test Technology (NMT) test site and test solution constitution.

Ion Species Forflux Testing	Test Position Site (µm)			Measuring Solutions Constitution	
	Pileorhiza	Rhizome	Mesophyll	Concentration	Constitution
Ca^{2+}	0 µm from the root apex	Cross-section	Cross-section	0.1 mM 0.1 mM	NH_4NO_3 $CaCl_2$
NH^{4+}	-	-	Cross-section	0.1 mM 0.1 mM	NH_4NO_3 $CaCl_2$
NO^{3-}	-	-	Cross-section	0.1 mM 0.1 mM	NH_4NO_3 $CaCl_2$

The system setup parameters in the experiment are as follows. Microsensors to measure Ca^{2+}, NH^{4+}, and NO^{3-} flux (Φ 4.5 ± 0.5 µm) were prepared by backfilling with electrolyte solution (100 mM $CaCl_2$, 100 mM NH_4Cl, and 10 mM KNO_3) up to approximately 1.0 cm from the tip. The micropipettes were front filled with 40–50 µm columns of selective liquid ion-exchange cocktails (Ca^{2+} LIX, XY-SJ-Ca, Xuyue (Beijing) Sci.&Tech. Co., Ltd., Beijing, China). An Ag/AgCl wire flux microsensor holder YG003-Y11 (Xuyue (Beijing) Sci.&Tech. Co., Ltd., Beijing, China) was inserted in the back of each flux microsensor to make electrical contact with the electrolyte solution. A flux microsensor with a Nernstian slope of >22 mV per decade was used in this study. The same flux microsensor was calibrated again according to the same procedure and standards after each test.

The Ca^{2+}, NH^{4+}, and NO^{3-} fluxes were calculated by Fick's law of diffusion as follows:

$$J = -D \cdot (dc/dx),$$

where dx (30 µm) is the distance the flux microsensor moved repeatedly from one point to another point perpendicular to the surfaces of samples at a frequency of ca. 0.3 Hz.

3.3.3. Changes of Hormone, Malondialdehyde (MDA), and Osmotic Adjustment Substance Content of *P. edulis* under Heterogeneous Drought Stress

The leaves of the clones [NC/ND] and clones [SC/SD] were separately determined by enzyme-linked immunosorbent assay (ELISA, Shanghai Enzyme-linked Biotechnology Co., Ltd., Shanghai, China). MDA, JA, ABA, Auxins (IAA), proline (Pro), and betaine (Bet) concentration was determined [46–48].

3.3.4. The Capacity of Antioxidant Defense Systems Activity of *P. edulis* under Heterogeneous Drought Stress

Leaves from the clonal ramets NC, ND, SC, and SD were collected, and Superoxide (SOD), peroxidase (POD), and catalase (CAT) activity in each sample was determined using the nitroblue tetrazolium (NBT) photochemical reduction, guaiacol, and theredox methods respectively [49,50].

3.3.5. Statistical Analysis

A two-factor repeated measures statistical analysis of variance was performed using SPSS 18.0 (SPSS Inc., Chicago, IL, USA) to calculate the studentized residual for four separate RWC values ($25 \pm 5\%$, $80 \pm 5\%$) and rhizome status (connected or disconnected), and to evaluate whether there were any abnormal values. Principal component analysis (PCA) maps were drawn using R software's ggplot2 package.

After carrying out the Shapiro–Wilk test, physiological indexes such as leaf hormone concentration, osmotic adjustment substance concentration and enzyme activity were found to approximate a normal distribution ($p > 0.05$). The results showed significant interactive effects between soil water content and rhizome status. The simple effect analysis (LSD, $p < 0.05$) of interactive effects between the connected disconnected rhizome groups was tested using syntax.

Significant differences are denoted with letters of the alphabet. At the same time, SPSS 18.0 was used to analyze the correlation between concentration of JA and ABA, concentration of Pro and MDA, and concentration of Bet and MDA.

4. Conclusions

This study shows that rhizome severance treatment on cloned ramets of *P. edulis* had significant effects on the water stress response. Under heterogeneous water stress, the rhizome that links cloned ramets of *P. edulis* plays an important role in regulating water physiology. LCSM was used to observe the structural characteristics of the stem and rhizome transverse section of *P. edulis*. It was found that the vascular bundle structures were similar in both plants. This is taken to mean that physiological integration provides a structural support of the clonal plant *P. edulis* in a heterogeneous water environment. We also used NMT to detect the Ca^{2+} flux in various *P. edulis* organs under heterogeneous water stress and found that the Ca^{2+} absorption ability in the connected rhizome was strong, and that the difference in flux between the two clonal ramets was small.

Combined with the related physiological and biochemical indices, this study found that the *P. edulis* rhizome plays an important part in the adaptation of the plant to drought stress, especially under heterogeneous water conditions where the sharing of water through the rhizome is possible. This facilitates the adaptability of the whole plant to drought stress conditions. Most physiological indexes were significantly lower in the SC treatment than in the SD treatment under the same soil moisture content.

Our findings reveal that the physiological changes caused by the cascade effect of the rhizome tends to be beneficial to the stress response. Compared with the control group with the disconnected rhizome, relevant physiological and biochemical indexes of the cloned ramets with a connected rhizome shifted towards adaptation to the stress. Under heterogeneous drought stress conditions, the benefits of the attached rhizome of clonal *P. edulis* ramets are obvious.

Author Contributions: Conceptualization, X.J., C.C. and S.F.; Data curation, X.J., G.L. and C.W.; Formal analysis, X.J., C.C., S.F., G.L., C.W. and B.C.; Funding acquisition, S.F.; Investigation, X.J. and B.C.; Methodology, X.J., C.C., S.F. and C.W.; Project administration, X.J., C.C. and S.F.; Resources, X.J., C.C., S.F. and G.L.; Software, X.J., C.W. and B.C.; Validation, X.J., C.C., G.L.; Visualization, X.J. and S.F.; Writing—original draft, X.J.; Writing—review & editing, X.J. All authors have read and agreed to the published version of the manuscript.

Funding: This study was supported by special research fund of International Centre for Bamboo and Rattan (No. 1632019005).

Acknowledgments: The authors acknowledge the financial support of the Foundation of International Centre for Bamboo and Rattan (No. 1632019005).

Conflicts of Interest: The authors declare no conflict of interest.

References

1. Carrão, H.; Naumann, G.; Barbosa, P. Global projections of drought hazard in a warming climate: A prime for disaster risk management. *Clim. Dyn.* **2018**, *50*, 2137–2155. [CrossRef]
2. Wu, S.; Hu, C.; Tan, Q.; Li, L.; Shi, K.; Zheng, Y. Drought stress tolerance mediated by zinc-induced antioxidative defense and osmotic adjustment in cotton (Gossypium Hirsutum). *Acta Physiol. Plant.* **2015**, *37*, 167. [CrossRef]
3. Haq, T.U.; Ali, A.; Nadeem, S.; Maqbool, M.M.; Ibrahim, M. Performance of canola cultivars under drought stress induced by withholding irrigation at different growth stages. *Soil Environ.* **2014**, *33*, 43–50.
4. Kaur, G.; Asthir, B. Molecular responses to drought stress in plants. *Biol. Plant.* **2016**, *2*, 201–209. [CrossRef]
5. Jiang, Z.H.; Fei, B.H.; Fan, S.H. Actively develop bamboo industry and vigorously promote the ecological civilization construction. *State Acad. For. Adm. J.* **2014**, *2*, 12–16.
6. Zhou, G.M.; Meng, C.F.; Jiang, P.K.; Xu, Q.F. Review of Carbon Fixation in Bamboo Forests in China. *Bot. Rev.* **2011**, *77*, 262–270. [CrossRef]
7. Bai, S.; Wang, Y.; Conant, R.T.; Zhou, G.; Xu, Y.; Wang, N.; Fang, F.; Chen, J. Can native clonal moso bamboo encroach on adjacent natural forest without human intervention? *Sci. Rep.* **2016**, *6*, 31504. [CrossRef] [PubMed]
8. Luo, F.L.; Xing, Y.P.; Wei, G.W.; Li, C.Y. Clonal integration facilitates spread of Paspalum paspaloides from terrestrial to cadmium contaminated aquatic habitats. *Plant Biol.* **2017**, *19*, 859–867. [CrossRef]
9. Czarnecka, B. Spatiotemporal Patterns of Genets and Ramets in a Population of Clonal Perennial Senecio rivularis: Plant features and habitat effects. *Ann. Bot. Fenn.* **2008**, *45*, 19–32. [CrossRef]
10. Dong, M. Clonal growth in plants in relation to resource heterogeneity: Foraging behavior. *Acta Bot. Sin.* **1996**, *38*, 828–835.
11. Poor, A.; Hershock, C.; Rosella, K.; Goldberg, D.E. Do Physiological Integration and Soil Heterogeneity Influence the Clonal Growth and Foraging of Schoenoplectus pungens? *Plant Ecol.* **2005**, *181*, 45–56. [CrossRef]
12. Sergior, R.; Susana, R.; Eduardo, D.L.P.; Helena, F. Physiological integration increases the survival and growth of the clonal invader *Carpobrotus edulis*. *Biol. Invasions* **2010**, *12*, 1815–1823.
13. Zhang, Y.; Zhang, Q.; Marek, S. Physiological Integration Ameliorates Negative Effects of Drought Stress in the Clonal Herb Fragaria orientalis. *PLoS ONE* **2012**, *7*, e44221. [CrossRef]
14. Herben, T. Physiological integration affects growth form and competitive ability in clonal plants. *Evol. Ecol.* **2004**, *18*, 493–520. [CrossRef]
15. Ito, R.; Miyafuji, H.; Kasuya, N. Rhizome and root anatomy of moso bamboo (*Phyllostachys pubescens*) observed with scanning electron microscopy. *J. Wood Sci.* **2015**, *61*, 431–437. [CrossRef]
16. Zhuang, M.H.; Li, Y.C.; Chen, S.L. Advances in the Researches of Bamboo Physiological Integration and Its Ecological Significance. *J. Bamboo Res.* **2011**, *30*, 1–9.
17. Jing, X.; Cai, C.J.; Fan, S.H.; Liu, G.L. Research Advances in Ecological Adaptability of Bamboo under Environmental Stresses. *World For. Res.* **2018**, *31*, 36–41.
18. Saitoh, T.; Nishiwaki, S.A. Importance of Physiological Integration of Dwarf Bamboo to Persistence in Forest Understorey: A Field Experiment. *J. Ecol.* **2002**, *90*, 78–85. [CrossRef]
19. Xui, S.N.; Liu, Y.C.; Liu, Y.H.; Chen, Z.L.; Li, Y.; Zhang, L.H. Physiological integration of growth and photosynthesis of *zoysia japonica* clonal ramets under nutrient heterogeneity. *J. Appl. Ecol.* **2018**, *29*, 811–817.
20. Qian, Y.Q.; Sun, Z.Y.; Han, L.; Ju, G.S. Photosynthate integration and regulation within clones of buffalograss under heterogeneous water supply. *Acta Ecol. Sin.* **2010**, *30*, 3966–3973.
21. Hu, J.J.; Chen, S.L.; Guo, Z.W.; Yang, Q.P.; Li, Y.C. Effects of Spacer Length on Water Physiological Integration of *Indocalamus decorus* Ramets under Heterogeneous Water Supply. *Acta Bot. Boreali Occident. Sin.* **2015**, *35*, 2532–2541.

22. Satchivi, N.M.; Stoller, E.W.; Wax, L.M.; Briskin, D.P. A Nonlinear Dynamic Simulation Model for Xenobiotic Transport and Whole Plant Allocation Following Foliar Application. I. Conceptual Foundation for Model Development. *Pestic. Biochem. Physiol.* **2000**, *68*, 67–84. [CrossRef]

23. Liu, Y.H.; Jia, H.S.; Gao, J. Review on researches of photoassimilates partitioning and its models. *Acta Ecol. Sin.* **2006**, *6*, 1981–1992.

24. Mh, C.D.C. Drought stress and reactive oxygen species: Production, scavenging and signaling. *Plant Signal. Behav.* **2008**, *3*, 156–165.

25. Sadiqov, S.T.; Akbulut, M.; Ehmedov, V. Role of Ca^{2+} in drought stress signaling in wheat seedlings. *Biochemistry* **2002**, *67*, 491–497. [PubMed]

26. Ahmad, P.; Bhardwaj, R.; Tuteja, N. Plant Signaling Under Abiotic Stress Environment. In *Environmental Adaptations and Stress Tolerance of Plants in the Era of Climate Change*; Springer: New York, NY, USA, 2012.

27. Roberts, A.W.; Haigler, C.H. Rise in chlorotetracycline fluorescence accompanies tracheary element differentiation in suspension cultures of Zinnia. *Protoplasm* **1998**, *152*, 37–45. [CrossRef]

28. Brauer, M.; Zhong, W.J.; Jelitto, T.; Schobert, C.; Sanders, D.; Komor, E. Free calcium ion concentration in the sieve-tube sap of Ricinus communis L. *Planta* **1998**, *206*, 103–107. [CrossRef]

29. Yu, F. *Studies on Developmental Biology of Phloem Ganglion in Phyllostachys Edulis*; Nanjing Forestry University: Nanjing, China, 2005.

30. Zhang, Q.X.; Lan, Y.P. Signal Transduction within Plants under Water Stress. *J. Shanxi Teach. Univ.* **2004**, *18*, 86–91.

31. Jacobsen, S.E.; Liu, F.; Jensen, C.R. Does root-sourced ABA play a role for regulation of stomata under drought in quinoa (Chenopodium quinoa Willd). *Sci. Hortic.* **2009**, *122*, 281–287. [CrossRef]

32. Kuromori, T.; Seo, M.; Shinozaki, K. ABA Transport and Plant Water Stress Responses. *Trends Plant Sci.* **2018**, *23*, 513–522. [CrossRef]

33. Ma, H.P.; Liu, Z.M. Transmission and distribution of ABA in seedlings and shoots of *Malus micromalus Makino* and their relationship with water. *Plant Physiol. J.* **1998**, *24*, 253–258.

34. Chen, J.S.; Lei, N.F.; Liu, Q. Defense signaling among interconnected ramets of a rhizomatous clonal plant, induced by jasmonic-acid application. *Acta Oecologica* **2011**, *37*, 355–360. [CrossRef]

35. Liu, Z.H.; Guo, X.; Gang, W.; Li, G.M. Drought stress and signal transduction of ABA. *Chin. Bull. Bot.* **2004**, *21*, 228–234.

36. Fujita, Y.; Fujita, M.; Satoh, R.; Maruyama, K.; Parvez, M.M.; Seki, M. Areb1 is a transcription activator of novel abre-dependent ABA signaling that enhances drought stress tolerance in Arabidopsis. *Plant Cell* **2005**, *17*, 3470. [CrossRef]

37. Li, C.J. Autoinhibition in polar transport of indoleacetic acid. *Acta Bot. Sin.* **1997**, *39*, 737–741.

38. Xu, S.C. The Roles of Ca^{2+}-Dependent and Calmodulin Stimulated Protein Kinase in Abscisic Acid-Induced Antioxidant Defense in Leaves of Maize Plants. Ph.D. Thesis, Nanjing Agricultural University, Najing, China, 2009.

39. Li, X.; Yue, H.; Huang, L.Q.; Guo, P.L. *Study on the Reaction Regular of Plant Antioxidant Enzymes under Environmental Stress*; Chinese Pharmaceutical Congress: Beijing, China, 2010.

40. Schuller, D.J.; Ban, N.; Huystee, R.B.; Mcpherson, A.; Poulos, T.L. The crystal structure of peanut peroxidase. *Structure* **1996**, *4*, 311. [CrossRef]

41. Xu, S.N.; Li, Y.; Chen, Z.L.; Zhong, L.H. Physiological integration of antioxidant enzymes and malondialdehyde in connected and disconnected clonal ramet under nutrient heterogeneity. *Pratacult. Sci.* **2018**, *35*, 13.

42. Qin, W.U.; Zhang, G.; Pei, B. Physiological and biochemical responses to different soil drought stress in three tree species. *Acta Ecol. Sin.* **2013**, *33*, 3648–3656. [CrossRef]

43. Wang, H.C. Effect of Light on Proline Synthesis under Drought Stress and Its Relationship with ABA. Ph.D. Thesis, Yangzhou University, Yangzhou, China, 2013.

44. An, H.; ShangGuan, Z.P. The nitrogen cycling of plants and it's physiological mechanism of root-zone environment. *Res. Soil Water Conserv.* **2006**, *13*, 83–85.

45. Xu, Y.; Sun, T.; Yin, L.P. Application of Non-invasive Microsensing System to Simultaneously Measure Both H^+, and O_2 Fluxes Around the Pollen Tube. *J. Integr. Plant Biol.* **2006**, *48*, 823–831. [CrossRef]

46. Xin, Z.Y.; Zhou, X.; Pilet, P.E. Level changes of jasmonic, abscisic, and indole-3yl-acetic acids in maize under desiccation stress. *J. Plant Physiol.* **1997**, *151*, 120–124. [CrossRef]

47. Engvall, E. Enzyme immunoassay ELISA and EMIT. *Methods Enzym.* **1980**, *70*, 419–439.

48. Shinagawa, A.; Suzuki, T.; Konosu, S. The Role of Free Amino Acids and Betaines in Intracellular Osmoregulation of Marine Sponges. *Nippon Suisan Gakkaishi* **1992**, *58*, 1717–1722. [CrossRef]
49. Lin, S.Y.; Wu, J.X. Discussion on the Determination of SOD by Tetrazolium Blue (NBT) Method. *Strait Pharm. J.* **2009**, *21*, 43–45.
50. Liu, S.Y.; An, W.; Xu, X.W.; Liu, H.Z. Determination of POD, CAT and PAL enzyme activity in leaves of Hippophae rhamnoides. *J. Jilin Agric. Univ.* **2009**, *31*, 584–586.

MDPI

St. Alban-Anlage 66

4052 Basel

Switzerland

Tel. +41 61 683 77 34

Fax +41 61 302 89 18

www.mdpi.com

Plants Editorial Office

E-mail: plants@mdpi.com

www.mdpi.com/journal/plants

9 783036 508306